Yun Wang

Dark Energy

Related Titles

Phillipps, S.
The Structure and Evolution of Galaxies
2005
ISBN 978-0-470-01585-8

Stiavelli, M. S.
From First Light to Reionization
The End of the Dark Ages

2009
ISBN 978-3-527-40705-7

Roos, M.
Introduction to Cosmology
2003
ISBN 978-0-470-84910-1

Shore, S. N.
The Tapestry of Modern Astrophysics
2003
ISBN 978-0-471-16816-4

Coles, P., Lucchin, F.
Cosmology
The Origin and Evolution of Cosmic Structure

2002
ISBN 978-0-471-48909-2

בס"ד

Dr. Chaim Yosef Mariategui-Levi

Yun Wang

Dark Energy

WILEY-VCH Verlag GmbH & Co. KGaA

The Author

Prof. Yun Wang
University of Oklahoma
Dept. of Physics & Astronomy
Norman, OK 73019
USA

All books published by **Wiley-VCH** are carefully produced. Nevertheless, authors, editors, and publisher do not warrant the information contained in these books, including this book, to be free of errors. Readers are advised to keep in mind that statements, data, illustrations, procedural details or other items may inadvertently be inaccurate.

Bibliographic information published by the Deutsche Nationalbibliothek
The Deutsche Nationalbibliothek lists this publication in the Deutsche Nationalbibliografie; detailed bibliographic data are available on the Internet at http://dnb.d-nb.de.

© 2010 WILEY-VCH Verlag GmbH & Co. KGaA, Weinheim

All rights reserved (including those of translation into other languages). No part of this book may be reproduced in any form – by photoprinting, microfilm, or any other means – nor transmitted or translated into a machine language without written permission from the publishers. Registered names, trademarks, etc. used in this book, even when not specifically marked as such, are not to be considered unprotected by law.

Composition   le-tex publishing services GmbH, Leipzig
Printing and Bookbinding   betz-druck GmbH, Darmstadt
Cover Design   Formgeber, Eppelheim

Printed in the U.S.A.

ISBN 978-3-527-40941-9

*For Sam, Tian, Chris,
and my parents*

Contents

Preface *XI*

1 The Dark Energy Problem *1*
1.1 Evidence for Cosmic Acceleration *1*
1.1.1 The Basic Cosmological Picture *1*
1.1.2 First Direct Observational Evidence for Cosmic Acceleration *4*
1.1.3 Current Observational Evidence for Cosmic Acceleration *8*
1.2 Fundamental Questions about Cosmic Acceleration *10*

2 The Basic Theoretical Framework *15*
2.1 Einstein's Equation *15*
2.2 Cosmological Background Evolution *16*
2.3 Cosmological Perturbations *18*
2.3.1 Cosmological Perturbations: Nonrelativistic Case *18*
2.3.2 Cosmological Perturbations: Generalized Case *22*
2.4 Framework for Interpreting Data *31*
2.4.1 Model-Independent Constraints *31*
2.4.2 Using the Fisher Matrix to Forecast Future Constraints *32*
2.4.3 Using the Markov Chain Monte Carlo Method in a Likelihood Analysis *32*
2.4.4 Self-Consistent Inclusion of Cosmic Microwave Background Anisotropy Data *33*

3 Models to Explain Cosmic Acceleration *35*
3.1 Dark Energy Models *35*
3.1.1 Quintessence, Phantom Field, and Chaplygin Gas *36*
3.1.2 Worked Example: PNGB Quintessence *39*
3.1.3 Worked Example: The Doomsday Model *40*
3.2 Modified Gravity Models *44*
3.2.1 $f(R)$ Gravity Models *44*
3.2.2 DPG Gravity Model *45*
3.2.3 The Cardassian Model *46*
3.3 A Cosmological Constant *47*

4 Observational Method I: Type Ia Supernovae as Dark Energy Probe *51*
4.1 Type Ia Supernovae as Distance Indicators *51*

Dark Energy. Yun Wang
Copyright © 2010 WILEY-VCH Verlag GmbH & Co. KGaA, Weinheim
ISBN: 978-3-527-40941-9

4.2	Possible Causes of Observational Diversity in SNe Ia 56
4.3	Supernova Rate 57
4.4	Systematic Effects 61
4.4.1	Extinction 62
4.4.2	K-Correction 64
4.4.3	Weak Lensing 66
4.4.4	Other Systematic Uncertainties of SNe Ia 71
4.5	Data Analysis Techniques 73
4.5.1	Light Curve Fitting 73
4.5.2	Flux-Averaging Analysis of SNe Ia 76
4.5.3	Uncorrelated Estimate of $H(z)$ 79
4.6	Forecast for Future SN Ia Surveys 83
4.7	Optimized Observations of SNe Ia 86
5	**Observational Method II: Galaxy Redshift Surveys as Dark Energy Probe** **91**
5.1	Baryon Acoustic Oscillations as Standard Ruler 91
5.2	BAO Observational Results 93
5.3	BAO Systematic Effects 97
5.3.1	Nonlinear Effects 98
5.3.2	Redshift-Space Distortions 99
5.3.3	Scale-Dependent Bias 101
5.4	BAO Data Analysis Techniques 103
5.4.1	Using the Galaxy Power Spectrum to Probe BAO 104
5.4.2	Using Two-Point Correlation Functions to Probe BAO 114
5.5	Future Prospects for BAO Measurements 119
5.6	Probing the Cosmic Growth Rate Using Redshift-Space Distortions 124
5.6.1	Measuring Redshift-Space Distortion Parameter β 124
5.6.2	Measuring the Bias Factor 128
5.6.3	Using $f_g(z)$ and $H(z)$ to Test Gravity 132
5.7	The Alcock–Paczynski Test 134
6	**Observational Method III: Weak Lensing as Dark Energy Probe** **135**
6.1	Weak Gravitational Lensing 135
6.2	Weak Lensing Observational Results 142
6.3	Systematics of Weak Lensing 147
6.3.1	Point Spread Function Correction 148
6.3.2	Other Systematic Uncertainties 154
6.4	Future Prospects for the Weak Lensing Method 157
6.5	The Geometric Weak Lensing Method 159
6.5.1	Linear Scaling and Off-Linear Scaling 160
6.5.2	Implementation of the Linear Scaling Geometric Method 161
7	**Observational Method IV: Clusters as Dark Energy Probe** **163**
7.1	Clusters and Cosmology 163
7.2	Cluster Abundance as a Dark Energy Probe 164

7.2.1 Theoretical Cluster Mass Function   *164*
7.2.2 Cluster Mass Estimates   *166*
7.2.3 Cluster Abundance Estimation   *172*
7.2.4 Cosmological Parameters Constraints   *175*
7.3 X-Ray Gas Mass Fraction as a Dark Energy Probe   *179*
7.4 Systematic Uncertainties and Their Mitigation   *182*

8 Other Observational Methods for Probing Dark Energy   *185*
8.1 Gamma Ray Bursts as Cosmological Probe   *185*
8.1.1 Calibration of GRBs   *185*
8.1.2 Model-Independent Distance Measurements from GRBs   *190*
8.1.3 Impact of GRBs on Dark Energy Constraints   *193*
8.1.4 Systematic Uncertainties   *195*
8.2 Cosmic Expansion History Derived from Old Passive Galaxies   *196*
8.3 Radio Galaxies as Cosmological Probe   *198*
8.4 Solar System Tests of General Relativity   *202*

9 Basic Instrumentation for Dark Energy Experiments   *205*
9.1 Telescope   *205*
9.2 NIR Detectors   *210*
9.3 Multiple-Object Spectroscopic Masks   *213*

10 Future Prospects for Probing Dark Energy   *219*
10.1 Designing the Optimal Dark Energy Experiment   *219*
10.2 Evaluating Dark Energy Experiments   *222*
10.3 Current Status and Future Prospects   *225*

References   *229*

Index   *239*

Preface

Solving the mystery of the observed cosmic acceleration is of fundamental importance. Dark energy research aims to determine whether the observed cosmic acceleration is due to a new energy component (dark energy) in the universe, or a modification of general relativity. The definitive results from this quest will revolutionize particle physics and cosmology.

Various observational projects (including some that are very ambitious) are currently being planned or envisioned to shed light on the nature of the observed cosmic acceleration. It is the perfect time to summarize and clarify the current state of the dark energy search, and examine the major observational techniques for probing dark energy, in order to foster the most fruitful and efficient approach to advancing the field.

Dark energy research is an extremely active field, with dozens of notable (and possibly significant) new papers coming out each month. Thus it is challenging to write a book on dark energy at this point. I have tried to include results from the public domain up to June 2009, if I consider them to be sufficiently significant and relevant to this book. This is no doubt a subjective choice, and reflects my own personal perspective on dark energy research.

I have tried to make this book as self-contained and pedagogical as possible, with emphasis on the basic physical pictures, and observational techniques that are most promising and have already been tested in practice. This book can be used as a textbook to train graduate students, and others who wish to enter the field of dark energy search. I have also aimed to make this book a useful reference to active researchers in the field (including myself), and others who are interested in dark energy.

Note that in the literature, "dark energy" and "cosmic acceleration" are often used interchangeably, although it is only correct if we already know that cosmic acceleration is not due to a modification of general relativity. However, it is cumbersome to refer to the various projects that aim to solve the mystery of cosmic acceleration as "cosmic acceleration projects", rather than the conventional "dark energy projects". In this book, I have made the distinction between "dark energy" and "cosmic acceleration" when the cause of the observed cosmic acceleration is being discussed, and used "dark energy probes" or "dark energy projects" in referring to the various observational techniques or projects that enable us to search for the cause of cosmic acceleration.

In order to maximize the accuracy of this book, I have sent key chapters from a draft of the book to experts in various areas for review. I am grateful to the following reviewers for invaluable comments and suggestions that helped improve the book:

Chapter 1–3 (Introduction, theoretical framework, and models):   *Richard Woodard, Robert Caldwell, Katherine Freese, James Fry*
Chapter 4 (Supernovae):   *David Branch, Peter Garnavich, Craig Wheeler*
Chapter 5 (Galaxy clustering):   *Carlton Baugh, Chris Blake, James Fry, Luigi Guzzo*
Chapter 6 (Weak lensing):   *Chris Hirata, Henk Hoekstra, David Wittman*
Chapter 7 (Clusters):   *Alexey Vikhlinin*
Chapter 9 (Instrumentation):   *Arlin Crotts, Massimo Robberto*

I taught a course on dark energy at the University of Oklahoma in spring 2009, using a draft of this book as the textbook. I thank the graduate students who took my course, in particular, Chia-Hsun (Albert) Chuang, Maddumage Don Hemantha, and Andre Lessa, for providing helpful feedback on the book.

Writing this book gave me a great opportunity to reflect on the field of dark energy research as a whole. I could not have done this without the love and support from my family.

University of Oklahoma, December 2009   *Yun Wang*

1
The Dark Energy Problem

The discovery that the expansion of the universe is accelerating was first made by Riess et al. (1998) and Perlmutter et al. (1999), with supporting evidence for this observation strengthening over time.

The cause for the observed acceleration is unknown, and is usually referred to as "the dark energy problem". It could be due to an unknown energy component in the universe (i.e., "dark energy"), or the modification of gravity as described by Einstein's general relativity (i.e., "modified gravity"). Solving the mystery of the observed cosmic acceleration is one of the most exciting challenges in cosmology today.

1.1
Evidence for Cosmic Acceleration

To understand the evidence for cosmic acceleration, we need to first introduce the basis of standard cosmology. We will then discuss the first and current evidence for cosmic acceleration.

1.1.1
The Basic Cosmological Picture

We live in an expanding universe, a fact first discovered by Hubble in 1929. Our physical universe can be described by the Robertson–Walker metric, the simplest metric that describes a homogeneous, isotropic, and expanding universe:

$$ds^2 = -c^2 dt^2 + a^2(t) \left[\frac{dr^2}{1 - \tilde{k} r^2} + r^2 d\theta^2 + r^2 \sin^2 \theta \, d\phi^2 \right] \tag{1.1}$$

where c is the speed of light, t is cosmic time, $a(t)$ is the cosmic scale factor, and $\tilde{k}$ is the curvature constant. The universe is flat for $\tilde{k} = 0$, open for $\tilde{k} < 0$ and closed for $\tilde{k} > 0$. The spatial location of an object is given by (r, θ, ϕ) in spherical coordinates.

Dark Energy. Yun Wang
Copyright © 2010 WILEY-VCH Verlag GmbH & Co. KGaA, Weinheim
ISBN: 978-3-527-40941-9

The physical wavelength of light emitted at time t is given by

$$\lambda_{\text{phys}} = a(t)\lambda, \tag{1.2}$$

where λ is the comoving wavelength. As the universe expands, $a(t)$ increases with time. Comoving quantities do not change with the expansion of the universe. The expansion of the universe leads to an increase in the observed wavelength (i.e., a redshift) of light from a distant source. The cosmological redshift is defined as

$$z \equiv \frac{1}{a(t)} - 1. \tag{1.3}$$

The redshift z is usually used as the indicator for cosmic time, because it can be measured for a given astrophysical object. If the light emitted by a distant object is stretched by a factor of $(1+z)$ in wavelength upon arrival at the observer, the object is said to be at a distance corresponding to redshift z.

The coordinate distance $r(z)$ from Eq. (1.1) gives the observer's *comoving distance* to an object located at redshift z. Our physical distance to the object is the *angular diameter distance* given by

$$d_A(z) \equiv a(t)r = \frac{r(z)}{1+z}. \tag{1.4}$$

If we know the intrinsic luminosity of an object, then measuring its apparent brightness allows us to infer our *luminosity distance* to the object

$$\left[\frac{d_L(z)}{10\,\text{pc}}\right]^2 = \frac{F_{\text{int}}}{F}, \tag{1.5}$$

where F is the observed flux from the object, and F_{int} is its "intrinsic flux", defined to be the flux from the object received by an observer located at a distance of 10 pc away from the object. In astronomical observations, magnitude is used as the unit for the observed flux. The magnitude difference between two objects with observed fluxes F_1 and F_2 is defined as

$$m_1 - m_2 \equiv 2.5 \log\left(\frac{F_2}{F_1}\right). \tag{1.6}$$

Thus, Eq. (1.5) becomes

$$m - M = 2.5 \log\left(\frac{F_{\text{int}}}{F}\right) = 5 \log\left[\frac{d_L(z)}{\text{Mpc}}\right] + 25, \tag{1.7}$$

where m and M are the apparent and absolute magnitudes of the object respectively, and $m-M$ is known as the *distance modulus*. Due to the redshifting of the light from the object, and the time dilation effect, the luminosity distance and the comoving distance to the object are related by

$$d_L(z) = (1+z)r(z). \tag{1.8}$$

The expansion rate of the universe at time t is known as the *Hubble parameter* $H(t)$, defined as

$$H(t) \equiv \frac{\dot{a}}{a} . \tag{1.9}$$

The Hubble parameter $H(t)$ and the cosmic scale factor $a(t)$ are functions of time (i.e., redshift) that depend on the composition of the universe, as well as the global spatial curvature of the universe. Setting $d^2s = 0$ and considering radial dependence only (i.e., considering the radial propagation of photons), Eq. (1.1) gives a relation between distance and redshift that depends on the Hubble parameter, which in turn depends on the composition and spatial curvature of the universe. For a flat universe, we have

$$d_L(z) = c(1+z) \int_0^z dz' \frac{1}{H(z')} , \quad \text{(flat universe)} . \tag{1.10}$$

In the standard cosmological model, the universe began in a very hot and very dense state, known as the *Big Bang*. It is likely that the universe went through a period of extremely rapid expansion (known as *inflation*) in the first tiny fraction of a second in the history of the universe. The universe was radiation dominated after the end of inflation, then became matter dominated at $z \sim 3000$.

For a universe consisting of matter, radiation, and a cosmological constant, the Hubble parameter $H(z)$ is

$$H^2(z) \equiv \left(\frac{\dot{a}}{a}\right)^2 = H_0^2 \left[\Omega_m(1+z)^3 + \Omega_r(1+z)^4 + \Omega_k(1+z)^2 + \Omega_\Lambda\right], \tag{1.11}$$

where the *Hubble constant* H_0 is defined as the value of the Hubble parameter today. The density fractions Ω_m, Ω_r, Ω_Λ, and Ω_k are defined by

$$\Omega_m \equiv \frac{\rho_m(t_0)}{\rho_c^0}, \quad \Omega_r \equiv \frac{\rho_{rad}(t_0)}{\rho_c^0}, \quad \Omega_\Lambda \equiv \frac{\rho_\Lambda}{\rho_c^0}, \quad \Omega_k \equiv -\frac{\tilde{k}}{H_0^2}, \tag{1.12}$$

where $\rho_m(t_0)$ and $\rho_{rad}(t_0)$ are the matter and radiation densities today, and ρ_Λ is the energy density due to a *cosmological constant* (also known as *vacuum energy*). The *critical density* ρ_c^0 is defined as

$$\rho_c^0 \equiv \frac{3H_0^2}{8\pi G} . \tag{1.13}$$

Requiring the consistency of Eq. (1.11) at $z = 0$, that is, $H(z=0) = H_0$, gives

$$\Omega_m + \Omega_r + \Omega_k + \Omega_\Lambda = 1 . \tag{1.14}$$

Note that $\Omega_r \ll \Omega_m$, thus the Ω_r term is usually omitted at $z \ll 3000$.

The matter-energy content of the universe is parametrized as an ideal fluid with density ρ and pressure p. Each component of the universe can be described by its equation of state w, defined as

$$w \equiv \frac{p}{\rho}. \tag{1.15}$$

Matter has $w = 0$. Radiation has $w = 1/3$. A cosmological constant corresponds to $w = -1$.

1.1.2
First Direct Observational Evidence for Cosmic Acceleration

Until about a decade ago, observational data favored a universe that is dominated by matter today. If we allow the universe to have an arbitrary spatial curvature (see Eq. (1.1)), and a possibly nonzero cosmological constant Λ (see Eq. (1.11)), then we can define a "deceleration parameter" of the cosmic expansion:

$$q_0 \equiv -\frac{\ddot{a}(t_0)/a(t_0)}{H_0^2} = \frac{\Omega_m}{2} - \Omega_\Lambda, \tag{1.16}$$

where $a(t_0)$ is the cosmic scale factor today, and we have used Eq. (1.11). Equation (1.16) shows clearly that a matter-dominated universe (with $\Omega_m > 2\Omega_\Lambda$) should be decelerating today.

For decades, astronomers tried to measure the cosmic "deceleration parameter" using Type Ia supernovae (SNe Ia) as cosmological distance indicators. SNe Ia can be calibrated to be good *standard candles*, with very small scatter in their intrinsic peak luminosity. Thus the measured apparent peak brightness of SNe Ia can be used to infer the distances to the SNe Ia. The observed spectra of the SNe Ia can be used to measure their redshifts. This yields an observed distance-redshift relation of SNe Ia that can be shown in a Hubble diagram. In a Hubble diagram, the distance modulus $m - M$ of a SN Ia, the difference between the observed apparent peak brightness of a SN Ia and its absolute peak brightness, is plotted as a function of redshift of the SN Ia. The measured distance modulus of a SN Ia gives a measurement of the luminosity distance $d_L(z)$ to the SN Ia at redshift z, see Eq. (1.7). This measured $d_L(z)$ can then be compared with the theoretical prediction (e.g., Eqs. (1.10) and (1.11)) to infer the values of cosmological parameters, and constrain q_0.

Unexpectedly, the quest to measure the cosmic "deceleration parameter" led to the discovery that the universe is accelerating today. This means that the universe is dominated by something that is not matter-like today, or general relativity does not give a complete description of the present universe. This discovery was made using the observed peak brightness of SNe Ia as cosmological distance indicators, independently by two teams of astronomers, Riess *et al.* (1998) and Perlmutter *et al.* (1999). Figure 1.1 shows the joint confidence contour for Ω_m and Ω_Λ from Riess *et al.* (1998) and Perlmutter *et al.* (1999). The discovery of cosmic acceleration was made at high statistical significance.

Figure 1.2 shows Hubble diagrams of SNe Ia from both teams. These show the observed distance-redshift relations, that is, the distance modulus $m - M$ versus

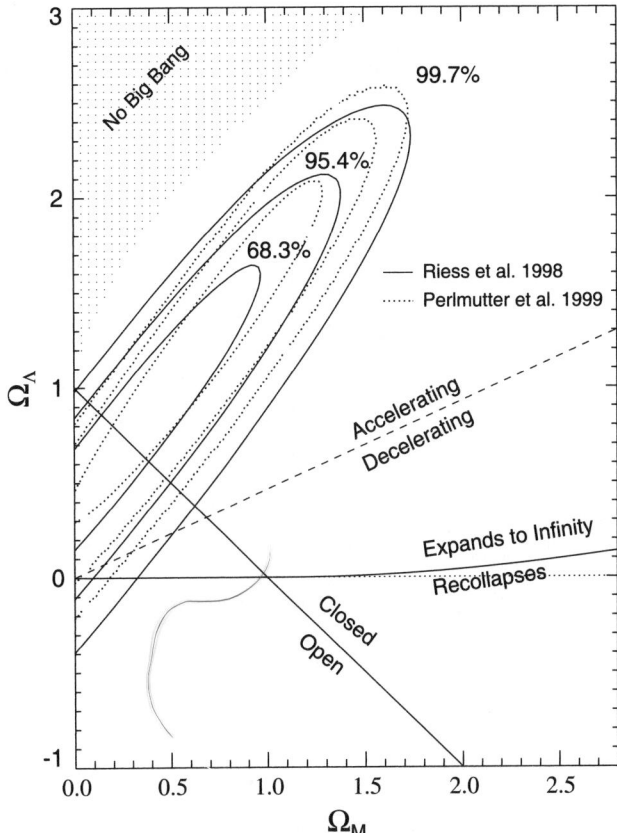

Figure 1.1 The discovery of cosmic acceleration by Riess *et al.* (1998) and Perlmutter *et al.* (1999) (Riess, 2000).

redshift z, compared to several theoretical models. It is difficult to see which cosmological model is favored. Figure 1.3 shows Hubble diagrams of flux-averaged data from Wang (2000a) using SNe Ia from both teams (Riess *et al.*, 1998; Perlmutter *et al.*, 1999). Flux averaging removes the weak gravitational lensing systematic effect of demagnification or magnification of SNe Ia due to the distribution of matter in the universe, since the total number of photons is unchanged by gravitational lensing (Wang, 2000a; Wang and Mukherjee, 2004). Flux averaging also makes data more transparent. It is interesting to note that flux averaging leads to slightly larger error ellipses that are shifted toward smaller Ω_Λ, making the first evidence for cosmic acceleration less strong than that suggested by Figure 1.1 (Wang, 2000a).

Another thing to note is that there were systematic differences in the data from the two teams, although both teams found cosmic acceleration using their own data and that of low-redshift measurements by Hamuy *et al.* (1996). Wang (2000a) compared the data of 18 SNe Ia published by both teams, see Figure 1.4. The error bars are the combined errors in the apparent B magnitude m_B^{eff} measured by Perlmutter

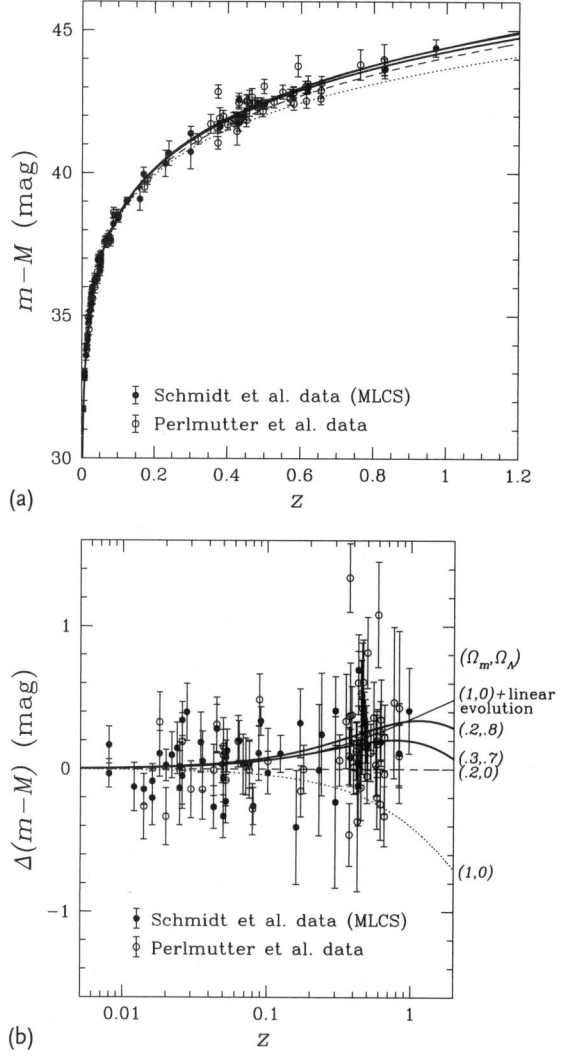

Figure 1.2 Supernova data from two independent teams, Riess *et al.* (1998), and Perlmutter *et al.* (1999) (from Wang, 2000a). Panel (b) is the same as panel (a), but with an open universe model ($\Omega_m = 0.2$, $\Omega_\Lambda = 0$) subtracted.

et al. (1999), and the distance modulus μ_0^{MLCS} measured by Riess *et al.* (1998). The difference of m_B^{eff} and μ_0^{MLCS} should be a constant (the SN Ia peak absolute magnitude) with zero scatter, if there were no differences in analysis techniques, and no internal dispersion in the SN Ia peak brightness. Wang (2000a) found a mean SN Ia peak absolute magnitude of $M_B = -19.33 \pm 0.25$. This scatter of 0.25 mag can be accounted for by the internal dispersion of each data set of about 0.20 mag in the

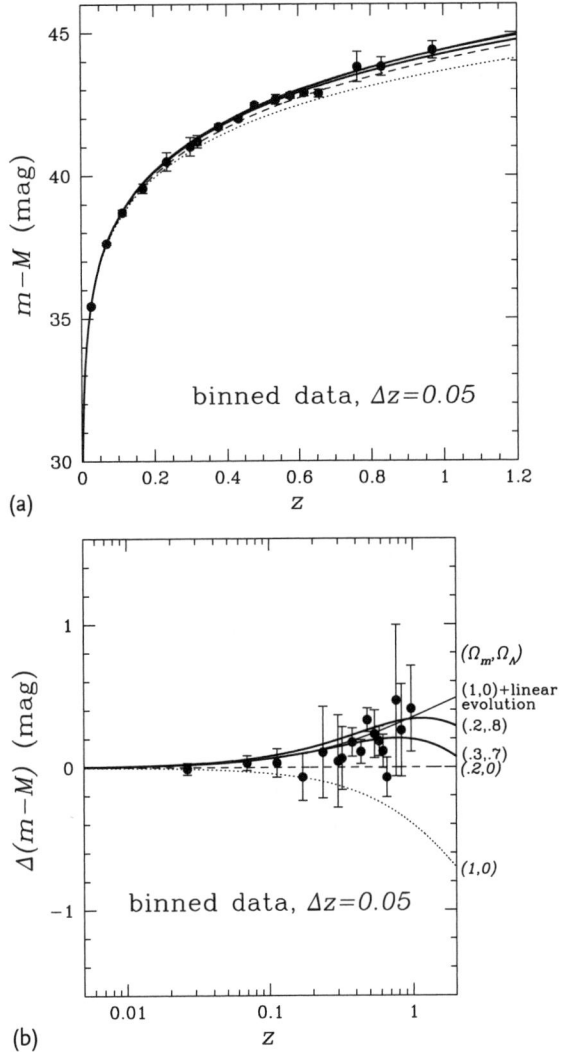

Figure 1.3 The data from Figure 1.2 flux-averaged (Wang, 2000a). Panel (b) is the same as panel (a), but with an open universe model ($\Omega_m = 0.2$, $\Omega_\Lambda = 0$) subtracted.

calibrated SN Ia peak absolute magnitudes, and an additional uncertainty of about 0.15 mag that is introduced by the difference in analysis techniques (Wang, 2000a).

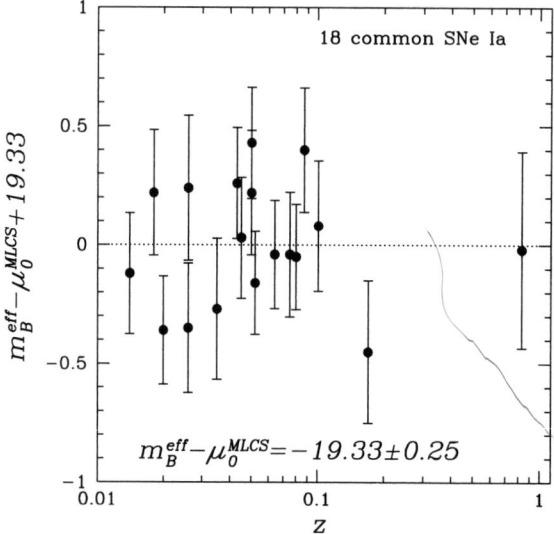

Figure 1.4 The difference between the apparent B magnitude m_B^{eff} measured by Perlmutter et al. (1999), and the distance modulus μ_0^{MLCS} measured by Riess et al. (1998) for the same 18 SNe Ia (from Wang, 2000a). The error bars are the combined errors in m_B^{eff} and μ_0^{MLCS}.

1.1.3
Current Observational Evidence for Cosmic Acceleration

The direct evidence for cosmic acceleration has strengthened over time, as a result of the observations of more SNe Ia, and an improvement in the analysis technique. The analysis by Wang (2000a) demonstrated the importance of analyzing all the SNe Ia using the same analysis technique. This was first done by Riess et al. (2004), who compiled a "gold" set of 157 SNe Ia. Most recently, Kowalski et al. (2008) compiled a "union" set of 307 SNe Ia in 2008, using data from Hamuy et al. (1996); Riess et al. (1998, 1999); Perlmutter et al. (1999); Tonry et al. (2003); Knop (2003); Krisciunas et al. (2004a,b); Barris et al. (2004); Jha et al. (2006); Astier et al. (2006); Riess et al. (2007), and Miknaitis et al. (2007).

Other observational data have provided strong indirect evidence for the existence of dark energy. Cosmic microwave background anisotropy data (CMB) have indicated that the global geometry of the universe is close to being flat, that is, $\Omega_{\text{tot}} \sim 1$ (de Bernardis et al., 2000). The observed abundance of galaxy clusters first revealed that we live in a low matter density universe ($\Omega_m \sim 0.2$–0.3) (Bahcall, Fan, and Cen, 1997). The CMB and galaxy cluster data together require the existence of dark energy that dominates the universe today.

To see the observational evidence for cosmic acceleration without assuming a cosmological constant, it is useful to measure the expansion history of the universe, the Hubble parameter $H(z)$, from data. Figure 1.5 shows the Hubble parameter $H(z)$, as well as $\dot{a}$, measured from a combination of current observation-

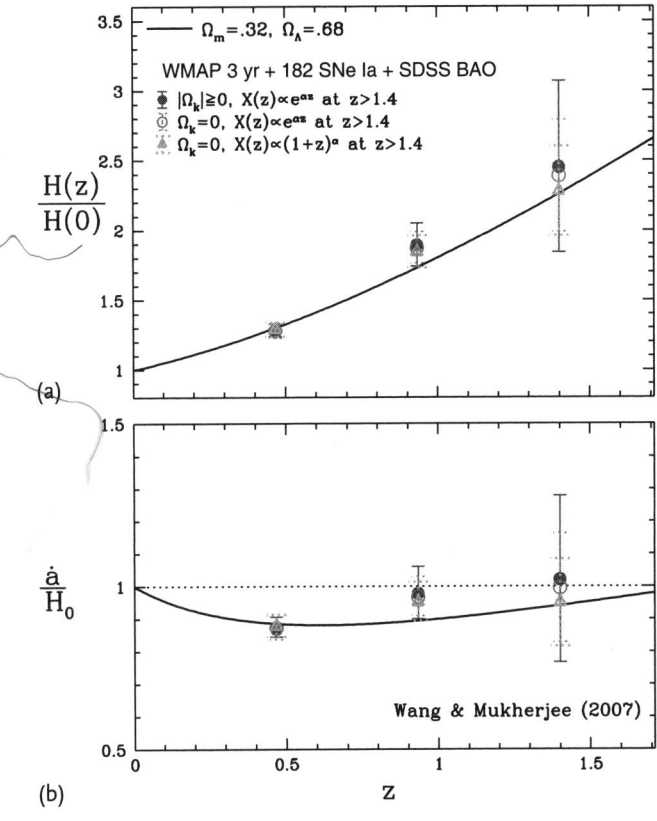

Figure 1.5 Expansion history of the universe measured from current data by Wang and Mukherjee (2007). Data used: CMB data from *WMAP* three-year observations (Spergel *et al.*, 2007); 182 SNe Ia (compiled by Riess *et al.* (2007), including data from the Hubble Space Telescope (HST) obtained by Riess *et al.* (2007), the Supernova Legacy Survey (SNLS) data obtained by Astier *et al.* (2006), as well as nearby SNe Ia); *SDSS* baryon acoustic oscillation measurement (Eisenstein *et al.*, 2005). Note that $X(z) \equiv \rho_X(z)/\rho_X(0)$ in the figure legends, with $\rho_X(z)$ denoting the dark energy density. Panels (a) and (b) use the same data but differ in y axis: $\dot{a} = H(z)a$.

al data by Wang and Mukherjee (2007): CMB data from *WMAP* 3 year observations (Spergel *et al.*, 2007); 182 SNe Ia (compiled by Riess *et al.* (2007), including data from the Hubble Space Telescope (HST) obtained by Riess *et al.* (2007), the Supernova Legacy Survey (SNLS) data obtained by Astier *et al.* (2006), as well as nearby SNe Ia); Sloan Digital Sky Survey (*SDSS*) baryon acoustic oscillation scale measurement (Eisenstein *et al.*, 2005). Clearly, the universe transitioned from cosmic deceleration (matter domination) to cosmic acceleration around $z \sim 0.5$.

The observed cosmic acceleration could be due to an unknown energy component (dark energy, e.g., Quintessence Models references; Linde (1987)), or a modification to general relativity (modified gravity, e.g., Modified Gravity Models references; Dvali, Gabadadze, and Porrati (2000); Freese and Lewis (2002)). The fol-

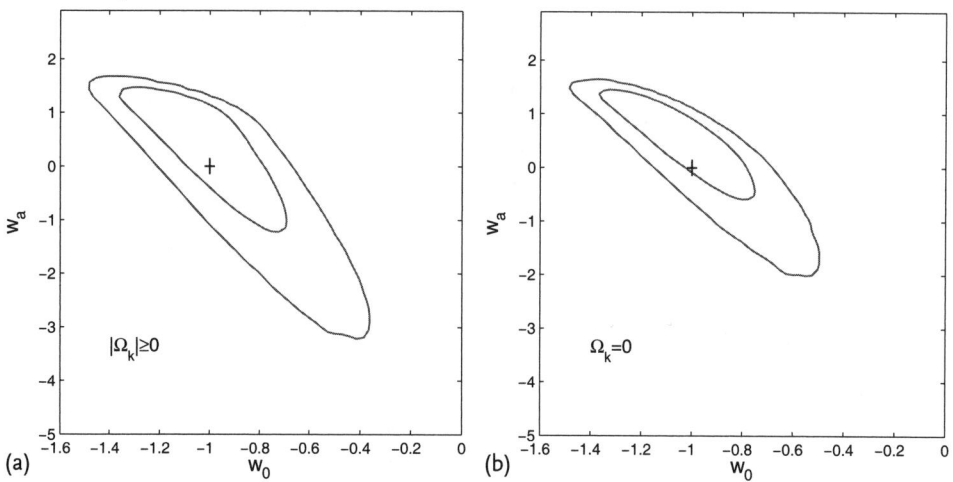

Figure 1.6 Constraints on the dark energy equation of state $w_X(a) = w_0 + w_a(1-a)$ obtained by Wang and Mukherjee (2007), using the same data as in Figure 1.5. A cosmological constant corresponds to $w_X(a) = -1$ (indicated by the cross in the figures). Panel (b) assumes a flat universe; panel (a) does not.

lowing references (Dark Energy Reviews; Copeland, Sami, and Tsujikawa, 2006; Caldwell and Kamionkowski, 2009) contain reviews with more complete lists of references of theoretical models. We discuss some of these models in Chapter 3.

The simplest explanation for the observed cosmic acceleration is that dark energy is a cosmological constant (although it is many orders smaller than expected based on known physics), and that gravity is not modified. Figure 1.6 shows constraints on the dark energy equation of state $w_X(a) = w_0 + w_a(1-a)$ (Chevallier and Polarski, 2001), obtained by Wang and Mukherjee (2007) using the same data as in Figure 1.5. Figures 1.5 and 1.6 show that a cosmological constant is consistent with current observational data, although uncertainties are large. Wang (2008a,c) found that this remains true from an analysis of more recent observational data. For complementary approaches to analyzing current data, see, for example, Wang and Tegmark (2005), and Current Data Results references.

1.2
Fundamental Questions about Cosmic Acceleration

Dark energy projects aim to solve the mystery of cosmic acceleration. In terms of observables, the **two fundamental questions** that need to be answered by dark energy searches are:

1. Is dark energy density constant in cosmic time?
2. Is gravity modified?

These questions can be answered by the precise and accurate measurement of the dark energy density $\rho_X(z)$ as a function of cosmic time (or the expansion history of the universe $H(z)$), and the growth history of cosmic large scale structure $f_g(z)$ from observational data. The answer to these questions will tell us whether cosmic acceleration is caused by dark energy or a modification of gravity, and if gravity is not modified, whether dark energy is a cosmological constant, or due to a dynamical field.

Dark energy is often parameterized by a linear equation of state (Chevallier and Polarski, 2001)

$$w_X(a) = w_0 + w_a(1-a) . \tag{1.17}$$

Because of our ignorance of the nature of dark energy, it is important to make model-independent constraints by measuring the dark energy density $\rho_X(z)$ (or the expansion history $H(z)$) as a free function of cosmic time (Wang and Garnavich, 2001; Tegmark, 2002; Daly and Djorgovski, 2003). Measuring $\rho_X(z)$ has advantages over measuring the dark energy equation of state $w_X(z)$ as a free function; $\rho_X(z)$ is more closely related to observables, hence is more tightly constrained for the same number of redshift bins used (Wang and Garnavich, 2001; Wang and Freese, 2006). Note that $\rho_X(z)$ is related to $w_X(z)$ as follows (Wang and Garnavich, 2001):

$$\frac{\rho_X(z)}{\rho_X(0)} = \exp\left\{ \int_0^z dz' \frac{3[1 + w_X(z')]}{1 + z'} \right\} . \tag{1.18}$$

Hence parametrizing dark energy with $w_X(z)$ implicitly assumes that $\rho_X(z)$ does not change sign in cosmic time. This precludes whole classes of dark energy models in which $\rho_X(z)$ becomes negative in the future ("Big Crunch" models, see Linde (1987); Wang et al. (2004) for an example) (Wang and Tegmark, 2004).

If the present cosmic acceleration is caused by dark energy, then

$$E(z) \equiv \frac{H(z)}{H_0} = \left[\Omega_m(1+z)^3 + \Omega_k(1+z)^2 + \Omega_X X(z)\right]^{1/2} , \tag{1.19}$$

which generalizes Eq. (1.11) by replacing Ω_Λ with $\Omega_X X(z)$, with the dark energy density function $X(z) \equiv \rho_X(z)/\rho_X(0)$. For a cosmological constant, $X(z) = 1$. Once $E(z)$ is specified, the evolution of matter density perturbations on large scales, $\delta_m^{(1)}(\mathbf{x}, t) = D_1(t)\delta_m(\mathbf{x})$, is determined by solving the following equation (assuming that dark energy perturbation $\delta_X = 0$):

$$D_1'' + 2E(z)D_1' - \frac{3}{2}\Omega_m(1+z)^3 D_1 = 0 , \tag{1.20}$$

where $D_1 = \delta_m^{(1)}(\mathbf{x},t)/\delta_m(\mathbf{x})$, and primes denote $d/d(H_0 t)$. Or, more conveniently:

$$a^2 E^2 D_1''(a) + \left[a^2 E \frac{dE}{da} + 3a E^2\right] D_1'(a) - \frac{3}{2}\Omega_m \frac{D_1}{a^3} = 0 , \tag{1.21}$$

where primes denote d/da. The usual initial condition is

$$D_1(a|a \to 0) = a .$$ (1.22)

The linear growth rate is defined as

$$f_g(z) \equiv \frac{d \ln D_1}{d \ln a} .$$ (1.23)

Note that we have assumed that dark energy and dark matter are separate (and that dark energy perturbations are negligible on scales of interest), which is true for the vast majority of dark energy models that have been studied in the literature. If dark energy and dark matter are coupled (a more complicated possibility), or if dark energy and dark matter are unified (unified dark matter models), Eq. (1.20) would need to be modified accordingly. Sandvik *et al.* (2004) found strong evidence for the separation of dark energy and dark matter by ruling out a broad class of so-called unified dark matter models. These models produce oscillations or exponential blowup of the dark matter power spectrum inconsistent with observations.

In the simplest alternatives to dark energy, the present cosmic acceleration is caused by a modification to general relativity. Such models can be tested by observational data, see for example, Modified Gravity (references with more details) for observational signatures of some modified gravity models. The only rigorously worked example is the DGP gravity model (Dvali, Gabadadze, and Porrati, 2000). The validity of the DGP model has been studied by Koyama (2007) and Song, Sawicki, and Hu (2007).

The DGP model can be described by a modified Friedmann equation:

$$H^2 - \frac{H}{r_0} = \frac{8\pi G \rho_m}{3} ,$$ (1.24)

where $\rho_m(z) = \rho_m(0)(1+z)^3$. Solving the above equation gives

$$E(z) = \frac{1}{2} \left\{ \frac{1}{H_0 r_0} + \sqrt{\frac{1}{(H_0 r_0)^2} + 4\Omega_m^0 (1+z)^3} \right\} ,$$ (1.25)

where Ω_m^0 and ρ_c^0 are defined by Eqs. (1.12) and (1.13), respectively. The added superscript "0" in Ω_m^0 denotes that this is the matter density fraction today in the DGP gravity model. Note that consistency at $z = 0$, $H(0) = H_0$ requires that

$$H_0 r_0 = \frac{1}{1 - \Omega_m^0} ,$$ (1.26)

thus the DGP gravity model is parametrized by a single parameter, Ω_m^0.

For DGP gravity, the evolution of matter density perturbations are modified; this is a hallmark of modified gravity models. The linear growth factor in the DGP gravity model is given by Lue, Scoccimarro, and Starkman (2004), and Lue (2006):

$$D_1'' + 2E(z) D_1' - \frac{3}{2} \Omega_m (1+z)^3 D_1 \left(1 + \frac{1}{3\alpha_{DGP}}\right) = 0 ,$$ (1.27)

where primes denote $d/d(H_0 t)$, and

$$\alpha_{\mathrm{DGP}} = 1 - 2Hr_0 \left(1 + \frac{\dot{H}}{3H^2}\right) = \frac{1 - 2H_0 r_0 + 2(H_0 r_0)^2}{1 - 2H_0 r_0} . \tag{1.28}$$

The dark energy model equivalent of the DGP gravity model is specified by requiring

$$\frac{8\pi G \rho_{\mathrm{de}}^{\mathrm{eff}}}{3} = \frac{H}{r_0} . \tag{1.29}$$

Equation (1.24) and the conservation of energy and momentum equation,

$$\dot{\rho}_{\mathrm{de}}^{\mathrm{eff}} + 3\left(\rho_{\mathrm{de}}^{\mathrm{eff}} + p_{\mathrm{de}}^{\mathrm{eff}}\right) H = 0 , \tag{1.30}$$

imply that (Lue, Scoccimarro, and Starkman, 2004; Lue, 2006)

$$w_{\mathrm{de}}^{\mathrm{eff}} = -\frac{1}{1 + \Omega_m(a)} , \tag{1.31}$$

where

$$\Omega_m(a) \equiv \frac{8\pi G \rho_m(z)}{3H^2} = \frac{\Omega_m^0 (1+z)^3}{E^2(z)} . \tag{1.32}$$

Note that

$$\Omega_m(a|a \to 0) = 1 , \quad w_{\mathrm{de}}^{\mathrm{eff}}(a|a \to 0) = -0.5 \tag{1.33}$$

$$\Omega_m(a|a \to 1) = \Omega_m^0 , \quad w_{\mathrm{de}}^{\mathrm{eff}}(a|a \to 1) = -\frac{1}{1 + \Omega_m^0} . \tag{1.34}$$

This means that the matter transfer function (which describes how the evolution of matter density perturbations depends on scale) for the dark energy model equivalent of a viable DGP gravity model ($\Omega_m^0 < 0.3$ and $w \leq -0.5$) is very close to that of the ΛCDM model at $k \gtrsim 0.001\,\mathrm{h\,Mpc^{-1}}$ (Ma et al., 1999).

It is easy and straightforward to integrate Eqs. (1.20) and (1.27) to obtain $D_1(a)$, and thus $f_g(z)$, for dark energy models and DGP gravity models, with the initial condition in Eq. (1.22), that is, $D_1(a|a \to 0) = a$ (which assumes that dark energy or modified gravity is negligible at sufficiently early times).

The measurement of $H(z)$ or $\rho_X(z)$ allows us to determine whether dark energy is a cosmological constant. The measurement of $f_g(z)$ allows us to determine whether gravity is modified. An ambitious SN Ia survey can provide measurement of $H(z)$ to a few percent in accuracy (Wang and Tegmark, 2005). An ambitious galaxy redshift survey can measure both $H(z)$ and $f_g(z)$ to better than a few percent in accuracy (Wang, 2008b). An ambitious weak lensing survey can measure $r(z)$ (which gives $H(z)$ in an integral form) to a few percent in accuracy and the growth factor $G(z)$ [$G(z) \propto D_1(t)$] to several percent in accuracy (Knox, Song, and Tyson,

2006). All these surveys are feasible within the next decade if appropriate resources are made available.

We will discuss each of the major observational methods for probing dark energy (i.e., determining the cause of the observed recent cosmic acceleration) in detail in Chapters 4–8, the key instrumentation for dark energy experiments in Chapter 9, and the future prospects for probing dark energy in Chapter 10.

2
The Basic Theoretical Framework

In order to extract constraints on dark energy from observational data, we need a basic theoretical framework that allows the analysis of data in a nearly model-independent manner. Here we lay out such a framework, with minimal variation from the standard cosmological framework.

The standard cosmological model is given by Einstein's equation, which relates geometry and the matter and energy contents of the universe, and the Robertson–Walker metric, which specifies the geometry of the universe.

In our discussion, the flat space-time metric $\eta_{\alpha\beta}$ is diagonal, with $\eta_{00} = -1$, $\eta_{11} = \eta_{22} = \eta_{33} = +1$. For convenience and simplicity, we omit the speed of light c from equations, except in the expressions for metrics and distances.

2.1
Einstein's Equation

Einstein's equation is

$$G_{\mu\nu} = 8\pi G\, T_{\mu\nu}. \tag{2.1}$$

$G_{\mu\nu}$ is the Einstein tensor which describes the properties of the space-time geometry:

$$G_{\mu\nu} = R_{\mu\nu} - \frac{1}{2} g_{\mu\nu} R, \tag{2.2}$$

with $g_{\mu\nu}$ denoting the metric tensor. The Ricci tensor $R_{\mu\nu}$ and the curvature scalar R are defined by

$$R_{\mu\nu} \equiv g^{\lambda\kappa} R_{\lambda\mu\kappa\nu} \tag{2.3}$$

$$R \equiv g^{\lambda\kappa} g^{\mu\nu} R_{\lambda\mu\kappa\nu} = g^{\mu\nu} R_{\mu\nu}, \tag{2.4}$$

where the inverse metric tensor $g^{\mu\nu}$ is defined by

$$g^{\mu\lambda} g_{\lambda\nu} = \delta^{\mu}_{\nu}, \tag{2.5}$$

Dark Energy. Yun Wang
Copyright © 2010 WILEY-VCH Verlag GmbH & Co. KGaA, Weinheim
ISBN: 978-3-527-40941-9

and the fully covariant curvature tensor is defined by

$$R_{\lambda\mu\kappa\nu} \equiv g_{\lambda\sigma} R^{\sigma}{}_{\mu\kappa\nu}, \tag{2.6}$$

with the Riemann–Christoffel curvature tensor defined as

$$R^{\lambda}{}_{\mu\nu\kappa} \equiv \frac{\partial \Gamma^{\lambda}{}_{\mu\kappa}}{\partial x^{\nu}} - \frac{\partial \Gamma^{\lambda}{}_{\mu\nu}}{\partial x^{\kappa}} + \Gamma^{\eta}{}_{\mu\kappa}\Gamma^{\lambda}{}_{\nu\eta} - \Gamma^{\eta}{}_{\mu\nu}\Gamma^{\lambda}{}_{\kappa\eta}. \tag{2.7}$$

The affine connection $\Gamma^{\sigma}_{\lambda\mu}$ is given by

$$\Gamma^{\sigma}{}_{\lambda\mu} = \frac{1}{2} g^{\nu\sigma} \left(\frac{\partial g_{\mu\nu}}{\partial x^{\lambda}} + \frac{\partial g_{\lambda\nu}}{\partial x^{\mu}} - \frac{\partial g_{\mu\lambda}}{\partial x^{\nu}} \right) \tag{2.8}$$

$T_{\mu\nu}$ is the energy-momentum tensor describing the matter and energy contents of the universe. For a perfect fluid (Weinberg, 1972; Ma and Bertschinger, 1995)

$$T^{\mu}{}_{\nu} = p g^{\mu}{}_{\nu} + (p + \rho) \mathcal{U}^{\mu} \mathcal{U}_{\nu}, \tag{2.9}$$

where $\mathcal{U}^{\mu}$ is the the fluid four-velocity, that is, the local value of $dx^{\nu}/d\tau$ for a comoving fluid element. The pressure p and energy density ρ are defined as being measured by an observer in a locally inertial frame that happens to be moving with the fluid at the instant of measurements, hence p and ρ are scalars.

2.2
Cosmological Background Evolution

The Robertson–Walker metric is the simplest metric that describes a homogeneous, isotropic, and expanding universe:

$$ds^2 = g^{(0)}_{\mu\nu} dx^{\mu} dx^{\nu} = -c^2 dt^2 + a^2(t) \left[\frac{dr^2}{1 - \tilde{k} r^2} + r^2 d\theta^2 + r^2 \sin^2\theta d\phi^2 \right] \tag{2.10}$$

where $a(t)$ is the cosmic scale factor, and $\tilde{k}$ is the curvature constant. The universe is flat for $\tilde{k} = 0$, open for $\tilde{k} < 0$ and closed for $\tilde{k} > 0$.

Given the Robertson–Walker metric, the 0–0 component of Einstein's equation leads to the Friedmann equation:

$$\left(\frac{\dot{a}}{a}\right)^2 + \frac{\tilde{k}}{a^2} = \frac{8\pi G \rho}{3}, \tag{2.11}$$

while the i–i component (together with the 0–0 component) of Einstein's equation leads to

$$\frac{\ddot{a}}{a} = -\frac{4\pi G}{3} (\rho + 3p). \tag{2.12}$$

Combining Eqs. (2.11) and (2.12) gives the conservation of stress energy:

$$d(\rho a^3) = -p\, d(a^3), \qquad (2.13)$$

which can be derived directly from $T^{\mu\nu}{}_{;\nu} = 0$ (see Eq. (2.76)). This implies that for matter ($p = 0$)

$$\rho_m \propto \frac{1}{a^3}, \qquad (2.14)$$

while for radiation ($p = \rho/3$)

$$\rho_r \propto \frac{1}{a^4}. \qquad (2.15)$$

Defining the equation of state

$$w \equiv \frac{p}{\rho}, \qquad (2.16)$$

Eq. (2.13) can be rewritten as

$$\frac{d\ln\rho}{da} = -\frac{3(1+w)}{a}, \qquad (2.17)$$

$$\rho(z) = \exp\left\{\int_0^z dz'\, \frac{3[1+w(z')]}{1+z'}\right\}. \qquad (2.18)$$

Note that for a cosmological constant, $\rho = $ const., and $w = -1$.

The minimal way to include dark energy in this framework is to add a new energy component with density $\rho_X(z)$. We can rewrite the Friedmann equation in terms of the Hubble parameter $H(z)$,

$$H^2(z) \equiv \left(\frac{\dot{a}}{a}\right)^2$$
$$= H_0^2\left[\Omega_m(1+z)^3 + \Omega_r(1+z)^4 + \Omega_k(1+z)^2 + \Omega_X X(z)\right] \qquad (2.19)$$

where the dark energy density function $X(z)$ is defined as

$$X(z) \equiv \frac{\rho_X(z)}{\rho_X(0)}. \qquad (2.20)$$

Requiring the consistency of Eq. (2.19) at $z = 0$, that is, $H(z=0) = H_0$, gives

$$\Omega_m + \Omega_r + \Omega_k + \Omega_X = 1. \qquad (2.21)$$

Note that $\Omega_r \ll \Omega_m$, thus the Ω_r term is usually omitted in dark energy studies, since dark energy is only important at late times.

Equation (2.19) can be easily manipulated to give the cosmic time and redshift relation:

$$t(z) = \int_z^\infty \frac{dz'}{(1+z')H(z')}, \qquad (2.22)$$

which gives the age of an object at redshift z.

Considering the radial, null geodesic of the Robertson–Walker metric

$$ds^2 = -c^2 dt^2 + a^2(t)\frac{dr^2}{1-\tilde{k}r^2} = 0 \qquad (2.23)$$

we find the comoving distance and redshift relation:

$$r(z) = cH_0^{-1}|\Omega_k|^{-1/2}\text{sinn}[|\Omega_k|^{1/2}\Gamma(z)], \qquad (2.24)$$

$$\Gamma(z) = \int_0^z \frac{dz'}{E(z')}, \quad E(z) = H(z)/H_0,$$

where $\text{sinn}(x) = \sin(x)$, x, $\sinh(x)$ for $\Omega_k < 0$, $\Omega_k = 0$, and $\Omega_k > 0$, respectively.

2.3
Cosmological Perturbations

In order to differentiate between dark energy and modified gravity, we need to know the growth rate of matter density perturbations in dark energy models, and compare it with the observed growth rate (see Section 1.2). We only study linear (small) perturbations here, because they are simpler in theory and easier to measure from observations. Note that nonlinear effects can be important both as a systematic effect, and as a source of cosmological information in most observational probes of dark energy.

If dark energy is not a cosmological constant, the rigorous treatment of cosmological perturbations requires a consistent application of general relativity, starting from suitably general forms for the metric and the energy-momentum tensor, and Einstein's equations. However, a nonrelativistic discussion has the benefit of being more physically intuitive, and leads to results that are applicable if dark energy is a cosmological constant. We will begin with the nonrelativistic discussion, and conclude with the general relativistic discussion.

2.3.1
Cosmological Perturbations: Nonrelativistic Case

To study the evolution of perturbations, we consider a universe filled with an ideal fluid, with mass density $\rho(r,t)$ and pressure $p(r,t)$. An element of the fluid has position and velocity

$$x = a(t)r(t) \tag{2.25}$$

$$v = \frac{dx}{dt}. \tag{2.26}$$

The time derivative along the trajectory of the element is

$$\frac{D}{Dt} = \frac{\partial}{\partial t} + \frac{dx}{dt}\frac{\partial}{\partial x} = \frac{\partial}{\partial t} + v \cdot \nabla. \tag{2.27}$$

Perturbations are introduced by

$$\rho(r, t) = \bar{\rho}(t) + \delta\rho(r, t) \tag{2.28}$$

$$p(r, t) = \bar{p}(t) + \delta p(r, t) \tag{2.29}$$

$$v = v_0 + \delta v(r, t) \tag{2.30}$$

$$\Phi_{gr}(r, t) = \Phi_0 + \Phi(r, t), \tag{2.31}$$

with

$$v_0 = H(t)x \tag{2.32}$$

$$\Phi_0 \equiv \frac{2\pi G \bar{\rho}(t)}{3} x^2. \tag{2.33}$$

The peculiar velocity δv and and the peculiar gravitational potential $\Phi(r, t)$ arise due to matter density perturbations $\delta\rho(r, t)$.

There are two equivalent approaches to deriving the equations satisfied by the cosmological perturbations: perturbing Einstein's equations (Peebles, 1993; Ma and Bertschinger, 1995; Bean and Dore, 2003; Weinberg, 2008), or perturbing the Eulerian equations describing a perfect fluid (Kolb and Turner, 1990; Peacock, 1999). The latter approach has the advantage of transparency, but is not fully valid in the general relativistic case. We will use the perturbative Eulerian equations approach for the nonrelativistic case, with $p \ll \rho$, that is, matter density perturbations.

The acceleration of the fluid element is given by the Euler equation:

$$\frac{Dv}{Dt} = -\frac{1}{\rho}\nabla p - \nabla \Phi_{gr}, \tag{2.34}$$

where Φ_{gr} is the gravitational potential which satisfies the Poisson equation:

$$\nabla^2 \Phi_{gr} = 4\pi G \rho. \tag{2.35}$$

Mass conservation leads to the continuity equation:

$$\frac{D\rho(r, t)}{Dt} = -\rho \nabla \cdot v \tag{2.36}$$

Note that the gradient operator

$$\nabla_i \equiv \frac{\partial}{\partial x^i} = \frac{1}{a(t)} \frac{\partial}{\partial r^i} = \frac{1}{a(t)} (\nabla_r)_i \,, \tag{2.37}$$

and that

$$\nabla \cdot \nabla |x|^n = n(n+1)|x|^{n-2} \,. \tag{2.38}$$

Defining the time derivative

$$\frac{d}{dt} \equiv \frac{\partial}{\partial t} + v_0 \cdot \nabla = \left.\frac{\partial}{\partial t}\right|_x + H(t) x \cdot \left.\frac{\partial}{\partial x}\right|_t \,, \tag{2.39}$$

we can write the equations satisfied by the perturbations to first order as

$$\nabla^2 \Phi(x, a) = 4\pi G \bar{\rho}(t) \delta(r, t) \,, \tag{2.40}$$

$$\frac{d\delta}{dt} = -\nabla \cdot \delta v \tag{2.41}$$

$$\frac{d\delta v}{dt} = -\nabla \Phi - \frac{\nabla \delta p}{\bar{\rho}(t)} - (\delta v \cdot \nabla) v_0 \,, \tag{2.42}$$

where

$$(\delta v \cdot \nabla) v_0 = H \delta v \,. \tag{2.43}$$

The matter density perturbation is defined by

$$\delta(x, a) \equiv \frac{\delta \rho}{\bar{\rho}} \,. \tag{2.44}$$

Since $x(t) = a(t) r(t)$, it is convenient to define

$$\delta v(t) = a(t) u(t) \,. \tag{2.45}$$

This leads to

$$\frac{du}{dt} + 2\frac{\dot{a}}{a} u = -\frac{\nabla \Phi}{a} - \frac{\nabla \delta p}{a \bar{\rho}} \tag{2.46}$$

$$\frac{d\delta}{dt} = -a \nabla \cdot u \,, \tag{2.47}$$

To cast the above equations in terms of comoving units, note that

$$df = \left.\frac{\partial f}{\partial t}\right|_r dt + \left.\frac{\partial f}{\partial r}\right|_t dr = \left.\frac{\partial f}{\partial t}\right|_x dt + \left.\frac{\partial f}{\partial x}\right|_t dx$$

$$= \left.\frac{\partial f}{\partial t}\right|_x dt + \left.\frac{\partial f}{\partial x}\right|_t \left(\left.\frac{\partial x}{\partial t}\right|_r dt + \left.\frac{\partial x}{\partial r}\right|_t dr\right), \tag{2.48}$$

thus

$$\left.\frac{\partial f}{\partial x}\right|_t = \frac{1}{a}\left.\frac{\partial f}{\partial r}\right|_t \qquad (2.49)$$

$$\left.\frac{\partial f}{\partial t}\right|_x = \left.\frac{\partial f}{\partial t}\right|_r - \frac{\dot{a}r}{a}\left.\frac{\partial f}{\partial r}\right|_t . \qquad (2.50)$$

Now Eq. (2.39) reduces to

$$\frac{d}{dt} = \left.\frac{\partial}{\partial t}\right|_x + \dot{a}r \cdot \left.\frac{\partial}{\partial x}\right|_t = \left.\frac{\partial}{\partial t}\right|_r .$$

Therefore Eqs. (2.46) and (2.47) reduce to

$$\dot{u} + 2\frac{\dot{a}}{a}u = -\frac{g}{a} - \frac{\nabla_r \delta p}{a^2 \bar{\rho}} \qquad (2.51)$$

$$\dot{\delta} = -\nabla_r \cdot u , \qquad (2.52)$$

where the dot denotes $(\partial/\partial t)|_r$, and the peculiar gravitational acceleration is

$$g \equiv \nabla \Phi = \frac{\nabla_r \Phi}{a} . \qquad (2.53)$$

Combining Eqs. (2.51) and (2.52) with Eq. (2.40), and considering $\delta \propto e^{-i\mathbf{k}\cdot\mathbf{r}}$ (where $\mathbf{k}$ is the comoving wavenumber), gives

$$\ddot{\delta} + 2\frac{\dot{a}}{a}\dot{\delta} = \delta\left(4\pi G \bar{\rho} - \frac{c_s^2 k^2}{a^2}\right), \qquad (2.54)$$

where the sound speed

$$c_s^2 \equiv \frac{\partial p}{\partial \rho} . \qquad (2.55)$$

For a flat universe with $\Omega_m = 1$, $p = 0$ ($w = 0$), $c_s^2 = 0$, and $\bar{\rho}(t) \propto a^{-3}$ (see Eq. (2.14)). Equation (2.11) gives

$$a(t) \propto t^{2/3} . \qquad (2.56)$$

It is easy to verify that Eq. (2.54) has a growing mode solution:

$$\delta_m(t) \propto a(t) \qquad (2.57)$$

Our universe is well-described by a model with $\Omega_m = 0.3$ and $\Omega_\Lambda = 0.7$. It is straightforward to solve Eq. (2.54) to find the factor of linear growth $D_1 \equiv \delta_m(x,t)/\delta_m(x)$, as described in Section 1.2. During the matter-dominated era ($z \lesssim 3000$), matter density perturbations grow more slowly than in an $\Omega_m = 1$ flat universe (for which $D_1(t) \propto a(t)$, see Eq. (2.57)).

During the epoch of radiation domination, the growth of matter density perturbations is suppressed due to radiation pressure. Matter density perturbations can only grow logarithmically during radiation domination (see, e.g., Dodelson (2003)).

2.3.2
Cosmological Perturbations: Generalized Case

Equation (2.10) is the background metric of the universe. It can be rewritten in a simpler form:

$$ds^2 = a^2(\tau)\left[-d\tau^2 + \gamma_{ij}dx^i dx^j\right], \tag{2.58}$$

where γ_{ij} is the three-metric for a space of constant spatial curvature $\tilde{k}$ given in Eq. (2.10), and τ is the *conformal time*, defined by

$$d\tau = \frac{dt}{a(t)}. \tag{2.59}$$

The introduction of metric perturbations depends on the choice of gauge. This gauge dependence arises from our freedom to choose a coordinate system to describe perturbations in general relativity, which in turn is due to the lack of a unique preferred coordinate system defined by the symmetry properties of the background.

Scalar perturbations can be written in the general form of

$$ds^2 = a^2(\tau)\left\{-(1+2\phi)d\tau^2 - 2B_{,i}dx^i d\tau + \left[(1-2\psi)\gamma_{ij} - 2E_{ij}\right]dx^i dx^j\right\}. \tag{2.60}$$

Choosing a gauge (i.e., a coordinate system) can make any two of the four functions ϕ, ψ, B, and E vanish. The simplest *gauge-invariant* linear combinations of these functions are (Mukhanov, Feldman, and Brandenberger, 1992; Mukhanov, 2005)

$$\Phi \equiv \phi - \frac{1}{a}\left[a\left(B - E'\right)\right]', \quad \Psi \equiv \psi + \frac{a'}{a}\left(B - E'\right). \tag{2.61}$$

The metric perturbations are related to the perturbations in the energy-momentum tensor, $T_{\mu\nu}$, through Einstein's equations. For a fluid with a small coordinate velocity $v^i \equiv dx^i/d\tau$, and an anisotropic shear perturbation $\Sigma^i{}_j$, the energy-momentum tensor to linear order in perturbations is given by

$$T^0{}_0 = -(\bar{\rho} + \delta\rho),$$

$$T^0{}_i = (\bar{\rho} + \bar{p})v_i = -T^i{}_0,$$

$$T^i{}_j = (\bar{p} + \delta p)\delta^i_j + \Sigma^i{}_j, \quad \Sigma^i{}_i = 0. \tag{2.62}$$

It is convenient to define

$$\theta \equiv \frac{ik^j \delta T^0{}_j}{\bar{\rho} + \bar{p}} = ik^j v_j, \tag{2.63}$$

$$\sigma \equiv -\frac{\left(\hat{k}_i \cdot \hat{k}_j - \delta_{ij}/3\right) \Sigma^i{}_j}{\bar{\rho} + \bar{p}}, \tag{2.64}$$

$$\hat{k} \equiv \frac{k}{k}.$$

Note that while Eq. (2.62) is valid in any gauge, the perturbations $\delta\rho$, δp, θ, and σ are dependent on the choice of the gauge.

In addition, we define

$$\mathcal{H} \equiv \frac{1}{a}\frac{\partial a}{\partial \tau} = \dot{a} \tag{2.65}$$

$$w \equiv \frac{\bar{p}}{\bar{\rho}} \tag{2.66}$$

$$c_s^2 \equiv \frac{\delta p}{\delta \rho}. \tag{2.67}$$

Substituting the metric from Eq. (2.60) into Einstein's equation, Eq. (2.1), and keeping the linear order perturbations, we can find the *gauge-invariant* perturbation equations (Mukhanov, 2005):

$$\Delta \Psi - 3\mathcal{H}\left(\Psi' + \mathcal{H}\Phi\right) = 4\pi G a^2 \delta T^0{}_0, \tag{2.68}$$

$$\left(\Psi' + \mathcal{H}\Phi\right)_{,i} = 4\pi G a^2 \delta T^0{}_i, \tag{2.69}$$

$$\left[\Psi'' + \mathcal{H}(2\Psi + \Phi)' + (2\mathcal{H}' + \mathcal{H}^2)\Phi + \frac{1}{2}\Delta(\Phi - \Psi)\right]\delta_{ij}$$
$$-\frac{1}{2}(\Phi - \Psi)_{,ij} = -4\pi G a^2 \delta T^i{}_j, \tag{2.70}$$

where the gauge-invariant metric perturbations Φ and Ψ are defined in Eq. (2.61).

Conformal Newtonian Gauge

The conformal Newtonian gauge (or the "longitudinal gauge") is defined by choosing $B = E = 0$ in the Eq. (2.60), with the perturbed Robertson–Walker metric given by

$$ds^2 = a^2(\tau)\left[-(1 + 2\phi)d\tau^2 + (1 - 2\psi)\gamma_{ij}dx^i dx^j\right]. \tag{2.71}$$

Note that the conformal Newtonian gauge is applicable only for scalar mode of the metric perturbations. The advantages of the conformal Newtonian gauge are that the metric tensor is diagonal, and that the metric perturbation ϕ plays the role of the gravitational potential in the Newtonian limit.

The perturbation equations in the conformal Newtonian gauge have the same form as Eqs. (2.68)–(2.70), with $\Phi = \phi$ and $\Psi = \psi$. For a mode with comoving wavenumber k,

$$k^2\psi + 3\mathcal{H}\left(\frac{\partial \psi}{\partial \tau} + \mathcal{H}\phi\right) = -4\pi G a^2 \delta\rho \tag{2.72}$$

$$k^2\left(\frac{\partial \psi}{\partial \tau} + \mathcal{H}\phi\right) = 4\pi G a^2 (\bar{\rho} + \bar{p})\theta \tag{2.73}$$

$$\frac{\partial^2 \psi}{\partial \tau^2} + \mathcal{H}\left(\frac{\partial \phi}{\partial \tau} + 2\frac{\partial \psi}{\partial \tau}\right) + \frac{3}{2}\mathcal{H}^2(1+w)\phi + \frac{k^2}{3}(\psi - \phi) = 4\pi G a^2 \delta p \tag{2.74}$$

$$k^2(\psi - \phi) = 12\pi G a^2 (\bar{\rho} + \bar{p})\sigma, \tag{2.75}$$

where we have used Eq. (2.62).

Conservation of energy-momentum

$$T^{\mu\nu}{}_{;\nu} \equiv \frac{\partial T^{\mu\nu}}{\partial x^\nu} + \Gamma^\mu_{\kappa\nu} T^{\kappa\nu} + \Gamma^\nu_{\kappa\nu} T^{\mu\kappa} = 0 \tag{2.76}$$

yields two equations (for $\mu = 0$ and $\mu = i$) in the conformal Newtonian gauge:

$$\frac{\partial \delta}{\partial \tau} = -(1+w)\left(\theta - 3\frac{\partial \psi}{\partial \tau}\right) - 3\mathcal{H}(c_s^2 - w)\delta \tag{2.77}$$

$$\frac{\partial \theta}{\partial \tau} = -\mathcal{H}(1 - 3c_s^2)\theta + \frac{c_s^2}{1+w}k^2\delta + k^2\phi - k^2\sigma. \tag{2.78}$$

Comparison of Eq. (2.72), Eq. (2.77), and Eq. (2.78) with Eq. (2.40), Eq. (2.46), and Eq. (2.47) show that a consistent treatment of the metric and energy-momentum tensor perturbations lead to additional terms compared to the perturbed Eulerian equations in the nonrelativistic case. It is difficult to manipulate the perturbation equations in the conformal Newtonian gauge to obtain a single perturbation equation for density perturbations. This is more easily done in the synchronous gauge.

Synchronous Gauge

The synchronous gauge corresponds to setting $\phi_{\text{syn}} = 0$ and $B_{\text{syn}} = 0$ in Eq. (2.60). The perturbed Robertson–Walker metric in the synchronous gauge is usually written as (Ma and Bertschinger, 1995)

$$ds^2 = a^2(\tau)\left[-d\tau^2 + (\delta_{ij} + h_{ij})dx^i dx^j\right], \tag{2.79}$$

where we have assumed a flat universe for simplicity. The metric perturbations h_{ij} can be decomposed into a trace part h and a traceless part η:

$$h_{ij}(\mathbf{x}, \tau) = \int d^3k \, e^{i\mathbf{k}\cdot\mathbf{x}} \left[\hat{k}_i \hat{k}_j h(\mathbf{k}, \tau) + \left(\hat{k}_i \hat{k}_j - \frac{1}{3}\delta_{ij}\right) 6\eta(\mathbf{k}, \tau)\right]. \tag{2.80}$$

In the synchronous gauge, the spatial hypersurfaces on which one defines the perturbations are orthogonal to constant-time hypersurfaces, and proper time corresponds to coordinate time. Thus this coordinate system is natural for freely falling observers or CDM particles. The main disadvantage of this gauge is that there are spurious gauge modes contained in the solutions to the equations for the density perturbations, which arise due to the residual gauge freedom (Ma and Bertschinger, 1995).

The perturbation equations in the synchronous gauge follow from Eqs. (2.68)–(2.70) with the substitutions (Mukhanov, 2005)

$$\Phi = \frac{1}{a}\left[a E'_{\text{syn}}\right]', \quad \Psi = \psi_{\text{syn}} - \frac{a'}{a} E'_{\text{syn}}, \tag{2.81}$$

and further relating ψ_{syn} and E_{syn} to h_{ij}. Or, they can be obtained directly by substituting the metric from Eq. (2.79) and the energy-momentum tensor from Eq. (2.62) into Einstein's equation, Eq. (2.1). Keeping the linear order perturbations, we find the perturbation equations for a mode with comoving wavenumber k in the synchronous gauge:

$$k^2 \eta - \frac{1}{2}\mathcal{H}\frac{\partial h}{\partial \tau} = -4\pi G a^2 \delta \rho \tag{2.82}$$

$$k^2 \frac{\partial \eta}{\partial \tau} = 4\pi G a^2 (\bar{\rho} + \bar{p}) \theta \tag{2.83}$$

$$\frac{\partial^2 h}{\partial \tau^2} + 2\mathcal{H}\frac{\partial h}{\partial \tau} - 2k^2 \eta = -24\pi G a^2 \delta p \tag{2.84}$$

$$\frac{\partial^2 h}{\partial \tau^2} + 6\frac{\partial^2 \eta}{\partial \tau^2} + 2\mathcal{H}\left(\frac{\partial h}{\partial \tau} + 6\frac{\partial \eta}{\partial \tau}\right) - 2k^2 \eta = -24\pi G a^2 (\bar{\rho} + \bar{p}) \sigma. \tag{2.85}$$

Conservation of energy-momentum, Eq. (2.76), yields two equations (for $\mu = 0$ and $\mu = i$) in the synchronous gauge:

$$\frac{\partial \delta}{\partial \tau} = -(1+w)\left(\theta + \frac{1}{2}\frac{\partial h}{\partial \tau}\right) - 3\mathcal{H}\left(c_s^2 - w\right)\delta \tag{2.86}$$

$$\frac{\partial \theta}{\partial \tau} = -\mathcal{H}\left(1 - 3c_s^2\right)\theta + \frac{c_s^2}{1+w}k^2\delta - k^2 \sigma. \tag{2.87}$$

Equations (2.82)–(2.85) and Eqs. (2.86)–(2.87) can be manipulated to find

$$\ddot{\delta} + \left[2 + 6\left(c_s^2 - w\right)\right]H\dot{\delta} + \mathcal{F}\delta = -3H(1+w)c_s^2\frac{\theta}{a^2}, \tag{2.88}$$

$$\mathcal{F} \equiv \frac{c_s^2 k^2}{a^2} - \frac{3}{2}H^2(1+3w)\left(1 - w + 2c_s^2\right)$$
$$+ 9H^2\left(c_s^2 - w\right)^2 + 3H\left(\dot{c}_s^2 - \dot{w}\right),$$

where $H^2 = 8\pi G \bar{\rho}/3$, and we have assumed zero shear ($\sigma = 0$) for simplicity. Note that Eq. (2.88) is only valid for a single fluid.

The density perturbations of a fluid in the synchronous gauge are only equal to the comoving density perturbations in the limit of zero fluid velocity (Hu, Spergel, and White, 1997). For a single fluid with $c_S^2 = w$ and $\theta = 0$, we find

$$\ddot{\delta} + 2H\dot{\delta} = \left[4\pi G\bar{\rho}(1+w)(1+3w) - \frac{c_S^2 k^2}{a^2}\right]\delta \,. \tag{2.89}$$

The above equation applies to pure matter ($w = 0$) and pure radiation ($w = 1/3$) fluids.

As a consequence of the synchronous gauge being the natural coordinate system for freely falling observers, the velocity perturbations vanish to lowest order in the synchronous gauge. This means that density perturbations equations derived in the synchronous gauge are in general good approximations to those for the comoving density perturbations.

Conversion from Synchronous Gauge to Conformal Newtonian Gauge

It is most convenient to solve the cosmological perturbation equations in the synchronous gauge. Thus it is useful to consider the conversion from synchronous gauge to conformal Newtonian gauge.

A general coordinate transformation,

$$x^\mu \to \tilde{x}^\mu = x^\mu + d^\mu \tag{2.90}$$

can be written as

$$\tilde{x}^0 = x^0 + \alpha(\mathbf{x}, \tau) \,,$$

$$\tilde{\mathbf{x}} = \mathbf{x} + \nabla\beta(\mathbf{x}, \tau) + \boldsymbol{\epsilon}(\mathbf{x}, \tau) \,, \tag{2.91}$$

where the vector $\mathbf{d}$ has been decomposed into a longitudinal component $\nabla\beta$ ($\nabla \times \nabla\beta = 0$) and a transverse component $\boldsymbol{\epsilon}$ ($\nabla \cdot \boldsymbol{\epsilon} = 0$). The metric tensor transforms as

$$\tilde{g}_{\mu\nu}(\tilde{x}) = \frac{\partial x^\sigma}{\partial \tilde{x}^\mu}\frac{\partial x^\rho}{\partial \tilde{x}^\nu} g_{\sigma\rho} \,. \tag{2.92}$$

Requiring that the line element ds^2 be invariant under the coordinate transformation from synchronous gauge to conformal Newtonian gauge, we find[1]

$$\alpha(\mathbf{x}, \tau) = \frac{\partial \beta(\mathbf{x}, \tau)}{\partial \tau} = \int d^3k\, e^{i\mathbf{k}\cdot\mathbf{x}} \frac{1}{2k^2}\left(\frac{\partial h}{\partial \tau} + 6\frac{\partial \eta}{\partial \tau}\right) \tag{2.93}$$

$$\beta(\mathbf{x}, \tau) = \int d^3k\, e^{i\mathbf{k}\cdot\mathbf{x}} \frac{1}{2k^2}\left[h(\mathbf{k}, \tau) + 6\eta(\mathbf{k}, \tau)\right] \tag{2.94}$$

$$\epsilon_i(\mathbf{x}, \tau) = \epsilon_i(\mathbf{x}) \,, \qquad \partial_i \epsilon_j + \partial_j \epsilon_i = 0 \,, \tag{2.95}$$

---

1) In general, $\alpha(x) = \partial\beta(\mathbf{x},\tau)/\partial\tau + \xi(\tau)$. Since $\xi(\tau)$ is an arbitrary function of time, and reflects the gauge freedom associated with a global redefinition of time with no physical significance, we set $\xi(\tau) = 0$.

and the conformal Newtonian gauge metric perturbations ϕ and ψ are related to the synchronous gauge metric perturbations h and η as follows (Ma and Bertschinger, 1995; Weinberg, 2008):

$$\phi = \frac{1}{2k^2}\left\{\frac{\partial^2 h}{\partial \tau^2} + 6\frac{\partial^2 \eta}{\partial \tau^2} + \mathcal{H}\left[\frac{\partial h}{\partial \tau} + 6\frac{\partial \eta}{\partial \tau}\right]\right\} \tag{2.96}$$

$$\psi = \eta(k,\tau) - \frac{\mathcal{H}}{2k^2}\left[\frac{\partial h}{\partial \tau} + 6\frac{\partial \eta}{\partial \tau}\right]. \tag{2.97}$$

The energy-momentum tensor transforms as

$$T^\mu{}_\nu(\tilde{x}) = \frac{\partial \tilde{x}^\mu}{\partial x^\lambda}\frac{\partial x^\rho}{\partial \tilde{x}^\nu} T^\lambda{}_\rho(x). \tag{2.98}$$

Thus the energy-momentum perturbations (including density perturbations δ) can be converted from the synchronous gauge to the conformal Newtonian gauge as follows (Ma and Bertschinger, 1995; Weinberg, 2008):

$$\delta_{\text{con}} = \delta_{\text{syn}} + \alpha\frac{1}{\bar{\rho}}\frac{\partial \bar{\rho}}{\partial \tau} \tag{2.99}$$

$$\theta_{\text{con}} = \theta_{\text{syn}} + \alpha k^2 \tag{2.100}$$

$$(\delta p)_{\text{con}} = (\delta p)_{\text{syn}} + \alpha\frac{\partial \bar{p}}{\partial \tau} \tag{2.101}$$

$$\sigma_{\text{con}} = \sigma_{\text{syn}}, \tag{2.102}$$

where α is given by Eq. (2.93).

Dark Energy Perturbation

Note that dark energy only affects matter clustering through the background evolution. If dark energy clusters on large scales, then the clustering of matter on those scales is modified. For simplicity, let us only consider matter and dark energy fluctuations:

$$\delta\rho = \delta\rho_{\text{m}} + \delta\rho_{\text{X}}, \quad \delta p = \delta p_{\text{X}}, \tag{2.103}$$

where $\delta\rho_{\text{X}}$ and δp_{X} denote the fluctuations in dark energy density and pressure. We have assumed that the sound speed for matter is zero, that is, $\delta p_{\text{m}} = 0$. We choose the synchronous gauge, in which $\theta_{\text{m}} = 0$ (Ma and Bertschinger, 1995). We parametrize dark energy with an equation of state w_{X} and sound speed $c_{\text{s,X}}^2$ defined as

$$c_{\text{s,X}}^2 \equiv \frac{\delta p_{\text{X}}}{\delta\rho_{\text{X}}}. \tag{2.104}$$

For simplicity, we assume zero dark energy shear, that is, $\sigma_{\text{X}} = 0$.

Most importantly, we will assume that matter and dark energy are *not* coupled. This means that there are no interactions between matter and dark energy. It does *not* mean that the evolution of matter and dark energy perturbations are not coupled.

The conservation of mean energy and momentum gives

$$\frac{\partial \bar{\rho}}{\partial \tau} = -3\mathcal{H}(\bar{p} + \bar{\rho}), \tag{2.105}$$

$$\bar{\rho} = \bar{\rho}_m + \bar{\rho}_X, \quad \bar{p} = \bar{p}_X. \tag{2.106}$$

Assuming that matter and dark energy are separate, we find

$$\frac{\partial \bar{\rho}_m}{\partial \tau} = -3\mathcal{H}\bar{\rho}_m, \quad \bar{\rho}_m \propto \frac{1}{a^3}, \tag{2.107}$$

$$\frac{\partial \bar{\rho}_X}{\partial \tau} = -3\mathcal{H}\bar{\rho}_X (1 + w_X). \tag{2.108}$$

Rewriting the equation for conservation of energy, Eq. (2.86), in terms of $\delta\rho$, and assuming that matter and dark energy are separate, we find

$$\frac{\partial (\delta \rho_m)}{\partial \tau} = -3\mathcal{H}(\delta \rho_m) - \frac{\bar{\rho}_m}{2}\frac{\partial h}{\partial \tau}, \tag{2.109}$$

$$\frac{\partial (\delta \rho_X)}{\partial \tau} = -3\mathcal{H}[(\delta \rho_X) + (\delta p_X)] - \bar{\rho}_X (1 + w_X)\left(\theta_X + \frac{1}{2}\frac{\partial h}{\partial \tau}\right). \tag{2.110}$$

We can use Eqs. (2.82) and (2.84) to eliminate the metric perturbation η, and relate the metric perturbation h to the density and pressure perturbations:

$$\frac{\partial^2 h}{\partial \tau^2} + \mathcal{H}\frac{\partial h}{\partial \tau} = -8\pi G a^2 (\delta\rho + 3\delta p). \tag{2.111}$$

We now define dimensionless matter and dark energy density perturbations:

$$\delta_m \equiv \frac{\delta \bar{\rho}_m}{\bar{\rho}_m}, \quad \delta_X \equiv \frac{\delta \bar{\rho}_X}{\bar{\rho}_X}. \tag{2.112}$$

Equation (2.109) now becomes

$$\frac{\partial \delta_m}{\partial \tau} = -\frac{1}{2}\frac{\partial h}{\partial \tau}. \tag{2.113}$$

Substituting the above equation into Eq. (2.111), we find

$$\frac{\partial^2 \delta_m}{\partial \tau^2} + \mathcal{H}\frac{\partial \delta_m}{\partial \tau} = 4\pi G a^2 \left[\bar{\rho}_m \delta_m + \bar{\rho}_X \delta_X (1 + 3c_{s,X}^2)\right], \tag{2.114}$$

or in the more familiar form

$$\ddot{\delta}_m + 2H\dot{\delta}_m = 4\pi G \left[\bar{\rho}_m \delta_m + \bar{\rho}_X \delta_X (1 + 3c_{s,X}^2)\right]. \tag{2.115}$$

Equations (2.109) and (2.110) gives

$$\frac{\partial \delta_X}{\partial \tau} + 3\mathcal{H}\delta_X \left(c_{s,X}^2 - w_X\right) = (1 + w_X)\left(\frac{\partial \delta_m}{\partial \tau} - \theta_X\right). \tag{2.116}$$

Combining the above equation with Eqs. (2.114) and (2.87) give

$$\frac{\partial^2 \delta_X}{\partial \tau^2} + \mathcal{H}\left[1 + 6\left(c_{s,X}^2 - w_X\right)\right]\frac{\partial \delta_X}{\partial \tau} + \mathcal{F}_X a^2 \delta_X$$
$$= (1 + w_X)\left(4\pi G a^2 \bar{\rho}_m \delta_m - 3\mathcal{H} c_{s,X}^2 \theta_X\right)$$

$$a^2 \mathcal{F}_X \equiv c_{s,X}^2 k^2 - 4\pi G a^2 \bar{\rho}_X (1 + w_X)(1 + 3c_{s,X}^2) + 3\left(c_{s,X}^2 - w_X\right) \cdot \frac{1}{a}\frac{\partial^2 a}{\partial \tau^2}$$
$$+ 9\mathcal{H}^2 \left(c_{s,X}^2 - w_X\right)^2 + 3\mathcal{H}\left(\frac{\partial c_{s,X}^2}{\partial \tau} - \frac{\partial w_X}{\partial \tau}\right),$$

(2.117)

or alternatively

$$\ddot{\delta}_X + H\left[2 + 6\left(c_{s,X}^2 - w_X\right)\right]\dot{\delta}_X + \mathcal{F}_X \delta_X$$
$$= (1 + w_X)\left(4\pi G \bar{\rho}_m \delta_m - \frac{3H c_{s,X}^2 \theta_X}{a^2}\right)$$

$$\mathcal{F}_X \equiv \frac{c_{s,X}^2 k^2}{a^2} - 4\pi G \bar{\rho}_X (1 + w_X)(1 + 3c_{s,X}^2) + 4\pi G \left(\bar{\rho} - 3\bar{p}\right)\left(c_{s,X}^2 - w_X\right)$$
$$+ 9H^2 \left(c_{s,X}^2 - w_X\right)^2 + 3H\frac{\partial}{\partial t}\left(c_{s,X}^2 - w_X\right).$$

(2.118)

Bean and Dore (2003) found that the $c_{s,X}^2 \theta$ term is always subdominant for all scales of interest, and can thus be omitted. This is as expected, since in the synchronous gauge, velocity perturbations vanish to the lowest order.

Late-Time Integrated Sachs–Wolfe (ISW) Effect

Using Eq. (2.96), we can express the gravitational potential ϕ in terms of the synchronous gauge metric perturbations h and η. Using Eqs. (2.85), (2.82), and (2.83), we find

$$\phi = -\frac{4\pi G a^2 \delta\rho}{k^2} - \frac{24\pi G \mathcal{H} a^2 (\bar{\rho} + \bar{p})\theta}{k^4}$$
$$= -\frac{4\pi G a^2 (\bar{\rho}_m \delta_m + \bar{\rho}_X \delta_X)}{k^2} - \frac{24\pi G \mathcal{H} a^2 (\bar{\rho}_X + \bar{p}_X)\theta_X}{k^4}.$$

(2.119)

Note that we have used (Ma and Bertschinger, 1995)

$$(\bar{\rho} + \bar{p})\theta = \sum_i (\bar{\rho}_i + \bar{p}_i)\theta_i.$$

(2.120)

If dark energy is a cosmological constant, or if dark energy is absent, Eq. (2.119) reduces to

$$\phi = -\frac{4\pi G a^2 \bar{\rho}_m \delta_m}{k^2} = -\frac{4\pi G \bar{\rho}_m^0}{k^2} \cdot \frac{\delta_m}{a}.$$

(2.121)

For a flat universe with $\Omega_m = 1$, $\delta_m \propto a$ (see Eq. (2.57)), and the gravitational potential ϕ is constant with time.

In the presence of dark energy, δ_m grows more slowly with cosmic time. This can be easily verified qualitatively using Eq. (2.115) for the case of $w_X = -1$ (a cosmological constant). Thus the presence of dark energy leads to the decay of the gravitational potential ϕ with time. This leads to a late-time integrated Sachs–Wolfe (ISW) effect in the cosmic microwave background (CMB) anisotropies when the universe evolves from the matter-dominated era to a dark energy-dominated era.

Metric Perturbations and Modification to Gravity

The gauge-invariant metric perturbations are equal, $\Phi = \Psi$, when the spatial part of the energy-momentum tensor is diagonal (see Eq. (2.70)). In the Newtonian gauge, this means $\phi = \psi$. In general, this need not be true if gravity is modified. The potential ψ is related to the peculiar acceleration $\mathbf{g}$ as follows:

$$\mathbf{g} = \nabla \psi \,. \tag{2.122}$$

Gravity can be modified by changing how the potential ϕ is related to matter density (assuming $\theta = 0$):

$$\nabla^2 \phi(\mathbf{x}, a) = 4\pi G \bar{\rho} Q(\mathbf{x}, a) \delta(\mathbf{x}, a) \,, \tag{2.123}$$

and by introducing an intrinsic shear stress:

$$\psi = \left[1 + S(\mathbf{x}, a)\right] \phi \tag{2.124}$$

and thus

$$\mathbf{g} = \nabla \left[(1 + S)\phi\right]. \tag{2.125}$$

As a concrete example, for the DGP gravity model, the evolution of matter density perturbations are modified, with the function Q and S in Eqs. (2.123) and (2.124) given by (Lue, Scoccimarro, and Starkman, 2004; Lue, 2006)

$$Q(a) = 1 - \frac{1}{3 a_{\text{DGP}}} \tag{2.126}$$

$$S(a) = \frac{2}{3 a_{\text{DGP}} - 1}, \tag{2.127}$$

where

$$a_{\text{DGP}} = 1 - 2 H r_0 \left(1 + \frac{\dot{H}}{3 H^2}\right) = \frac{1 - 2 H_0 r_0 + 2(H_0 r_0)^2}{1 - 2 H_0 r_0}. \tag{2.128}$$

We do not discuss matter density perturbations in modified gravity models in more detail in this book, since there are no compelling models of modified gravity at present.

A robust approach is to measure the growth rate of cosmic large scale structure $f_g(z)$ from observational data, and compare it with the prediction of dark energy models. If we detect a statistically significant deviation in this comparison, then we know that gravity is modified, and the measured deviation itself can provide guidance to building the correct modified gravity model. Henceforth, we will focus on the search for dark energy, which will also enable us to test gravity.

2.4
Framework for Interpreting Data

2.4.1
Model-Independent Constraints

The use of Type Ia supernovae (SNe Ia), galaxy redshift surveys, weak lensing, and galaxy clusters in probing dark energy are recognized by the community as the most promising methods for the dark energy search. Each of these methods will be discussed in detail in this book. Cosmic microwave background anisotropy (CMB) data and independent measurements of H_0 are required to break the degeneracy between dark energy and cosmological parameters (see e.g. Wang and Mukherjee (2007)), and hence are also important in the search.

As described in the previous section, there are two fundamental questions that need to be answered by the search for dark energy: (1) Is dark energy density constant in cosmic time? (2) Is gravity modified? These questions can be answered by the precise and accurate measurement of the dark energy density $\rho_X(z)$ as a function of cosmic time (or the expansion history of the universe, $H(z)$, see Eq. (1.19)), and the growth history of cosmic large scale structure, $f_g(z)$ (see Eqs. (1.20) and (1.23)), from observational data.

The measurement of $H(z)$ or $\rho_X(z)$ allows us to determine whether dark energy is a cosmological constant. The measurement of $f_g(z)$ allows us to determine whether gravity is modified.

Different observational methods probe $r(z)$ (which is related to $H(z)$ in an integral form), $H(z)$, and $f_g(z)$ in different ways. SNe Ia probe the luminosity distance, $d_L(z) = (1+z)r(z)$. The baryon acoustic oscillations (BAO) measured from a galaxy redshift survey probe the angular diameter distance, $D_A(z) = r(z)/(1+z)$, and $H(z)$. The redshift distortions measured from a galaxy redshift survey probe $f_g(z)$. A weak lensing survey probes a combination of $H(z)$ and $G(z)$ in an integral form (note that $G(z) \propto \delta_m(z)$, and $f_g(z) = d\ln G(z)/d\ln a$). Clearly, multiple probes allow a number of cross-checks to be made to ensure the robustness of the dark energy constraints.

In particular, one can convert each set of data into a measurement of $r(z)$, or $H(z)$, in order to compare all the different methods on the same footing in a model-independent manner, and to search for systematic effects.

Since the early work by Wang and Garnavich (2001) advocating model-independent constraints on dark energy, much progress has been made in that direction.

Sahni and Starobinsky (2006) reviewed the large body of work by multiple groups on the model-independent reconstruction of dark energy properties.

2.4.2
Using the Fisher Matrix to Forecast Future Constraints

The simplest and widely used method to forecast constraints from future observations is to use the Fisher matrix formalism. The Fisher information matrix of a given set of parameters, s, approximately quantifies the amount of information on s that we "expect" to get from our future data. The Fisher matrix can be written as

$$F_{ij} = -\frac{\partial^2 \ln L}{\partial s_i \partial s_j}, \tag{2.129}$$

where L is the likelihood function, the expected probability distribution of the observables given parameters s.

The Cramér-Rao inequality (Kendall and Stuart, 1969) states that no unbiased method can measure the i-th parameter with standard deviation less than $1/\sqrt{F_{ii}}$ if other parameters are known, and less than $\sqrt{(F^{-1})_{ii}}$ if other parameters are estimated from the data as well. Note that the derivatives in Eq. (2.129) are calculated assuming that the cosmological parameters are given by an a priori model, and thus the errors on the parameters are somewhat dependent on the assumed model. It is straightforward to apply Eq. (2.129). For Gaussian distributed measurements, $L \propto \exp(-\chi^2/2)$.

2.4.3
Using the Markov Chain Monte Carlo Method in a Likelihood Analysis

Note that while a Fisher matrix analysis can estimate the most optimistic errors for a given experiment, it cannot determine the most likely values of the estimated parameters. These require a likelihood analysis, which must be performed in analyzing real or realistically simulated data. In such a likelihood analysis, we need to find the high confidence region of our parameter space given the data. Traditionally, this is done by using a grid of parameter values that span the allowed parameter space.

The Markov Chain Monte Carlo (MCMC) method uses the Metropolis algorithm:

1. Choose a candidate set of parameters, s_*, at random from a proposal distribution.
2. Accept the candidate set of parameters with probability $A(s, s_*)$; otherwise, reject it.

For Gaussian distributed observables, the acceptance function

$$A(s, s_*) = \min\left\{1, \exp\left[-\chi^2(s_*) + \chi^2(s)\right]\right\}. \tag{2.130}$$

For sufficient sampling, the true probability density functions (pdf) are recovered. See Neil (1993) for a review of MCMC.

MCMC is much faster than the grid method, and scales linearly with the number of parameters. It gives a smooth pdf for estimated parameters since they receive a contribution from all MCMC samples (each with a random set of parameter values). Public software containing MCMC is available (see Lewis and Bridle (2002)).

Wang et al. (2004) showed the difference between the pdf derived from MCMC, and that from a Fisher matrix approximation. Even when setting the inverse of the Fisher matrix to the covariance matrix derived from MCMC, there are significant differences. MCMC gives the actual, usually asymmetric, pdf of the model parameters given the data, while the Fisher matrix approach by definition gives symmetric error bars. This provides some insight on the limit of the Fisher matrix error forecasts of future projects. For future projects that are sufficiently known in detail, an MCMC simulation can provide a robust and realistic forecast of its performance in constraining models.

2.4.4
Self-Consistent Inclusion of Cosmic Microwave Background Anisotropy Data

Observational bounds on dark energy depend on our assumptions about the curvature of the universe. Cosmic microwave background (CMB) anisotropy data provide the most important ingredient in constraining cosmic curvature.

CMB data give us the comoving distance to the photon-decoupling surface $r(z_*)$, and the comoving sound horizon at the photo-decoupling epoch (Eisenstein and Hu, 1998; Page et al., 2003):

$$r_s(z_*) = \int_0^{t_*} \frac{c_s dt}{a} = c H_0^{-1} \int_{z_*}^{\infty} dz \frac{c_s}{E(z)} ,$$

$$= c H_0^{-1} \int_0^{a_*} \frac{da}{\sqrt{3(1 + \overline{R}_b a) a^4 E^2(z)}} , \qquad (2.131)$$

where a is the cosmic scale factor, $a_* = 1/(1 + z_*)$, and $a^4 E^2(z) = \Omega_m(a + a_{eq}) + \Omega_k a^2 + \Omega_X X(z) a^4$, with $a_{eq} = \Omega_{rad}/\Omega_m = 1/(1 + z_{eq})$, and $z_{eq} = 2.5 \times 10^4 \Omega_m h^2 (T_{CMB}/2.7 \text{ K})^{-4}$. The sound speed is $c_s = 1/\sqrt{3(1 + \overline{R}_b a)}$, with $\overline{R}_b a = 3\rho_b/(4\rho_\gamma)$, $\overline{R}_b = 31\,500\, \Omega_b h^2 (T_{CMB}/2.7 \text{ K})^{-4}$. Four-year COBE data give $T_{CMB} = 2.728 \pm 0.004$ K (95% C.L.) (Fixsen, 1996). However, when using the CMB bounds derived by Komatsu et al. (2009), we should take $T_{CMB} = 2.725$.

Wang and Mukherjee (2007) showed that the CMB shift parameters

$$R \equiv \sqrt{\Omega_m H_0^2} r(z_*) , \quad l_a \equiv \pi r(z_*)/r_s(z_*) , \qquad (2.132)$$

together with $\Omega_b h^2$, provide an efficient summary of CMB data as far as dark energy constraints go. Note that R is the scaled distance to the photon-decoupling sur-

Table 2.1 Inverse covariance matrix for the extended WMAP distance priors.

	$l_A(z_*)$	$R(z_*)$	z_*	$100\Omega_b h^2$
$l_a(z_*)$	31.001	−5015.642	183.903	2337.977
$R(z_*)$		876 807.166	−32 046.750	−403 818.837
z_*			1175.054	14 812.579
$100\Omega_b h^2$				187 191.186

Table 2.2 Covariance matrix for (R, l_a, Ω_b).

	$R(z_*)$	$l_a(z_*)$	$\Omega_b h^2$
$R(z_*)$	3.67435E−04	1.81817E−03	−2.02061E−06
$l_a(z_*)$		0.73178	−3.16089E−04
$\Omega_b h^2$			3.55483E−07

face, while l_a is the angular scale of the sound horizon at photon-decoupling (Page et al., 2003).

We can use the covariance matrix of $[R(z_*), l_a(z_*), \Omega_b h^2]$ from the five-year WMAP data, with z_* given by fitting formulae from Hu and Sugiyama (1996):

$$z_* = 1048 \left[1 + 0.00124(\Omega_b h^2)^{-0.738}\right] \left[1 + g_1(\Omega_m h^2)^{g_2}\right], \tag{2.133}$$

where

$$g_1 = \frac{0.0783(\Omega_b h^2)^{-0.238}}{1 + 39.5(\Omega_b h^2)^{0.763}} \tag{2.134}$$

$$g_2 = \frac{0.560}{1 + 21.1(\Omega_b h^2)^{1.81}}. \tag{2.135}$$

CMB data are included in a combined analysis by adding the following term to the χ^2 of a given model with $p_1 = R(z_*)$, $p_2 = l_a(z_*)$, and $p_3 = \Omega_b h^2$:

$$\chi^2_{\text{CMB}} = \Delta p_i \left[\text{Cov}^{-1}(p_i, p_j)\right] \Delta p_j, \quad \Delta p_i = p_i - p_i^{\text{data}}, \tag{2.136}$$

where p_i^{data} are the maximum likelihood values.

Table 2.1 (from Komatsu et al. (2009)) gives the inverse covariance matrix for $(l_a(z_*), R(z_*), z_*, 100\Omega_b h^2)$. Wang (2008a) showed that using $(l_a(z_*), R(z_*), z_*)$ is essentially identical to using $(R, l_a, \Omega_b h^2)$. The five-year WMAP maximum likelihood values of $(l_a(z_*), R(z_*), z_*, 100\,\Omega_b h^2)$ are: $l_a(z_*) = 302.10$, $R(z_*) = 1.710$, $z_* = 1090.04$, and $100\,\Omega_b h^2 = 2.2765$. Table 2.2 gives the covariance matrix of $(R, l_a, \Omega_b h^2)$.

3
Models to Explain Cosmic Acceleration

Modeling the observed cosmic acceleration is one of the most active areas of research in theoretical cosmology today. The references found in Dark Energy Reviews and Copeland, Sami, and Tsujikawa (2006) contain in-depth reviews of the current status of cosmic acceleration models. A useful way to categorize the models is according to how they modify Einstein's equation, which relates geometry and the matter and energy contents of the universe. Models that modify the right hand side of Einstein's equation, Eq. (2.1), are referred to as "dark energy models". Models that modify the left hand side of Einstein's equation are referred to as "modified gravity models".

Note that in principle, dark energy and modified gravity models can give any cosmic expansion history $H(t)$ by design. For example, Woodard (2005) showed how one can obtain the potential $V(\phi)$ of the dark energy scalar field, or the form of the modified gravity model $f(R)$ in terms of any $H(t)$. This means that the measurement of *both* the cosmic expansion history and the growth rate of large scale structure are needed to differentiate the various models (see Section 1.2).

In this chapter we briefly discuss some cosmic acceleration models. For a basic discussion of density perturbations in dark energy models, see Section 2.3.2. For a detailed discussion of density perturbations in both dark energy and modified gravity models, the reader is referred to Copeland, Sami, and Tsujikawa (2006), Boehmer *et al.* (2008), and Caldera-Cabral, Maartens, and Urena-Lopez (2008).

3.1
Dark Energy Models

Dark energy models do not modify geometry. Thus Eqs. (2.10), (2.11), (2.12), and (2.17) are valid. We introduce a few popular models in Section 3.1.1, and present two worked examples in Sections 3.1.2–3.1.3.

3.1.1
Quintessence, Phantom Field, and Chaplygin Gas

Quintessence

The leading class of new energy component models is quintessence (Quintessence Models references). These models introduce a scalar field ϕ that is minimally coupled to gravity, essentially the same class of models that have been studied to obtain cosmic inflation (required to solve the horizon problem, flatness problem, and the structure formation problem of the old "standard cosmology" model). Quintessence is given by the action

$$S = \int d^4x \sqrt{-g} \left[-\frac{1}{2}(\nabla\phi)^2 - V(\phi) \right], \tag{3.1}$$

where $(\nabla\phi)^2 = g^{\mu\nu}\partial_\mu\phi\partial_\nu\phi$, and $V(\phi)$ is the potential of the field. The energy momentum tensor for quintessence is

$$\begin{aligned} T_{\mu\nu} &= -\frac{2}{\sqrt{-g}} \frac{\delta S}{\delta g^{\mu\nu}} \\ &= \partial_\mu\phi\partial_\nu\phi - g_{\mu\nu}\left[\frac{1}{2}g^{\alpha\beta}\partial_\alpha\phi\partial_\beta\phi + V(\phi)\right]. \end{aligned} \tag{3.2}$$

For a flat universe described by the Robertson–Walker metric, the energy density and pressure of the scalar field are given by

$$\rho = -T_0^0 = \frac{1}{2}\dot\phi^2 + V(\phi) \tag{3.3}$$

$$p = \sum_i T^i{}_i = \frac{1}{2}\dot\phi^2 - V(\phi). \tag{3.4}$$

Thus for quintessence models,

$$H^2 = \frac{8\pi G}{3}\left[\frac{1}{2}\dot\phi^2 + V(\phi)\right], \tag{3.5}$$

$$\frac{\ddot a}{a} = -\frac{8\pi G}{3}[\dot\phi^2 - V(\phi)]. \tag{3.6}$$

Equation (3.6) shows that cosmic acceleration occurs for $\dot\phi^2 < V(\phi)$, that is, if the potential is flat enough so that the scalar field rolls very slowly (same as the requirement for the inflaton potential).

The equation of state for the scalar field ϕ is

$$w_\phi = \frac{p}{\rho} = \frac{\dot\phi^2 - 2V(\phi)}{\dot\phi^2 + 2V(\phi)}, \tag{3.7}$$

which satisfies

$$-1 \leq w_\phi \leq 1. \tag{3.8}$$

The conservation of energy-momentum, Eq. (2.18), implies that

$$1 \le \frac{\rho_\phi(z)}{\rho_\phi(0)} \le (1+z)^6 . \tag{3.9}$$

Note that if we require a power-law cosmic expansion

$$a(t) \propto t^\alpha , \tag{3.10}$$

where $\alpha > 1$ to ensure accelerated cosmic expansion, Eqs. (3.5) and (3.6) give

$$\frac{\ddot{a}}{a} - H^2 = \dot{H} = -\frac{8\pi G \dot{\phi}^2}{3} . \tag{3.11}$$

Thus we find

$$\phi = \left(\frac{3\alpha}{8\pi G}\right)^{1/2} \ln t + \text{const} . \tag{3.12}$$

and

$$V(\phi) = \frac{3}{8\pi G}\left(H^2 + \frac{\dot{H}}{2}\right) \tag{3.13}$$

$$= \frac{3}{8\pi G} \cdot \frac{\alpha(\alpha - 1/2)}{t^2}$$

$$= V_0 \exp\left[-4\left(\frac{2\pi}{3\alpha}\right)^{1/2} \frac{\phi}{m_{Pl}}\right], \tag{3.14}$$

where we have defined the Planck mass

$$m_{Pl} \equiv \frac{1}{\sqrt{G}} . \tag{3.15}$$

In addition to giving rise to accelerated cosmic expansion, exponential potentials lead to cosmological scaling solutions, in which $\rho_\phi \propto \rho_m$ (Copeland, Liddle, and Wands, 1998).

Phantom Field

Since current observational data indicate that $w_X < -1$ is allowed at 68% confidence level, a new class of scalar field models, known as the "phantom field" models, have been introduced. The phantom field can have $w_X < -1$ since they have a negative kinetic energy (Caldwell, 2002), which could be motivated by S-brane construction in string theory (Phantom Models references). For phantom fields,

$$S = \int d^4x \sqrt{-g}\left[\frac{1}{2}(\nabla\phi)^2 - V(\phi)\right], \tag{3.16}$$

and

$$w_\phi = \frac{p}{\rho} = \frac{\dot{\phi}^2 + 2V(\phi)}{\dot{\phi}^2 - 2V(\phi)}, \tag{3.17}$$

thus $w_\phi < -1$ for $\dot{\phi}^2 < 2V(\phi)$.

Phantom fields are generally plagued by ultraviolet quantum instabilities. Because the energy density of a phantom field is unbounded from below, the vaccum becomes unstable against the production of ghosts and positive energy fields (Carroll, Hoffman, and Trodden, 2003). If phantom models are only low-energy effective theories with an upper limit (cutoff) in energy, they are still strongly constrained by existing observational data. Assuming that the phantom field interacts only gravitationally, the cutoff energy has to be smaller than 3 MeV for consistency with the cosmic gamma ray background (Cline, Jeon, and Moore, 2004).

Chaplygin Gas

Another popular class of dark energy models is the Chaplygin gas (Kamenshchik, Moschella, and Pasquier, 2001), which is a fluid that can lead to cosmic acceleration at late times. In its simplest form, the Chaplygin gas has the equation of state

$$p = -\frac{A}{\rho}, \tag{3.18}$$

where A is a positive constant. This equation of state was first introduced by Chaplygin (1904) to study the lifting force on a plane wing in aerodynamics. Chaplygin gas can be motivated by string theory (Bordemann and Hoppe, 1993) and supersymmetry (Hoppe, 1993; Jackiw and Polychronakos, 2000).

The generalized Chaplygin gas has an equation of state

$$p = -\frac{A}{\rho^\alpha}. \tag{3.19}$$

Equations (3.19) and (2.17) give

$$\rho = \left[A + \frac{B}{a^{3(1+\alpha)}} \right]^{1/(1+\alpha)}, \tag{3.20}$$

where B is a constant. The equation of state of generalized Chaplygin gas is

$$w = \frac{p}{\rho} = -\frac{A}{\rho^{\alpha+1}} = -\frac{1}{1 + (B/A)(1+z)^{3(1+\alpha)}}. \tag{3.21}$$

The generalized Chaplygin gas behaves as matter at early times:

$$\rho \propto a^{-3}, \quad w \simeq 0, \quad \text{for} \quad a \ll 1 \tag{3.22}$$

and as a cosmological constant at late times:

$$\rho \simeq \text{const.}, \quad w \simeq -1, \quad \text{for} \quad a \gg \left(\frac{B}{A}\right)^{1/[3(1+\alpha)]}. \tag{3.23}$$

The original Chaplygin gas model corresponds to $\alpha = 1$. A cosmological constant corresponds to $\alpha = 0$ and $B = 0$.

Chaplygin gas provides an interesting possibility for the unification of dark energy and dark matter. However, the Chaplygin gas models are severely constrained by current observational data. Sandvik *et al.* (2004) showed that Chaplygin gas leads to the exponential blowup of the matter power spectrum in the absence of CDM, thus ruling out Chaplygin gas as a CDM substitute. Bean and Dore (2003) and Barreiro, Bertolami, and Torres (2008) showed that current observational data favor the generalized Chaplygin gas that is similar to a cosmological constant.

3.1.2
Worked Example: PNGB Quintessence

One of the best motivated quintessence models is the PNGB quintessence, the pseudo-Nambu–Goldstone-boson quintessence model of dark energy (Kaloper and Sorbo, 2006). The potential of the PNGB field can be well-approximated by

$$V(\phi) = M^4 \left[\cos\left(\frac{\phi}{f}\right) + 1 \right]. \tag{3.24}$$

In order for the theory to be stable and finite, the decay constant f is required to be smaller than the Planck mass m_{Pl} (Kaloper and Sorbo, 2006; Abrahamse *et al.*, 2008), defined as

$$m_{\text{Pl}} \equiv \frac{1}{\sqrt{8\pi G}}. \tag{3.25}$$

The equations of motion for the PNGB field are

$$\ddot{\phi} + 3H(a)\dot{\phi} + \frac{\partial V}{\partial \phi} = 0, \tag{3.26}$$

$$H^2 = \frac{1}{3m_{\text{Pl}}^2} \left(\frac{\rho_r^0}{a^4} + \frac{\rho_m^0}{a^3} + \rho_\phi + \frac{\rho_k^0}{a^2} \right). \tag{3.27}$$

The energy and pressure of the PNGB field are

$$\rho_\phi = \frac{1}{2}\dot{\phi}^2 + V(\phi), \quad p_\phi = \frac{1}{2}\dot{\phi}^2 - V(\phi). \tag{3.28}$$

The initial value of the field, ϕ_I, can be taken to be between 0 and πf, since the potential is periodic and symmetric. The initial value of $\dot{\phi}$ can be taken to be zero, since nonzero values of $\dot{\phi}$ are expected to have been quickly damped by the rapid expansion of the early universe (Abrahamse *et al.*, 2008).

Abrahamse *et al.* (2008) showed that a suitable choice of the parameter set in the PNGB model is: $\{V_I, \phi_I/f, f\}$. V_I is the initial value of the potential, $V_I \equiv V(\phi_I)$. V_I and f can be made dimensionless by using m_{Pl}^4 and m_{Pl}, respectively, as the unit.

Once the initial conditions are specified, the equations of motion can be solved to find the expansion history $H(z)$, and the dark energy equation of state

$$w_\phi \equiv \frac{p_\phi}{\rho_\phi} = \frac{\dot{\phi}^2/2 - V(\phi)}{\dot{\phi}^2/2 + V(\phi)}. \tag{3.29}$$

Given $H(z)$, Eq. (1.20) can be solved to obtain the growth factor of matter density perturbations, $G(z) \propto D_1(z)$.

Since the PNGB potential is very flat, and the PNGB field starts rolling from rest, the dark energy equation of state from the PNGB field is very close to $w = -1$ (a cosmological constant). In particular, $w_\phi = -1$ at high z, and only deviates very slightly from $w_\phi = -1$ at $z = 0$ (see Abrahamse et al., 2008). Not surprisingly, it is extremely difficult to differentiate between PNGB quintessence and a cosmological constant. A very ambitious future observational project is required to rule out a cosmological constant by $> 3\sigma$ (Abrahamse et al., 2008).

3.1.3
Worked Example: The Doomsday Model

We now present another worked example to give readers a detailed picture of the process of making observable predictions for a scalar field dark energy model, and constraining it with available data (Wang et al., 2004). The doomsday model presented here was one of the first models of dark energy (Linde, 1987). In this model the cosmological constant was replaced by the energy density of a slowly varying scalar field ϕ with the linear effective potential

$$V(\phi) = V_0(1 + \alpha \phi). \tag{3.30}$$

Note that $\alpha = 0$ corresponds to a cosmological constant.

For simplicity, we assume a flat universe, consistent with observational data. Note that Eqs. (3.3) and (3.4) give the density and pressure of the ϕ field. The total density and pressure are

$$\rho = \rho_m + \rho_\phi, \quad p = p_\phi. \tag{3.31}$$

The Hubble parameter is

$$H^2(a) \equiv \left(\frac{\dot{a}}{a}\right)^2 = \frac{8\pi G}{3}\left[\frac{\rho_m^0}{a^3} + \frac{\dot{\phi}^2}{2} + V_0(1 + \alpha\phi)\right], \tag{3.32}$$

where a is the cosmic scale factor, and ρ_m^0 is the matter density today.

The equations of motion are

$$\ddot{\phi} + 3H(a)\dot{\phi} + \frac{\partial V}{\partial \phi} = 0, \tag{3.33}$$

$$\frac{\ddot{a}}{a} = \frac{8\pi G}{3}\left[V(\phi) - \dot{\phi}^2 - \frac{1}{2}\frac{\rho_m^0}{a^3}\right], \tag{3.34}$$

where the dots denote derivatives with respect to the cosmic time t.

It is not straightforward to solve the equations of motion for a given set of (Ω_m, α, H_0), since Ω_m depends on H_0, which in turn is only known once we have solved the equations for $a(t)$ and found its derivative at $a = 1$.

3.1 Dark Energy Models

We rewrite the equations of motion as

$$\frac{d^2\overline{\phi}}{d\tau^2} + 3E_*(z)\frac{d\overline{\phi}}{d\tau} + \overline{\alpha} = 0, \qquad (3.35)$$

$$\frac{1}{a}\frac{d^2 a}{d\tau^2} = 1 + \overline{\alpha}\overline{\phi} - \left(\frac{d\overline{\phi}}{d\tau}\right)^2 - \frac{1}{2}\frac{\Omega_m^*}{a^3}, \qquad (3.36)$$

where

$$E_*^2(z) \equiv \left(\frac{1}{a}\frac{da}{d\tau}\right)^2 = \frac{\Omega_m^*}{a^3} + \frac{1}{2}\left(\frac{d\overline{\phi}}{d\tau}\right)^2 + 1 + \overline{\alpha}\overline{\phi}, \qquad (3.37)$$

$$\tau \equiv \sqrt{\frac{8\pi G V_0}{3}}t, \quad \overline{\phi} \equiv \sqrt{\frac{8\pi G}{3}}\phi, \quad \overline{\alpha} \equiv \sqrt{\frac{3}{8\pi G}}\alpha, \quad \Omega_m^* \equiv \frac{\rho_m^0}{V_0}. \qquad (3.38)$$

Note that the equations of motion (Eqs. (3.35)–(3.36)) only depend on $(\Omega_m^*, \overline{\alpha})$, with no dependence on H_0.

At $\tau \to 0$, $a \to 0$, we take $\phi = d\phi/dt = 0$ (Kallosh et al., 2003). Evolving the equations of motion to $a = 1$ ($z = 0$) gives us

$$H_0 = \sqrt{\frac{8\pi G V_0}{3}}E_*(z=0), \quad \Omega_m = \frac{\Omega_m^*}{E_*^2(z=0)}. \qquad (3.39)$$

Figure 3.1 shows the cosmic scale factor $a(t)$ as a function of cosmic time t for the doomsday model for $\alpha = 0, 0.71, 0.76, 0.86, 1.13$. Figure 3.2 shows the dark energy equation of state for the same five cases. Figure 3.3 shows the ratio of the collapse time from today, t_c, and the age of the universe today, t_0, as a function of the linear potential parameter α.

In the linear model of Eq. (3.30), the expansion rate of the universe depends on the parameters (ρ_m^0, V_0, α) (see Eq. (3.32)). We assume uniform priors for ρ_m^0 and

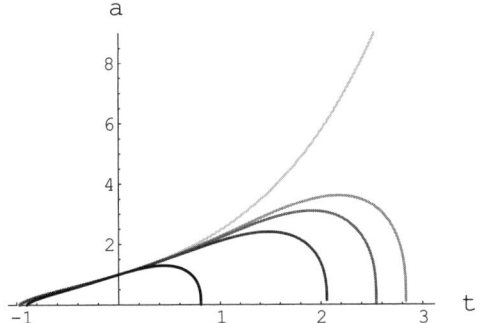

Figure 3.1 The cosmic scale factor $a(t)$ as a function of cosmic time t for the doomsday model with $\alpha = 0, 0.71, 0.76, 0.86, 1.13$ (Kallosh et al., 2003). Note that for $\alpha = 0$ (a cosmological constant), a increases indefinitely with increasing t.

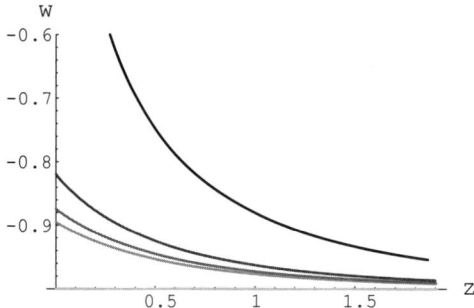

Figure 3.2 The dark energy equation of state for the doomsday model with $\alpha = 0, 0.71, 0.76, 0.86, 1.13$ (Kallosh et al., 2003). Note that $\alpha = 0$ corresponds to a cosmological constant, with $w_X(z) = -1$, and coincides with the x axis in this figure.

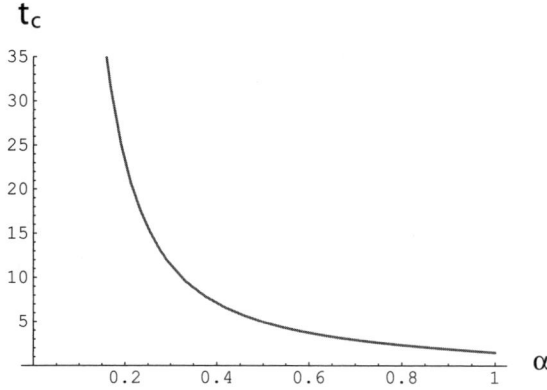

Figure 3.3 The ratio of the collapse time from today, t_c, and the age of the universe today, t_0, as a function of the linear potential parameter α (Kallosh et al., 2003).

V_0. It can be shown that choosing a uniform prior for V_0 is equivalent to choosing a uniform prior for the initial value of the scalar field ϕ after inflation, which is the standard assumption made in Linde (1987) and Garriga, Linde, and Vilenkin (2004). For α, we can consider three different priors (Garriga, Linde, and Vilenkin, 2004):

$$P(\alpha) \propto 1 \; ; \quad P(\alpha) \propto \alpha^{-0.5} \; ; \quad P(\alpha) \propto \alpha \tag{3.40}$$

which correspond to $n = 2, 3, 3/2$ in Eq. (29) of Garriga, Linde, and Vilenkin (2004).

Using Bayes' theorem, we can take the prior in α into consideration by replacing the usual χ^2 with

$$\tilde{\chi}^2 = \chi^2 - 2 \ln P(\alpha) \,. \tag{3.41}$$

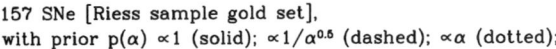
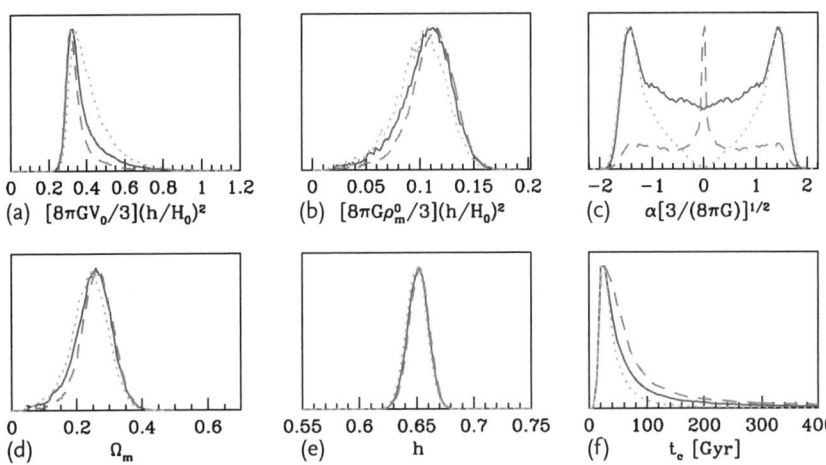

Figure 3.4 The constraints on the linear model parameters from the Riess et al. (2004) gold sample of 157 SNe Ia, for different priors on α (Wang et al., 2004). (a)–(c) show the probability distribution functions of the set of independent parameters V_0, ρ_m^0, and α. (d)–(f) show derived parameters Ω_m, h, and the time to collapse from today t_c.

Once we find $\phi(a)$ and $a(t)$, we can find $H(z)$ (see Eq. (3.32)), and make predictions and compare with observational data in a Markov Chain Monte Carlo (MCMC) likelihood analysis (see Section 2.4.3).

Using the data from the Riess et al. (2004) gold sample of 157 SNe Ia (flux-averaged with $\Delta z = 0.05$), the CMB data from WMAP one-year observations (Bennett et al., 2008), CBI (Pearson et al., 2003), ACBAR (Kuo et al., 2004), and the 2dF measurement of $f_g(z)$ at $z = 0.35$ (Verde et al., 2002; Hawkins et al., 2003), Wang et al. (2004) found that the collapse time of the universe is $t_c > 42$ (24) Gyr from today at 68% (95%) confidence.

Figure 3.4 shows the constraints on the linear model parameters from the Riess et al. (2004) gold sample of 157 SNe Ia, with different priors on α from Eq. (3.40): $P(\alpha) \propto 1$ (solid line); $P(\alpha) \propto \alpha^{-0.5}$ (dashed line); and $P(\alpha) \propto \alpha$ (dotted line). The first row (a–c) shows the probability distribution functions of the set of independent parameters V_0, ρ_m^0, and α. The second row (d–f) shows derived parameters Ω_m, h, and the time to collapse from today t_c.

The doomsday model only begins to significantly deviate from a cosmological constant at late times (see Figure 3.2). Therefore, it is most strongly constrained by data at $z \lesssim 0.5$ (Wang et al., 2004). This differs from many dark energy models which differ significantly from a cosmological constant at intermediate and high redshifts ($1 \lesssim z \lesssim 5$), and resemble a cosmological constant at low redshifts. Thus both high z and low z data will be important to constrain viable dark energy models.

3.2
Modified Gravity Models

Modified gravity models are alternative theories of gravity; they modify general relativity. These models explain the observed recent cosmic acceleration by modifying the left hand side of Einstein's equation (Eq. (2.1)). The most studied models are the $f(R)$ gravity models and the DGP gravity model.

3.2.1
$f(R)$ Gravity Models

The $f(R)$ gravity theories can be recast as scalar-tensor theories, which appear as low-energy limits of string theories (Caldwell and Kamionkowski, 2009). In the $f(R)$ gravity models (see for example, Woodard (2005) for a review), the action is

$$S = \int d^4x \sqrt{-g} f(R), \qquad (3.42)$$

where $f(R)$ is an arbitrary function of R. These models are interesting since they encompass both recent cosmic acceleration models and inflation models. Low curvature (or late time) modifications, such as $f(R) = A/R + ...$, can explain the observed recent cosmic acceleration. High curvature (or early time) modifications, such as $f(R) = AR^2 + ...$, can lead to cosmic inflation (Starobinsky, 1980; Wang, 1990).

Varying Eq. (3.42) with respect to the metric tensor gives the field equations with

$$G_{\mu\nu} = \left(\frac{\partial f}{\partial R}\right)^{-1} \left[\frac{1}{2} g_{\mu\nu}\left(f - \frac{\partial f}{\partial R} R\right) + \nabla_\mu \nabla_\nu \frac{\partial f}{\partial R} - \Box\left(\frac{\partial f}{\partial R}\right) g_{\mu\nu}\right], \qquad (3.43)$$

where $G_{\mu\nu}$ is the Einstein tensor.

The field equations for the $f(R)$ models are usually solved in the Einstein frame, and the solutions are then transformed back to the original frame (known as the Jordan frame). In the Jordan frame, the metric measures physical distances and times. In the Einstein frame, the gravitational kinetic term is the Ricci scalar R, that is, gravity is described by general relativity. To go from the Jordan frame to the Einstein frame, we make a conformal transformation,

$$g^{(E)}_{\mu\nu} = e^{2\omega} g_{\mu\nu}, \qquad (3.44)$$

where "E" denotes the metric in the Einstein frame. We can choose the conformal factor of the form

$$2\omega = \ln\left(2\kappa^2 \left|\frac{\partial f}{\partial R}\right|\right), \qquad (3.45)$$

where $\kappa^2 = 8\pi G = 8\pi m_{pl}^{-2} = m_{pl}^{-2}$. Consequently ω behaves like a scalar field ϕ, defined by

$$\kappa \phi \equiv \sqrt{6}\omega = \frac{\sqrt{6}}{2} \ln\left(2\kappa^2 \left|\frac{\partial f}{\partial R}\right|\right). \qquad (3.46)$$

Then the action in the Einstein frame is given by

$$S_E = \int d^4x \sqrt{-g_E}\, \mathcal{L}, \tag{3.47}$$

with Lagrangian density

$$\mathcal{L} = \frac{1}{2\kappa^2} R(g_E) - \frac{1}{2}(\nabla_E \phi)^2 - U(\phi), \tag{3.48}$$

and potential

$$U(\phi) = (\text{sign})e^{-\frac{2\sqrt{6}}{3}\kappa\phi} \left[\frac{(\text{sign})}{2\kappa^2} \operatorname{Re}^{\frac{\sqrt{6}}{3}\kappa\phi} - f \right], \tag{3.49}$$

where $(\text{sign}) = (\partial f/\partial R)/|\partial f/\partial R|$.

Amendola, Polarski, and Tsujikawa (2006) pointed out that in all $f(R)$ theories that behave as a power of R at large or small R, the scale factor during the matter-dominated stage evolves as $a \propto t^{1/2}$ instead of $a \propto t^{2/3}$, except for Einstein gravity. This leads to disagreement with CMB and galaxy redshift survey data, which establish a matter-dominated epoch in the early universe prior to the current epoch of cosmic acceleration. For the $f(R)$ models to provide a viable explanation for all current observational data, they would need to allow a matter-dominated epoch to exist before the late-time acceleration.

There is a lot of freedom in constructing $f(R)$ models. De Felice, Mukherjee, and Wang (2008) showed that one can place constraints on $f(R)$ models using observational data without assuming an explicit form for the functions. They used a general form of $f(R)$ with a valid Taylor expansion up to second order in R about redshift zero. The coefficients of this expansion can be reconstructed via the cosmic expansion history measured using current cosmological observations. These are the quantities of interest for theoretical considerations relating to ghosts and instabilities. De Felice, Mukherjee, and Wang (2008) found that current data provide interesting constraints on the coefficients.

Capozziello and Salzano (2009) provided a review of the latest results on observational tests of $f(R)$ models. The next-generation dark energy surveys should shrink the allowed parameter space for $f(R)$ models quite dramatically. In particular, the precise and accurate measurement of the growth rate of large scale structure $f_g(z)$ will allow us to differentiate between the $f(R)$ models and dark energy models that give the same cosmic expansion history $H(z)$.

3.2.2
DPG Gravity Model

The DGP gravity model (Dvali, Gabadadze, and Porrati, 2000) can be described by the action for a brane embedded in a five-dimensional Minkowski bulk:

$$S = -\frac{M_5^3}{2} \int d^5 X \sqrt{-g}\, R_5 - \frac{m_{pl}^2}{2} \int d^4x \sqrt{-h}\, R_4 + \int d^4x \sqrt{-h}\, \mathcal{L}_m + S_{GH}, \tag{3.50}$$

where g_{ab} is the metric in the bulk and $h_{\mu\nu}$ is the induced metric on the brane. $\mathcal{L}_m$ is the matter Lagrangian confined to the brane. The second term containing the four-dimensional Ricci scalar on the brane is unique to the DGP model, and such a term can be induced by quantum effects in the matter sector on the brane. The last term S_{GH} is a Gibbons–Hawking boundary term necessary for the consistency of the variational procedure and leads to the Israel junction conditions (Copeland, Sami, and Tsujikawa, 2006). The DGP model arises from braneworld theory, and provides a new line of theoretical investigation by linking cosmic acceleration to the effects of an unseen extra dimension at very large distances. Note that there are tachyonic instabilities in the self-accelerating branch of solutions in the DGP model (Charmousis et al., 2006). Efforts to solve this problem include the "cascading DGP" model, which is a higher codimension generalization of the DGP scenario (de Rham et al., 2008).

The characteristic length scale for DGP gravity is given by

$$r_0 = \frac{m_{pl}^2}{2 M_5^3} , \qquad (3.51)$$

where m_{pl} is the four-dimensional Planck mass, and M_5 is its counter part in the five-dimensional bulk. For length scales much smaller than r_0, gravity manifests itself as a four-dimensional theory. At large distances, gravity leaks into the bulk, making the higher dimensional effects important. The weak-field gravitational potential behaves as

$$\Phi \sim \begin{cases} r^{-1} & \text{for } r < r_0 , \\ r^{-2} & \text{for } r > r_0 . \end{cases} \qquad (3.52)$$

If the brane is described by the Robertson–Walker metric (Eq. (2.10)) with zero curvature, the DGP model leads to the modified Friedmann equation given in Eq. (1.24). Equations (1.24–1.27) can be used to make predictions of the DGP model, and derive constraints on the model in a likelihood analysis using observational data.

Current observational data already tightly constrain the DGP model. Depending on which sample of SNe Ia is used together with the SDSS baryon acoustic oscillation scale measurement (Eisenstein et al., 2005), the DGP model is either allowed at the 95% C.L. (Alam and Sahni, 2006), or excluded at 99% C.L. (Fairbairn and Goobar, 2006). Clearly, future data with smaller systematic uncertainties will be required to definitively constrain the DGP model. See also the discussion in Section 1.2.

3.2.3
The Cardassian Model

In the Cardassian model (Freese and Lewis, 2002), cosmic acceleration arises from modifications to the Friedmann equation:

$$H^2 = g(\rho_m) , \qquad (3.53)$$

where $g(\rho_m)$ is a function of the energy density, ρ_m contains only matter and radiation (no vacuum energy). The function $g(\rho_m)$ returns to the usual $8\pi\rho_m/(3m_{pl}^2)$ during the early history of the universe, but takes a different form that drives an accelerated expansion after a redshift $z \sim 1$. In the Cardassian model, the universe is flat, matter dominated, and accelerating.

There is no unique four-dimensional or higher-dimensional theory that gives rise to the Cardassian model. Such modifications to the Friedmann equation may arise, for example, as a consequence of our observable universe living as a three-dimensional brane in a higher dimensional universe (Chung and Freese, 2000). Alternatively, Gondolo and Freese (2003) noticed that nonstandard physics in the geometry (left hand side of Einstein's equations) could equivalently be treated as a nonstandard fluid (right hand side of Einstein's equations). Thus, such a Friedmann equation may arise if there is dark matter with self-interactions characterized by negative pressure (Gondolo and Freese, 2003).

The original version of the Cardassian model invoked the addition of a simple power-law term: $H^2 = \frac{8\pi G\rho}{3}\rho + B\rho^n$ with $n < 2/3$ and B, a constant determined by data. However, the generalized Cardassian model allows any modification of the form of Eq. (3.53). For example, one possibility that allows direct comparison to data is

$$H^2 = \frac{8\pi G\rho_m}{3}\left[1 + \left(\frac{\rho_{Card}}{\rho_m}\right)^{q(1-n)}\right]^{1/q}, \qquad (3.54)$$

where $n < 2/3$ and $q > 0$, and ρ_{Card} is a characteristic constant energy density. The original power-law Cardassian model corresponds to $q = 1$. This model is also known as the "modified polytropic Cardassian" (MP Cardassian). The name "modified polytropic" arises in the context of treating the right hand side of Eq. (3.54) as a single fluid; then the relationship between energy density and pressure is roughly polytropic (see Gondolo and Freese, 2003).

The distance-redshift relation predictions of generalized Cardassian models can be very different from generic quintessence models, and can be differentiated with future observational data (Wang et al., 2003). The fluid interpretation of the Cardassian model has difficulties with large scale structure formation, similar to the generalized Chaplygin gas model. The structure formation in other possible interpretations of the Cardassian model has not yet been studied.

3.3
A Cosmological Constant

The cosmological constant remains a viable candidate for dark energy. If future observational data increase the evidence for a cosmological constant, we will need to once again seriously face the cosmological constant that has puzzled theoretical particle physicists for several decades. The fact that the cosmological constant is vanishingly small, yet not zero, may make deriving it from first principles easier. A fundamental symmetry may require the cosmological constant to be precisely zero, and a perturbative effect may give it a very small value.[2]

The mystery of dark energy may be a reflection of our incomplete or incorrect understanding of gravity (see for example, Padmanabhan, 2008). Quantum gravitational corrections may hold the key to solving this mystery. An interesting possibility for modifying gravity is the fully nonlocal effective action that results from quantum gravitational corrections (Woodard, 2005). Potentially more interesting is the possibility of very strong infrared effects from the epoch of primordial inflation (Tsamis and Woodard, 1995; Martineau and Brandenberger, 2005).

Padmanabhan (2008) pointed out that no other approach really alleviates the difficulties faced by the cosmological constant. This is because in all other attempts to model dark energy, one still has to explain why the bulk cosmological constant (treated as a low-energy parameter in the action principle) is zero. Until the theory is made invariant under the shifting of the Lagrangian by a constant, one cannot obtain a satisfactory solution to the cosmological constant problem. However, this is impossible in any generally covariant theory with the conventional low-energy matter action, if the metric is varied in the action to obtain the field equations.

To see explicitly how the cosmological constant problem arises in general relativity, note that in general relativity, the gravitational field equations are obtained by varying the action

$$S = \int \sqrt{-g}\, d^4x \mathcal{L}_{tot} ,\qquad(3.55)$$

with respect to the metric $g_{\mu\nu}$. The Lagrangian

$$\mathcal{L}_{tot} = \mathcal{L}_{grav}(g) + \mathcal{L}_{matter}(g, \phi) .\qquad(3.56)$$

$\mathcal{L}_{grav}(g)$ is the gravitational Lagrangian dependent on the metric and its derivatives. $\mathcal{L}_{matter}(g, \phi)$ is the matter Lagrangian which depends on both the metric and the matter fields ϕ.

Adding a constant λ_m to the matter Lagrangian induces the change

$$\mathcal{L}_{matter} \rightarrow \mathcal{L}_{matter} + \lambda_m ,\qquad(3.57)$$

this leaves the equations of motion for matter invariant at scales below supersymmetry-breaking (which occurred in the early universe). Thus Eq. (3.57) is a symmetry of the matter sector. But gravity, according to general relativity, breaks this symmetry, as clearly indicated by Einstein's equation, Eq. (2.1), which is *not* invariant under the corresponding shift of

$$T^a{}_b \rightarrow T^a{}_b + \rho_\Lambda \delta^a_b .\qquad(3.58)$$

---

2) Alternatively, if quantum fields exist in extra compact dimensions, they will give rise to a quantum vacuum or Casimir energy. That vacuum energy will manifest itself as a cosmological constant. Milton et al. (2001) showed that SN Ia and CMB data place a lower bound on the size of the extra dimensions, while laboratory constraints on deviations from Newton's law place an upper limit. The allowed region is so small as to suggest that either extra compact dimensions do not exist, or their properties are about to be tightly constrained by experimental data.

One possible solution is to treat the metric *not* as the fundamental dynamical degrees of freedom (as in general relativity), but as a coarse grained description of the space-time at macroscopic scales, similar to the density of a solid (which has no meaning at atomic scales) (Alternative Gravity Theory references). It is interesting to note that the metric is emergent in string theory, which supports the idea that the metric might not be a fundamental dynamical variable.

Padmanabhan (2008) reviewed an alternative perspective in which gravity arises as an emergent, long-wavelength phenomenon and can be described in terms of an effective theory using an action associated with normalized vectors n^a in the space-time (instead of the metric tensor $g_{\mu\nu}$). This action is explicitly invariant under the shift of the energy momentum tensor $T_{ab} \to T_{ab} + \Lambda g_{ab}$ and any bulk cosmological constant can be gauged away. This is because, as a consequence of the condition $n_a n^a = 1$ on the dynamical variables, there is a "gauge freedom" that allows an arbitrary integration constant to appear in the theory, which can absorb the bulk cosmological constant.

Once the bulk value of the cosmological constant is eliminated, the observed dark energy could be fluctuations in the vaccum energy in the correct theory of quantum gravity. We are still a long way from having such a theory.

4
Observational Method I:
Type Ia Supernovae as Dark Energy Probe

4.1
Type Ia Supernovae as Distance Indicators

The use of Type Ia supernovae (SNe Ia) is the best-established method for probing dark energy, since this is the method through which cosmic acceleration was discovered (see Riess et al. (1998) and Perlmutter et al. (1999)). The unique advantage of this method is that it is independent of the clustering of matter, and can provide a robust measurement of $H(z)$ through the measured luminosity distance as a function of redshift

$$d_L(z) = (1+z)r(z), \tag{4.1}$$

where the comoving distance $r(z)$ from the observer to redshift z is given by Eq. (2.24).

It is now well-established that a SN Ia is a thermonuclear explosion that completely destroys a carbon/oxygen white dwarf very near the Chandrasekhar limit of $1.4 M_\odot$ for a stellar composition with the number of free electrons per nucleon $Y_e = \frac{1}{2}$. The sub-Chandrasekhar-mass white dwarf models do not reproduce the light curves and spectra of observed SNe Ia (Höflich and Khokhlov, 1996; Nugent et al., 1997). A super-Chandrasekhar-mass white dwarf would produce a super-luminous SN Ia, such as the SNe Ia found by Howell et al. (2006) and Hicken et al. (2007); these can be easily separated from the normal SNe Ia used for cosmology.[3] The uniformity in the explosion mass is the reason normal SNe Ia are so uniform in peak luminosity.

The properties of a SN Ia depend on the properties of the star that exploded. One common feature of SNe Ia is that their brightness depends on the production of ^{56}Ni during the explosion. ^{56}Ni (with a half-life of 6.1 days) decays into ^{56}Co (with a half-life of 77.7 days), which then decays into stable ^{56}Fe. These decays are accompanied by the emission of gamma-rays and positrons, which eventually become the SN light.

---

3) The spectra and light curves of the super-luminous SN 2006gz show very strong signatures of unburned carbon (Hicken et al., 2007), while the spectra of normal SNe Ia show low carbon abundance (Marion et al., 2006).

Dark Energy. Yun Wang
Copyright © 2010 WILEY-VCH Verlag GmbH & Co. KGaA, Weinheim
ISBN: 978-3-527-40941-9

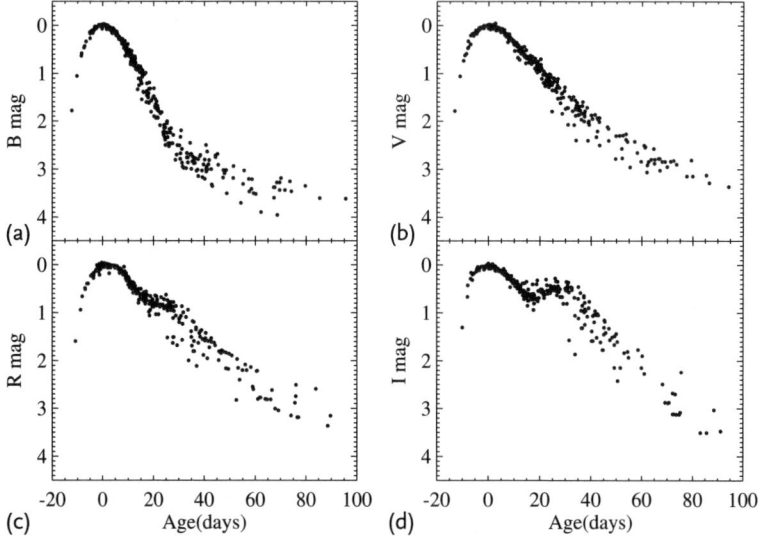

Figure 4.1 Composite B (a), V (b), R (c), and I (d) light curves of 22 SNe Ia (Riess *et al.*, 1999). The light curves were normalized in time and brightness to the initial peak, including a correction for the $1 + z$ time dilation and a K-correction.

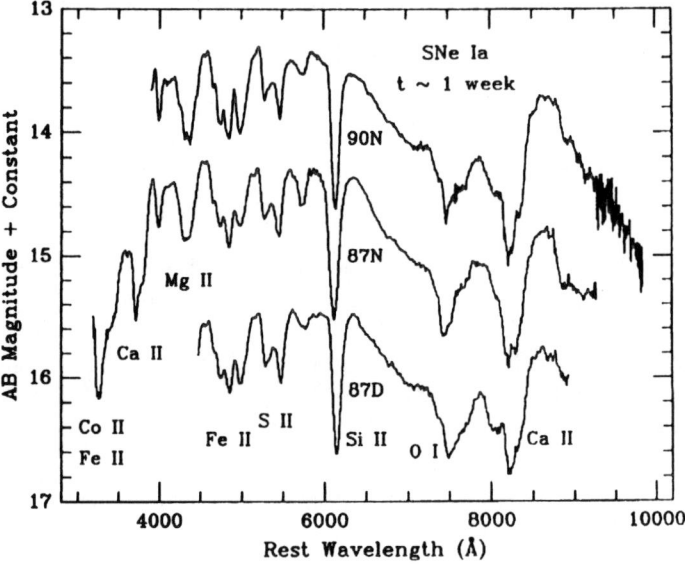

Figure 4.2 Spectra of three nearby SNe Ia, SN 1987D, SN 1987N, and SN 1990N, about one week after maximum (Filippenko, 1997).

There is an inhomogeneity in SN Ia light curves and spectra. Figure 4.1 shows the composite B, V, R, and I light curves of 22 SNe Ia, normalized in time and brightness to the initial peak, including a correction for the $1 + z$ time dilation and

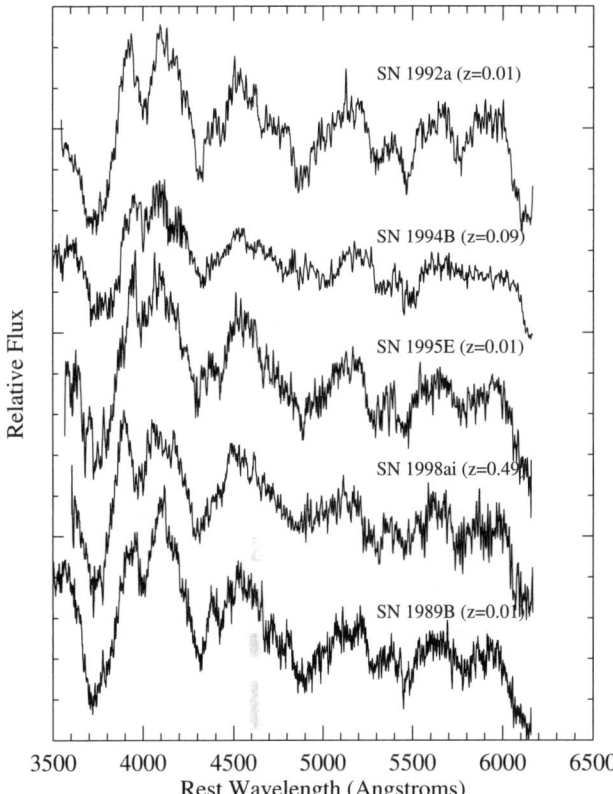

Figure 4.3 Spectra of four nearby SNe Ia and SN 1998ai at $z = 0.49$ (Riess et al., 1998). The spectra of the nearby SNe Ia were resampled and convolved with Gaussian noise to match the quality of the of SN 1998ai.

a K-correction. Figure 4.2 shows the spectra of three nearby SNe Ia (Filippenko, 1997). Some of the spectral lines are labeled with line identification that indicates the atom that causes the absorption. Figure 4.3 shows the spectra of four nearby SNe Ia and SN 1998ai at $z = 0.49$. There are physical reasons behind the inhomogeneities in SNe Ia; these are discussed in Section 4.2.

The first challenge to overcome when using SNe Ia as cosmological standard candles is properly incorporating the intrinsic scatter in SN Ia peak luminosity. The usual calibration of SNe Ia reduces the intrinsic scatter in SN Ia peak luminosity (Hubble diagram dispersion) to about 0.16 mag (Phillips, 1993; Riess, Press, and Kirshner, 1995). The calibration techniques used so far are based on one observable parameter, the light curve width, which can be parametrized either as Δm_{15} (decline in magnitudes for a SN Ia in the first 15 days after B-band maximum (Phillips, 1993, 1999), or a stretch factor (which linearly scales the time axis (Goldhaber et al., 2001)). The correlation between light curve width (as measured by light curve decline rate) and peak luminosity was first proposed by Pskovskii (1977), and supported by the work of Branch (1981). However, due to the paucity of

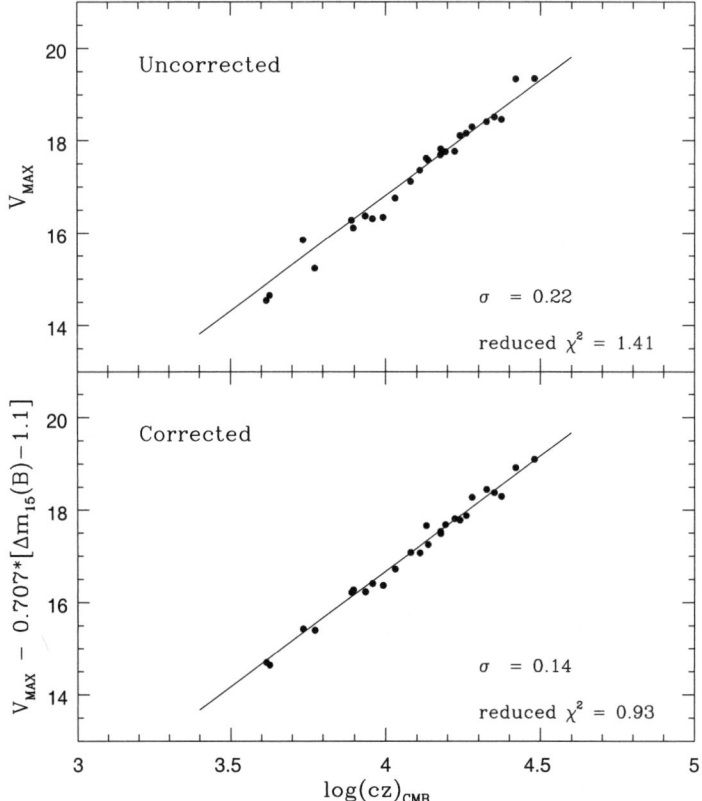

Figure 4.4 Hubble diagrams showing 26 SNe Ia with $B_{max} - V_{max} \leq 0.20$ from the Calan/Tololo sample compiled by Hamuy et al. (1996). This sample provided half of the data for the discovery of the cosmic acceleration in 1998 (Riess et al., 1998). The solid lines indicate Hubble's law; perfect standard candles (with $\sigma = 0$) fall on these lines.

well-observed SNe Ia, this relation was not firmly established until 1993 by Phillips (1993, 1999).

Figure 4.4 shows how correcting the SN peak brightness (measured by the V-band magnitude at peak brightness, V_{MAX}) using Δm_{15} improves the homogeneity of nearby SNe Ia (Hamuy et al., 1996). The dramatic reduction in the dispersion of the Hubble diagram was achieved by using the calibration relation (derived from the same sample of SNe Ia)

$$V_{MAX} - 0.707(\pm 0.150)\left[\Delta m_{15}(B) - 1.1\right] = 5\log(cz) - 3.329(\pm 0.031), \quad (4.2)$$

where the second term on the left corrects the observed magnitudes of each SN Ia to the equivalent magnitudes of an event with $\Delta m_{15}(B) = 1.1$ mag.

It is critical to use the same light curve-fitting technique to derive distances to the SNe Ia used to constrain cosmology; this removes the significant uncertainty introduced by the difference in fitting techniques (Wang, 2000a). Riess et al. (2007)

compiled a homogeneous sample of SNe Ia. The most recent homogeneous sample of SNe Ia was compiled by Kowalski *et al.* (2008).

The Pskovskii–Branch–Phillips (PBP) relation between SN Ia light curve width and peak luminosity can be understood as follows. The light curve width is associated with the amount of ^{56}Ni produced in the SN Ia explosion, which in turn depends on when the carbon burning makes the transition from turbulent deflagration to a supersonic detonation. An earlier transition produces less ^{56}Ni, which leads to a cooler explosion, and results in a less luminous SN Ia. On the other hand, the cooler explosion corresponds to lower opacity, which results in a faster decline of the SN Ia brightness (Wheeler, 2003). A more complex explanation arises from detailed radiative transfer calculations of Chandrasekhar-mass SN Ia models, which showed that the faster B-band decline rate of dimmer SNe Ia reflects their faster ionization evolution as a result of their cooler explosions (Kasen and Woosley, 2007).

There may be additional physical parameters associated with SN Ia light curves (see, e.g., Wang and Hall (2008)) or spectra (see, e.g., Bongard *et al.* (2006), Branch, Dang, and Baron (2009)) that can further improve the calibration of SNe Ia. The light curve fitting technique SALT (see Section 4.5.1) uses both the SN Ia light curve width and the SN Ia color to minimize the scatter of SNe Ia on the Hubble diagram. Note, however, SN Ia color has contributions from both the intrinsic color of SN Ia and dust extinction, thus is not a purely physical parameter.

The key to the efficient use of SNe Ia for probing dark energy is to obtain the largest possible unbiased sample of SNe Ia ranging from nearby to the greatest distances from the observer (Wang and Lovelace, 2001). This can be achieved by an ultra-deep survey of the same areas in the sky every few days for at least one year (Wang, 2000b). Figure 4.5 shows the comparison of an ultra-deep supernova

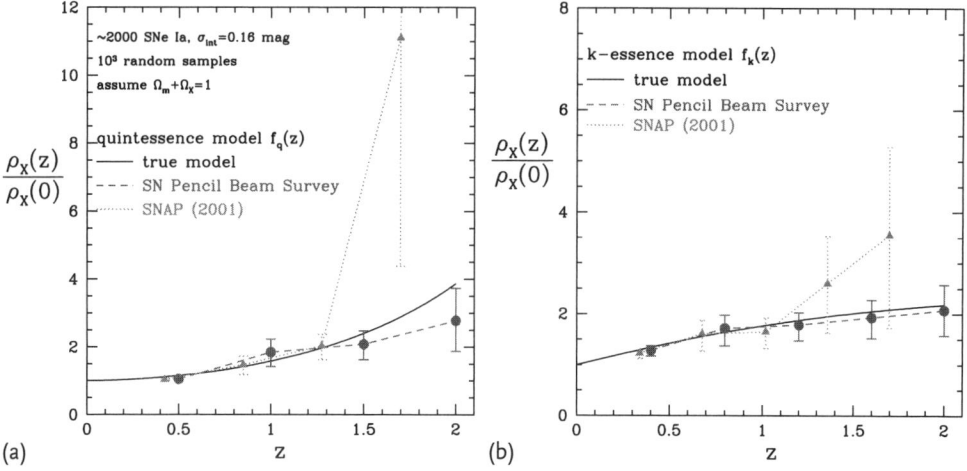

Figure 4.5 The comparison of an ultra-deep supernova survey (Wang, 2000b) with a much shallower survey in the reconstruction of the dark energy density $\rho_X(z)$ as a free function of cosmic time (Wang and Lovelace, 2001). Panels (a) and (b) use the same simulated data, but assume different fiducial dark energy models.

survey (Wang, 2000b) with a much shallower survey. Clearly, a sufficiently deep supernova survey is required to reconstruct the dark energy density $\rho_X(z)$ as a free function of cosmic time (i.e., to measure $H(z)$ precisely).

4.2
Possible Causes of Observational Diversity in SNe Ia

A SN Ia is a thermonuclear explosion of a C+O white dwarf (WD) that accretes matter from a nondegenerate companion star until it approaches the Chandrasekhar mass. By the time an SN Ia is detected, it has settled into homologous expansion, with velocity proportional to radius. The observed properties of a SN Ia depend on the basic explosion characteristics: (a) the density-velocity distribution, and the associated total kinetic energy E_k, and (b) the composition-velocity structure, including the key parameter M_{Ni} (the mass of ^{56}Ni produced).

Normal SNe Ia have highly but not perfectly homogeneous observational properties, and various kinds of peculiar SNe Ia have been found. Possible causes of the observational diversity include the following (Branch, 2009):

1. *The composition structure of the pre-SN WD.* The C/O ratio is expected to depend on the main-sequence mass of the WD progenitor, and to be a complicated function of radius. This is because different C/O ratios result from different He-burning phases in the WD progenitor, as well as from He-burning in the accreted outer layers (Dominguez, Höflich, and Straniero, 2001). The metallicity (e.g., the iron mass fraction inherited from the progenitor of the WD) is expected to affect M_{Ni} and other aspects of explosive nucleosynthesis, and it directly affects the line blocking in the spectrum. Hoeflich, Wheeler, and Thielemann (1998) studied the influence of the initial composition of the exploding WD on the nucleosynthesis, light curves, and spectra of SNe Ia.
2. *The density structure of the pre-SN WD, and the associated binding energy.* The explosion kinetic energy is the difference between the nuclear energy and the WD binding energy. The pre-SN WD density structure is expected to depend on the initial WD mass (before accretion from the companion) and on the time evolution of the accretion rate. A lower accretion rate allows more time for cooling, so the WD must reach a higher central density in order to ignite carbon (Nomoto, 1982).
3. *Rotation.* The WD accretes not only mass but also angular momentum, which spins up the WD, breaks spherical symmetry, affects the density structure, and may allow accretion to super-Chandrasekhar mass before ignition (Uenishi, Nomoto, and Hachisu, 2003; Yoon and Langer, 2002).
4. *The WD magnetic field.* If sufficiently strong, this may significantly affect the explosion (Ghezzi et al., 2004).
5. *Stochasticity.* The development of the initial nuclear flame may depend on extremely fine details of the state of the matter near the center of the WD (which

could involve convection) – details at a level that is impossible to predict (Sorokina and Blinnikov, 2000).
6. *Double-degenerate events* (the merger of binary WDs), if they can produce SNe Ia. The pre-SN merger products will in principle have a distribution of total masses, up to twice the Chandrasekhar mass.
7. *Asymmetries in the ejecta*, including the hole in the ejecta caused by the presence of the companion star (Marietta, Burrows, and Fryxell, 2000). These asymmetries will affect the observed colors of SNe Ia and their deduced reddening (Wheeler, 2009). There is some spectroscopic and spectropolarimetric evidence for global (shape) asymmetries, and for local (clumpy) asymmetries (see Wang et al. (2003); Thomas et al. (2004); and Kasen et al. (2003, 2004)). For a review on spectropolarimetry of supernovae, see Wang and Wheeler (2003).
8. *Detached high-velocity features*, recently recognized to be common in SN Ia spectra (Wang et al., 2003; Branch et al., 2004; Thomas et al., 2004; Mazzali et al., 2005). These may be a signal of interaction between the SN ejecta and a circumstellar medium (Gerardy et al., 2004), in which case the density, velocity, composition, and asymmetry of the circumstellar matter opens up many possibilities for (usually subtle) diversity. Whatever the cause of the detached high-velocity features, it is clear that if we see them in the SN spectrum then they also affect the SN photometry. More generally, we cannot understand the SN Ia photometric diversity without understanding the spectroscopic diversity.
9. *Parent–galaxy extinction*, not just the amount but also the exact shape of the extinction curve, which is not universal (see, e.g., Branch and Tammann (1992); Wang, Baade, and Patat (2007); and Kessler et al. (2009)).

To shed light on these possible effects, we have to stick to the empirical path, e.g., the approach described by Branch et al. (2001). They argued that two-parameter (light curve width and B–V color) luminosity corrections completely standardize SN Ia luminosities within current observational errors, so that with good data distances can be obtained to within 3% or better. From existing data, with its errors, it is not possible to tell how much better we can improve distance measurements. Much larger sets of SN Ia photometric and spectroscopic data are key to advancing our understanding of SN Ia physics, and exploring the systematic floor of SNe Ia as cosmological distance indicators.

4.3
Supernova Rate

To forecast the impact of future surveys of SNe Ia on dark energy constraints, we need to estimate the number of SNe Ia such surveys expect to observe. This requires knowing the rate of SNe Ia.

The rate of SNe Ia depends on the SN Ia explosion time scale, the time separating the explosion and the formation of the progenitor star. This time scale is usually re-

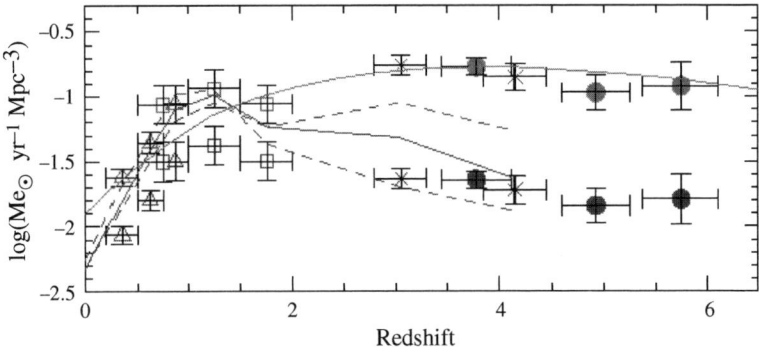

Figure 4.6 The average star formation density of UV-bright star-forming galaxies as a function of redshift (Giavalisco et al., 2004). The GOODS points (solid circles) have been obtained from the specific luminosity density using the conversion factor by Madau, Pozzetti, and Dickinson (1998). The other points are from Steidel et al. (1999) after conversion to the same cosmological model. The lower points are as observed, the upper points have been corrected for dust obscuration using the procedure proposed by Adelberger and Steidel (2000). The upper solid curve is from semi-analytical models (Somerville, Primack, and Faber, 2001).

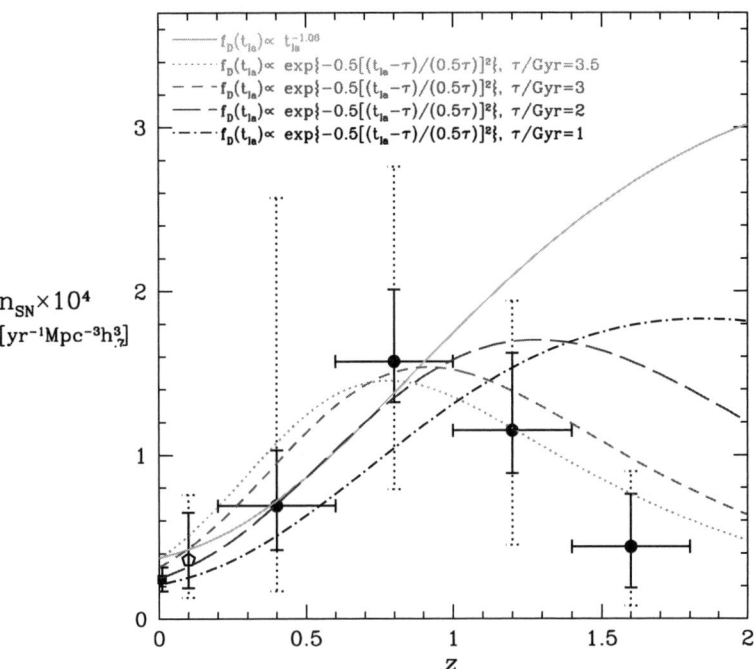

Figure 4.7 SN Ia rate as function of redshift for different assumptions about the SN Ia delay time. The data points are from Cappellaro, Evans, and Turatto (1999), Hardin et al. (2000), and Dahlen et al. (2004). The solid error bars indicate the statistical 68.3% confidence intervals. The dotted error bars include systematic uncertainties (Dahlen et al., 2004).

ferred to as the SN Ia delay time. Our knowledge of the rate of SNe Ia has improved dramatically, since Oemler and Tinsley (1979) recognized a class of short-lived SNe Ia, and Mannucci, Della Valle, and Panagia (2006) pointed out correctly that SNe Ia have a wide range of delay time (from $t < 0.1$ Gyr to $t > 10$ Gyr).

The SN Ia delay time distribution can be converted into the SN Ia rate using an extinction-corrected star-formation rate (SFR)[4]:

$$\mathrm{SNR_{Ia}}(t) = \nu \int_{t_F}^{t} \mathrm{SFR}(t') \cdot f_D(t-t') dt' , \qquad (4.3)$$

where t is the age of the universe at redshift z, t_F is the time when the first stars were formed ($z_F = 10$ is usually set for computational convenience), ν is the number of SNe Ia per formed solar mass, and $f_D(t-t')$ is the SN Ia delay time distribution function. Note that $\int \mathrm{SNR_{Ia}}(z) dz \equiv \int \mathrm{SNR_{Ia}}(t) dt$.

Figure 4.6 shows the measured SFR (Giavalisco et al., 2004). The measured and extinction-corrected SFR can be fitted by (Strolger et al., 2004)

$$\mathrm{SFR}(t) = a \left[t^b e^{-t/c} + d e^{d(t-t_0)/c} \right] ,$$

$$a = 0.182 , \quad b = 1.26 , \quad c = 1.865 , \quad d = 0.071 , \qquad (4.4)$$

where SFR is in units of $M_\odot \, \mathrm{yr}^{-1} \, \mathrm{Mpc}^{-3} h_{70}$ (with $h_{70} \equiv H_0/70 \, \mathrm{km \, s^{-1} \, Mpc^{-1}}$), and t is given in gigayears.

Figure 4.7 shows the SN Ia rate as a function of redshift for different assumptions about the SN Ia delay time. The data points are from Cappellaro, Evans, and Turatto (1999), Hardin et al. (2000), and Dahlen et al. (2004).

Most of the model curves in Figure 4.7 assume the distribution in the SN Ia delay time to be a Gaussian (Dahlen et al., 2004):

$$f_D(t_{\mathrm{Ia}}) \propto \exp\left[-\frac{(t_{\mathrm{Ia}} - \tau)^2}{2\sigma^2}\right] , \quad \sigma = 0.5\tau \qquad (4.5)$$

for $\tau = 3.5, 3, 2, 1$ Gyr.

The delay time distribution of SNe Ia was recently measured directly by Totani et al. (2008), using the faint variable objects detected in the Subaru/XMM-Newton Deep Survey (SXDS) down to $i' \sim 25.5$:

$$f_D(t_{\mathrm{Ia}}) = f_D(1 \, \mathrm{Gyr}) \left(\frac{t_{\mathrm{Ia}}}{1 \, \mathrm{Gyr}}\right)^\alpha$$

$$f_D(1 \, \mathrm{Gyr}) = 0.53^{+0.12}_{-0.11} \, \mathrm{century}^{-1} \left(10^{10} L_{K,0,\odot}\right)^{-1}$$

$$\alpha = -1.06^{+0.15}_{-0.15} , \qquad (4.6)$$

---

[4] In the "two-component model" (Scannapieco and Bildsten, 2005), the SNe Ia rate consists of a prompt piece that is proportional to the SFR, and an extended piece that is proportional to the total stellar mass. This is essentially a two-component model of the delay time distribution $f_D(t-t')$: one component has a peak at time $t - t' = 0$ and is zero at all other times (i.e., very short delay time), and the other has $f_D = $ const. with time (i.e., long delay times).

where $L_{K,0}$ is the K-band luminosity. Thus the delay time of SNe Ia is continuously distributed with $t_{Ia} = 0.1 - 8 \, \text{Gyr}$ (see Figure 4.8). Figure 4.7 shows that assuming a Gaussian distribution in the SN Ia delay times significantly underestimates the SN Ia rate at $z \gtrsim 2$.

Our knowledge of SN Ia progenitors is rather limited at present. Two scenarios are possible: the single-degenerate (SD) and double-degenerate (DD) progenitor models. In the SD model, SNe Ia result from binaries consisting of a white dwarf and a nondegenerate companion (main sequence or red giant star). In the DD model, SNe Ia result from binaries consisting of a white dwarf and a degenerate companion. Both the single-degenerate (SD) and double-degenerate (DD) progenitor models for SNe Ia can explain the observed delay time distribution. Figure 4.9 shows that the DD scenario fits the observed SN Ia time-delay distribution.

In the SD model, if the mass range of the companion star to the white dwarf were too narrow, its delay time distribution would be too limited around the companion's main-sequence lifetime to be consistent with the observed delay time distribution. However, Hachisu, Kato, and Nomoto (2008) found that a SD model that consists of two channels, (white dwarf + red giant) and (white dwarf + main-sequence star), fits the observed SN Ia delay time distribution as well (see Figure 4.10). In these

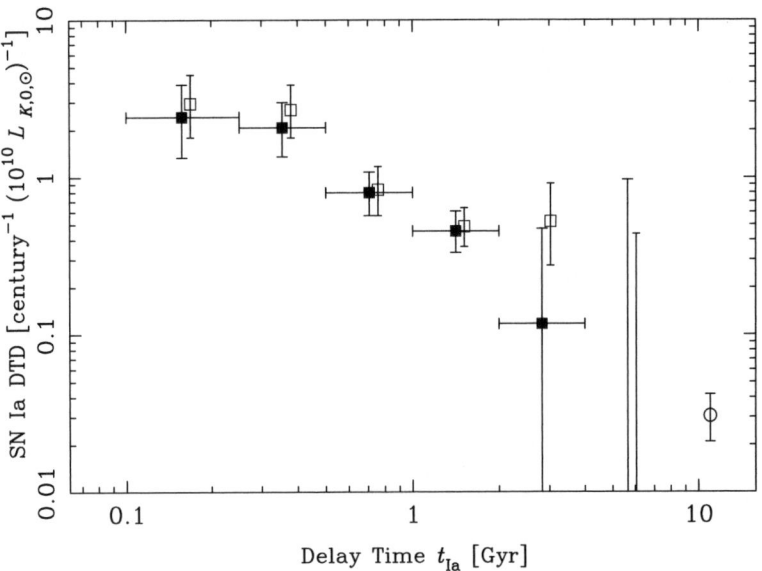

Figure 4.8 The observed SNe Ia delay time distribution, per unit delay time t_{Ia} [century^{-1}] for a single starburst population whose total K-band luminosity is $10^{10} L_{K,\odot}$ at the age of 11 Gyr (Totani et al., 2008). The filled squares are the final observational estimates by Totani et al. (2008) based on their baseline analysis, and the error bars are statistical 1σ errors. The open squares are the same but using a simpler method to estimate the delay time. The time bins are the same as those for the filled squares, but the open squares are slightly shifted in time in this plot for presentation. The open circle is delay time distribution inferred from the SN Ia rate in elliptical galaxies in the local universe by Mannucci et al. (2005).

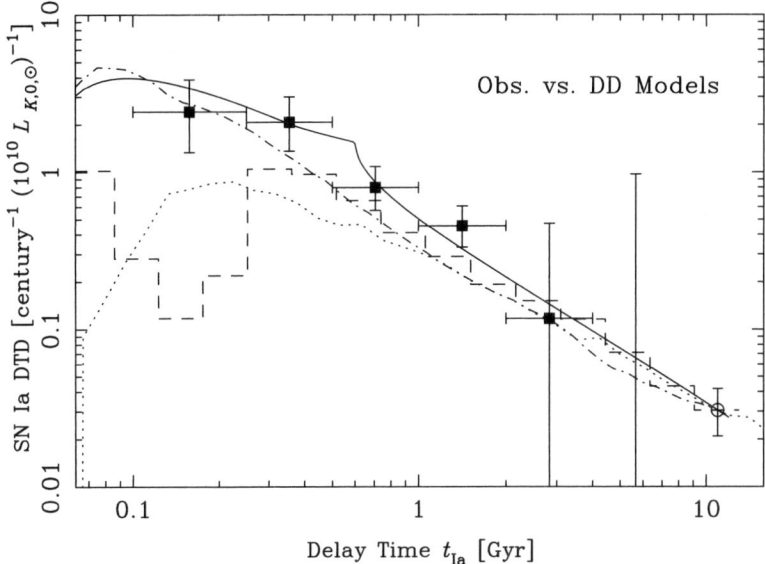

Figure 4.9 The observed SN Ia DTD $f_D(t_{Ia})$ compared with the theoretical predictions based on the DD scenario by Ruiz-Lapuente & Canal (1998, dotted), Yungelson & Livio (2000, dot-dashed), Greggio (2005, solid), and Belczynski *et al.* (2005, dashed) (Totani *et al.*, 2008). The data points are the same as those in Figure 4.8. The model curves are normalized by the DTD data at $t_{Ia} = 11$ Gyr, but the normalization is arbitrary and only the DTD shapes should be compared between the data and models.

channels, the companion stars have a mass range of ~ 0.9–3 $M_\odot$ (WD + RG) and $\sim 2 - 6$ $M_\odot$ (WD + MS). The combined mass range is wide enough to yield the featureless delay time distribution.

4.4
Systematic Effects

Calibration of SN Ia photometry is a critical issue for the use of SNe Ia as cosmological standard candles, and it sets the floor for systematic uncertainties for SNe Ia. It is anticipated that 1–2% absolute calibration is required to tie in space observations with low redshift, ground-based data, and an intra-band calibration accuracy of 0.5–1% is required to control overall systematic uncertainties (Phillips *et al.*, 2006). Requirements on photometric calibration are the strongest driver for the instrumentation design for a SN Ia survey. Since photometric calibration uncertainties are fixed for a given experiment (with its own specific photometric calibration instrumentation and strategy), we will not discuss them in detail here.

4.4.1
Extinction

The extinction by normal dust can be corrected using multi-band imaging data. The extinction by host galaxy dust is usually small, and it can be minimized using multi-band photometry. The extinction by the Milky Way can be minimized by choosing the observing fields to be near the ecliptic poles.

Recent data show that the apparent dust extinction of SNe Ia is very different from the typical extinction law due to Milky Way dust, possibly due to the mixing

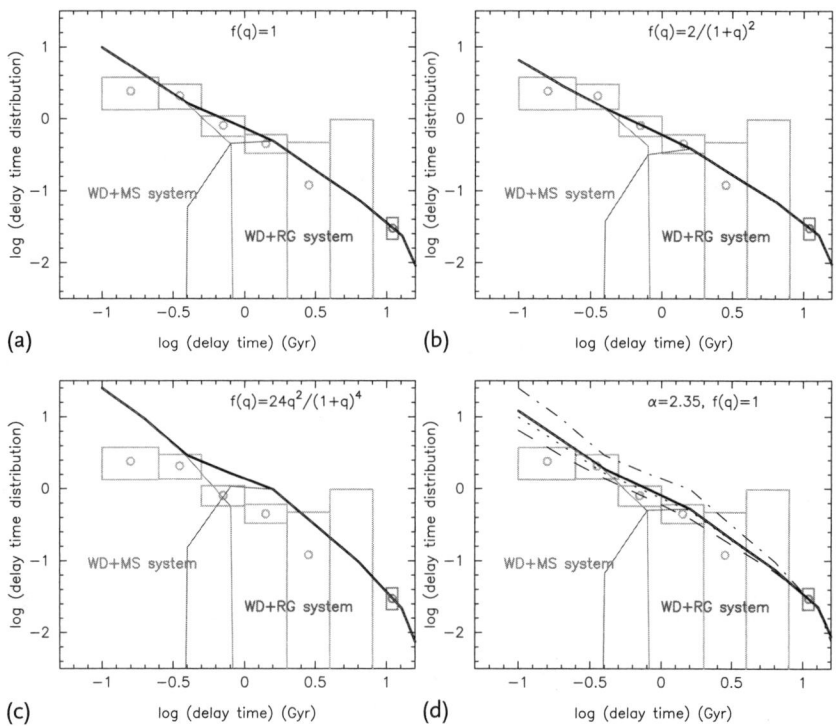

Figure 4.10 The delay time distributions (DTDs) for single-degenerate (SD) models of SNe Ia (Hachisu, Kato, and Nomoto, 2008). The ordinate is the DTD in units of per century and per $10^{10} L_{K,\odot}$. Open circles with an open box are the observational DTD taken from Totani et al. (2008) (for <10 Gyr) and Mannucci et al. (2005) (at 11 Gyr). Each open box indicates a one-sigma error of each measurement. Throughout all figures from (a) to (d), thick solid lines indicate the total DTD coming both from the WD + MS and WD + RG systems. Each contribution from each channel is separately shown by thin solid lines. The assumed binary mass ratio distribution $f(q)$ and the initial mass function power α are given in the figure legends in each panel.

of intrinsic SN Ia color variation with dust extinction, or variations in the properties of dust (Conley et al., 2007; Wang, Baade, and Patat, 2007).

The key to minimizing the systematic effect due to dust extinction is to observe SNe Ia in the near infrared (NIR), since dust extinction decreases with wavelength. Cardelli, Clayton, and Mathis (1989) found that the mean extinction ratio at wavelength λ and the V band can be fitted to the form

$$\left\langle \frac{A_\lambda}{A_V} \right\rangle = a(x) + \frac{b(x)}{R_V}, \qquad (4.7)$$

where A_λ denotes the extinction in a given band λ (in units of magnitudes), and

$$x \equiv \frac{1\,\mu m}{\lambda}, \qquad R_V \equiv \frac{A_V}{E(B-V)}, \qquad (4.8)$$

with $E(B-V)$ denoting the color excess

$$E(B-V) = (B-V)_{\text{observed}} - (B-V)_{\text{intrinsic}}, \qquad (4.9)$$

where $(B-V)$ is the color index; B and V denote the apparent magnitudes in B and V bands. Figure 4.11 shows the mean extinction ratio A_λ/A_V as a function of λ for

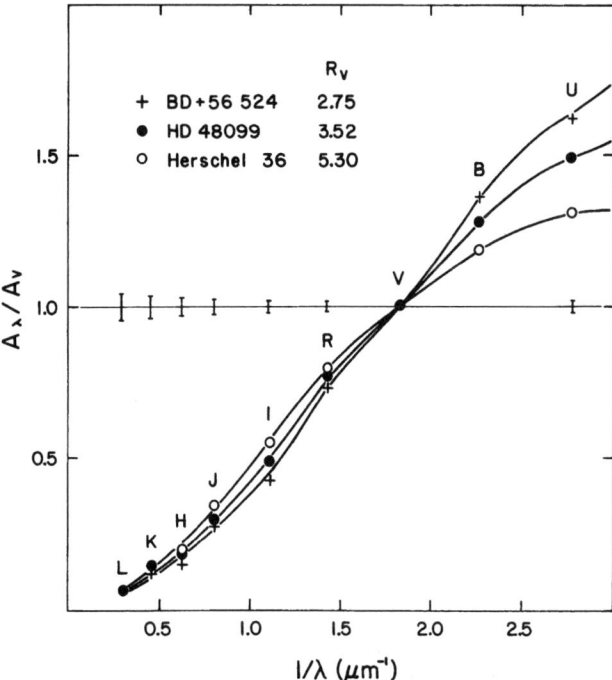

Figure 4.11 The mean extinction ratio A_λ/A_V as a function of λ for three different R_V values, based on NIR and optical data for three stars (Cardelli, Clayton, and Mathis, 1989). Clearly, at shorter wavelengths, the extinction is larger and sensitive to the value of R_V.

three different R_V values. Clearly, at shorter wavelengths, the extinction is larger and more sensitive to the value of R_V.

A feasible supernova survey from space can obtain restframe J band light curves for all the SNe Ia with $0 \lesssim z \lesssim 2$ (Wang et al., 2004; Crotts et al., 2005; Phillips et al., 2006). Since $A_J/A_V = 0.282$ (Cardelli, Clayton, and Mathis, 1989), such a survey is much less sensitive to dust extinction compared to surveys that obtain restframe SN Ia light curves at optical wavelengths. The overall effect of extinction can be reduced to less than 1%.

4.4.2
K-Correction

In order to obtain correct cosmological constraints from SN Ia data, the photometry of the SNe Ia at different redshifts in different color bands must be mapped onto a consistent restframe band. Much of the current SN Ia observations have been processed by mapping to the restframe B band. Figure 4.12 gives an example of how an observed filter can differ from the restframe B band.

For a high-redshift supernova, the light of a restframe filter band is redshifted to a longer wavelength, and is observed through an observed filter band that overlaps, but is not identical to, the redshifted restframe filter band (see Figure 4.12). The apparent magnitude in the observed filter, y, is related to the absolute magnitude in the restframe filter, x as follows (Hogg et al., 2002):

$$m_y = M_x + \mu_0 + K_{xy}, \tag{4.10}$$

where μ_0 is the distance modulus

$$\mu_0 = 5 \log \left(\frac{d_L(z)}{10 \, \text{pc}} \right). \tag{4.11}$$

The cross-filter K-correction (Kim, Goobar, and Perlmutter, 1996), K_{xy}, allows one to transform the magnitude in the observed filter, y, to the magnitude in the restframe filter, x:

$$K_{xy}(t, z, \mathbf{p}) = -2.5 \log \left(\frac{\int \lambda T_x(\lambda) Z(\lambda) d\lambda}{\int \lambda T_y(\lambda) Z(\lambda) d\lambda} \right)$$
$$+ 2.5 \log \left(\frac{\int \lambda T_x(\lambda) S(\lambda, t, \mathbf{p}) d\lambda}{\int \lambda T_y(\lambda) \left[S(\lambda/(1+z), t, \mathbf{p}) /(1+z) \right] d\lambda} \right), \tag{4.12}$$

where t is the epoch, z is the redshift, and $\mathbf{p}$ is a vector of parameters that affect the broadband colors of the SN Ia (for example, light curve shape and reddening). The right hand side of Eq. (4.12) contains integration over the wavelength λ. In the integrands, T_x and T_y denote the effective transmission of the x and y filter bands, respectively, Z denotes the spectral energy distribution (SED) for which the x–y color is precisely known, and S denotes the SED of the SN Ia. Note that the normalization of the SED is arbitrary. Equation (4.12) relates the SN Ia fluxes in a restframe filter x and a de-redshifted observed filter y.

In order to minimize the systematic errors caused by an assumed SN Ia SED, Hsiao et al. (2007) outlined a prescription for deriving a mean spectral template time series using a library of $\sim$600 observed spectra of $\sim$100 SNe Ia from heterogeneous sources. They removed the effects of broadband colors and measured the remaining uncertainties in the K-correction associated with the diversity in spectral features, and presented a template spectroscopic sequence near maximum light for further improvement on the K-correction estimate (Hsiao et al., 2007).

We can use a set of restframe template spectra to build a restframe SED for the entire SN Ia wavelength range. Spectral features of SNe Ia evolve rapidly around maximum light ($t = 0$) and slow down past $t = 30$ days. It is important to use small epoch bins near maximum light (Hsiao et al., 2007).

With a very large sample of SNe Ia with at least several spectra per SN Ia at different epochs, we can calibrate and improve the template SED, and reduce the

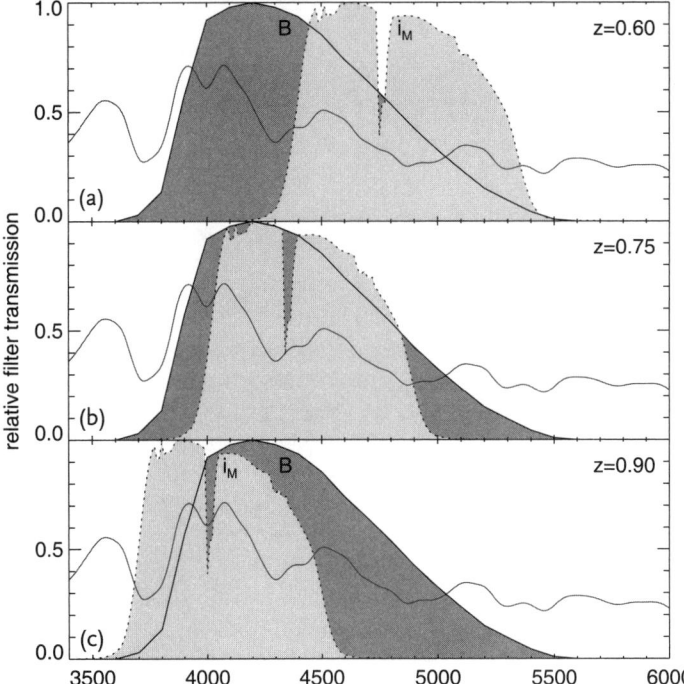

Figure 4.12 The pairing of observed filter i_M and restframe filter B at redshifts 0.6 (a), 0.75 (b) and 0.9 (c) (Hsiao et al., 2007). The solid curves are the transmission of B band, and the dotted curves are the transmission of the deredshifted i_M band. A typical SN Ia spectrum is shown in the background. The filter bands are misaligned at $z = 0.6$ (a) and $z = 0.9$ (c). The K-corrections at these redshifts depend heavily on the assumed spectral template (the faint continuous line extending over the entire wavelength range in all three panels).

uncertainty in K-corrections to around 0.01 mag (Wang et al., 2004; Crotts et al., 2005; Phillips et al., 2006).

4.4.3
Weak Lensing

The bending of the light from SNe Ia by intervening matter will modify the observed brightness of SNe Ia. The weak lensing amplification of SNe Ia by cosmic large scale structure can be modeled by a universal probability distribution function for weak lensing amplification based on the measured matter power spectrum (Wang, Holz, and Munshi, 2002). The effect of weak lensing on the SN Ia data can be minimized through flux averaging, which is discussed in Section 4.5.2.

Due to the deflection of light by density fluctuations along the line of sight, a source (at redshift z_s) will be magnified by a factor $\mu \simeq 1 + 2\kappa$ (the weak lensing limit), where the convergence κ is given by (Bernardeau, Van Waerbeke, and Mellier, 1997; Kaiser, 1998)

$$\kappa = \frac{3}{2}\Omega_m \int_0^{\chi_s} d\chi\, w(\chi,\chi_s)\delta(\chi), \qquad (4.13)$$

with

$$d\chi = \frac{cH_0^{-1}dz}{E(z)},$$

$$w(\chi,\chi_s) = \frac{H_0^2}{c^2}\frac{\mathcal{D}(\chi)\mathcal{D}(\chi_s - \chi)}{\mathcal{D}(\chi_s)}(1+z),$$

$$\mathcal{D}(\chi) = \frac{cH_0^{-1}}{\sqrt{|\Omega_k|}}\text{sinn}\left(\sqrt{|\Omega_k|}\chi\right),$$

and where "sinn" is defined as sinh if $\Omega_k > 0$, and sin if $\Omega_k < 0$. If $\Omega_k = 0$, the sinn and the Ω_k disappear. The density contrast $\delta \equiv (\rho_m - \bar{\rho}_m)/\bar{\rho}_m$. Since $\rho_m \geq 0$, there exists a minimum value of the convergence. Note, however, the minimum convergence is *not* given by setting $\delta = -1$ (i.e., $\rho_m = 0$) in Eq. (4.13), but by requiring that

$$\mu_{\min} = \frac{1}{(1-\tilde{\kappa}_{\min})^2}. \qquad (4.14)$$

The minimum magnification can be found by considering the alternative definition of magnification in terms of angular diameter distances (Schneider, Ehlers, and Falco, 1992)

$$\mu \equiv \left|\frac{D_A(\tilde{\alpha}=1)}{D_A(\tilde{\alpha})}\right|^2, \qquad (4.15)$$

where $\tilde{\alpha}$ is the mass-fraction $\tilde{\alpha}$ of the smoothly distributed matter in the universe, and $D_A(z|\tilde{\alpha})$ is the Dyer–Roeder distance. For a given redshift z, the largest possible distance for light bundles which have not passed through a caustic is given by the Dyer–Roeder distance, which satisfies the Dyer–Roeder equation (Dyer and Roeder, 1973; Kantowski, 1998):

$$g(z)\frac{d}{dz}\left[g(z)\frac{dD_A}{dz}\right] + \frac{3}{2}\tilde{\alpha}\Omega_m(1+z)^5 D_A = 0,$$

$$D_A(z=0) = 0, \quad \left.\frac{dD_A}{dz}\right|_{z=0} = \frac{c}{H_0},$$

$$g(z) \equiv (1+z)^2 E(z). \tag{4.16}$$

Since $D_A(z|\tilde{\alpha})$ increases with decreasing $\tilde{\alpha}$, and $\tilde{\alpha} \geq 0$, the minimum magnification is given by

$$\mu_{min} = \left|\frac{D_A(\tilde{\alpha}=1)}{D_A(\tilde{\alpha}=0)}\right|^2 \tag{4.17}$$

and the minimum convergence is given by

$$\tilde{\kappa}_{min}(z) \equiv 1 - \frac{D_A(\tilde{\alpha}=0|z)}{D_A(\tilde{\alpha}=1|z)}$$

$$= -\frac{3}{2}\frac{\Omega_m}{cH_0^{-1}} \int_0^z dz' \frac{(1+z')^2}{E(z')} \frac{r(z')}{r(z)}(1+z)[\lambda(z) - \lambda(z')], \tag{4.18}$$

using $D_A(\tilde{\alpha}=1|z) = r(z)/(1+z)$ (Wang, 2000a). Note that $\tilde{\kappa}_{min}(z) < 0$. The affine parameter is

$$\lambda(z) = cH_0^{-1} \int_0^z \frac{dz'}{(1+z')^2 E(z')}. \tag{4.19}$$

The observed flux from a SN Ia is

$$f = \mu L_{int}, \tag{4.20}$$

where L_{int} is the intrinsic brightness of the SN Ia, and μ is the magnification due to intervening matter. Note that μ and L_{int} are statistically independent. The probability density distribution (pdf) of the product of two statistically independent variables can be found given the pdf of each variable (see, e.g., Lupton (1993)).

We find that the pdf of the observed flux f is given by (Wang, 2005)

$$p(f) = \int_0^{L_{int}^{max}} \frac{dL_{int}}{L_{int}} p_L(L_{int}) p\left(\frac{f}{L_{int}}\right), \tag{4.21}$$

where $p_L(L_{int})$ is the pdf of the intrinsic peak brightness of SNe Ia, $p(\mu)$ is the pdf of the magnification of SNe Ia. The upper limit of the integration, $L_{int}^{max} = f/\mu_{min}$, results from $\mu = f/L_{int} \geq \mu_{min}$.

A definitive measurement of $p_L(L_{\text{int}})$ will require a much greater number of well-measured SNe Ia at low z than is available at present. Since $p_L(L_{\text{int}})$ is not sufficiently well-determined at present, we present results for two different $p_L(L_{\text{int}})$: Gaussian in flux and Gaussian in magnitude.

The $p(\mu)$ can be computed numerically using cosmological volume N-body simulations (Wambsganss et al., 1997; Barber et al., 2000; Premadi et al., 2001; Vale and White, 2003). We can derive the $p(\mu)$ for an arbitrary cosmological model by using the universal probability distribution function (UPDF) of weak lensing amplification (Wang, 1999; Wang, Holz, and Munshi, 2002), with the corrected definition of the minimum convergence in Eq. (4.18) (Wang, Tenbarge, and Fleshman, 2003).

We now present an improved UPDF (compared with the original UPDF (Wang, Holz, and Munshi, 2002)), with the corrected minimum convergence and extended to high magnifications. The numerical simulation data of $p(\mu)$ are converted to the modified UPDF, pdf of the reduced convergence η,

$$P(\eta) = \frac{1}{1+\eta^2} \exp\left[-\left(\frac{\eta - \eta_{\text{peak}}}{w_\eta \eta^q}\right)^2\right], \tag{4.22}$$

where

$$\eta \equiv 1 + \frac{\mu - 1}{|\mu_{\text{min}} - 1|}. \tag{4.23}$$

The parameters of the UPDF, η_{peak}, w_η, and q are only functions of the variance of η, ξ_η, which absorbs all the cosmological model dependence. The functions $\eta_{\text{peak}}(\xi_\eta)$, $w_\eta(\xi_\eta)$, and $q(\xi_\eta)$ are extracted from the numerical simulation data.

For an arbitrary cosmological model, one can readily compute ξ_η (Valageas, 2000)

$$\xi_\eta = \int_0^{\chi_s} d\chi \left(\frac{w}{F_s}\right)^2 I_\mu(\chi), \tag{4.24}$$

with

$$F_s = \int_0^{\chi_s} d\chi\, w(\chi, \chi_s), \quad I_\mu(z) = \pi \int_0^\infty \frac{dk}{k} \frac{\Delta^2(k,z)}{k} W^2(\mathcal{D}k\theta_0),$$

$$\Delta^2(k,z) = 4\pi k^3 P_m(k,z), \quad W(\mathcal{D}k\theta_0) = \frac{2J_1(\mathcal{D}k\theta_0)}{\mathcal{D}k\theta_0}, \tag{4.25}$$

where k is the wavenumber, $P_m(k,z)$ is the matter power spectrum, θ_0 is the smoothing angle, and J_1 is the Bessel function of order 1.

Given ξ_η, the UPDF can be computed using Eq. (4.22). Since $\mu = 1 + |\mu_{\text{min}} - 1|(\eta - 1)$, we find (Wang, Holz, and Munshi, 2002)

$$P(\mu) = \frac{P(\eta|\xi_\eta)}{|\mu_{\text{min}} - 1|}. \tag{4.26}$$

Using data from Barber (2003), we find the following fitting formulae:

$$\eta_{\text{peak}}(\xi_\eta) \simeq 0.56465 + 0.78717x - 0.61943x^2 \tag{4.27}$$

$$w(\xi_\eta) \simeq 0.46824 - 0.22612x + 0.14831x^2 \tag{4.28}$$

$$q(\xi_\eta) \simeq 1.09206 - 0.10658x + 0.27844x^2 , \tag{4.29}$$

where

$$x(\xi_\eta) = \frac{0.04}{\left[(\sqrt{\xi_\eta} - 0.049)/2.2951\right]^{1/2} - 0.3174} . \tag{4.30}$$

Equations (4.27)–(4.29) are valid for $\sqrt{\xi_\eta(z)} \lesssim 0.4$. For a flat universe with $\Omega_m = 0.3$ and $\Omega_\Lambda = 0.7$, we found $x(\xi_\eta) = z/5$, and

$$\sigma_\eta \equiv \sqrt{\xi_\eta} \simeq 0.28031 + \frac{1.45701}{5z} + \frac{2.29508}{(5z)^2} \tag{4.31}$$

$$\sigma_{\text{lens}} \equiv \sigma_\mu \simeq 0.00311 + 0.43433\left(\frac{z}{5}\right) - 0.23759\left(\frac{z}{5}\right)^2 . \tag{4.32}$$

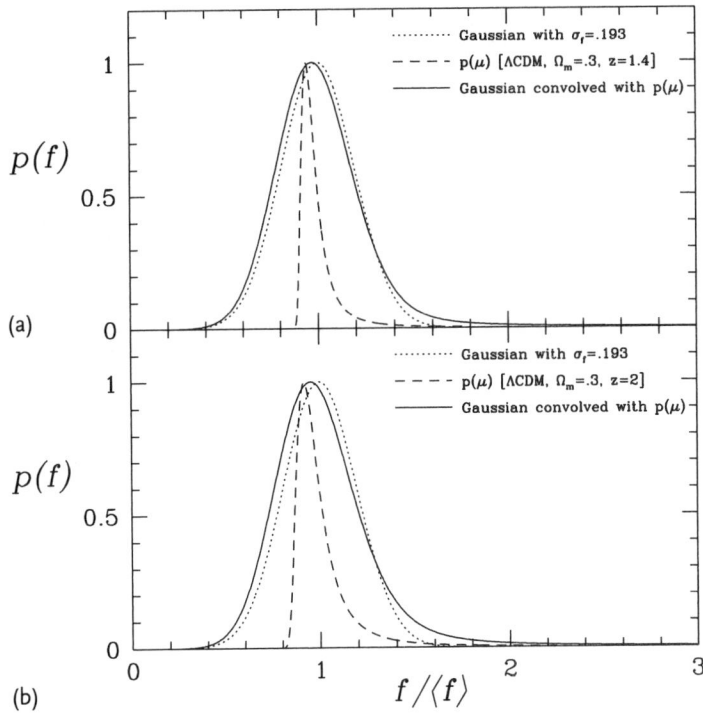

Figure 4.13 Prediction of the observed flux distributions of SNe Ia for magnification distribution $p(\mu)$ given by a ΛCDM model ($\Omega_m = 0.3$, $\Omega_\Lambda = 0.7$) at $z = 1.4$ (a) and $z = 2$ (b), respectively (Wang, 2005). It is assumed that the intrinsic peak brightness distribution, $p_L(L_{\text{int}})$, is Gaussian with a rms variance of 0.193 (in units of the mean flux).

The total uncertainty in each SN Ia data point is $\sqrt{\sigma_{\mathrm{int}}^2 + \sigma_{\mathrm{lens}}(z)^2}$, with $\sigma_{\mathrm{lens}}(z)$ given by Eq. (4.32).

Figure 4.13 shows the prediction of the observed flux distributions of SNe Ia, with magnification distribution $p(\mu)$ given by a ΛCDM model ($\Omega_m = 0.3$, $\Omega_\Lambda = 0.7$) at $z = 1.4$ (a) and $z = 2$ (b), respectively. We have assumed that the intrinsic peak brightness distribution, $p_L(L_{\mathrm{int}})$, is Gaussian in flux with a rms variance of 0.193 (in units of the mean flux, and chosen to be the same as the $0.02 \leq z \leq 0.1$ subset of the Riess et al. sample (Riess et al., 2004)). Figure 4.14 is the same as Figure 4.13, except here we have assumed that $p_L(L_{\mathrm{int}})$ is Gaussian in magnitude, with a rms variance of 0.213 mag (chosen to be the same as the $0.02 \leq z \leq 0.1$ subset of the Riess et al. sample (Riess et al., 2004)).

Clearly, there are two signatures of the weak lensing of SNe Ia in the observed brightness distribution of SNe Ia. The first signature is the presence of a non-Gaussian tail at the bright end, which is due to the high magnification tail of the magnification distribution. The second signature is the slight shift of the peak toward the faint end (compared to the pdf of the intrinsic SN Ia peak brightness), which is due to $p(\mu)$ peaking at $\mu < 1$ (demagnification) because the universe is mostly empty. As the redshift of the observed SNe Ia increases, the non-Gaussian

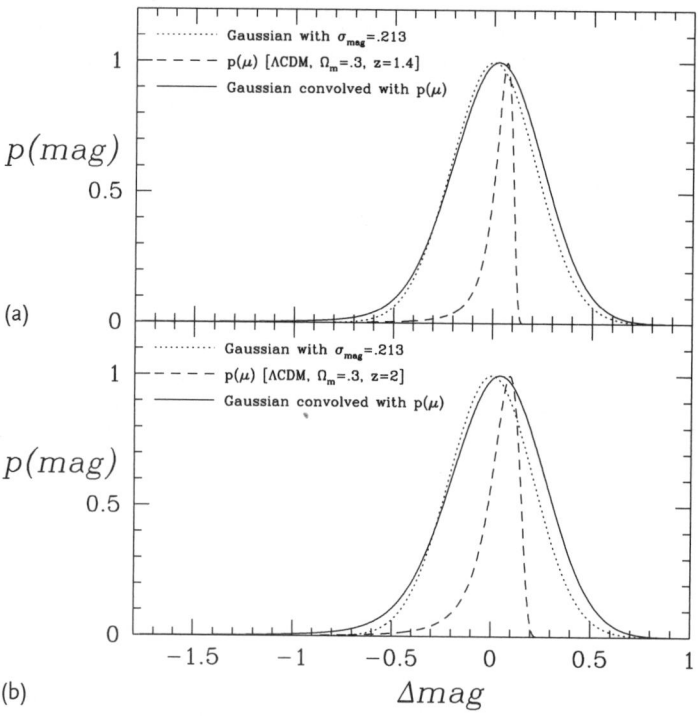

Figure 4.14 Same as Figure 4.13, except here it is assumed that the intrinsic peak brightness distribution, $p_L(L_{\mathrm{int}})$, is Gaussian in magnitude, with a rms variance of 0.213 mag (Wang, 2005). (a) $z = 1.4$ (b) $z = 2$.

tail at the bright end will grow larger, while the peak will shift further toward the faint end (see Figures 4.13 and 4.14).

If the distribution of the intrinsic SN Ia peak brightness is Gaussian in flux, the dominant signature of weak lensing is the presence of the high magnification tail in flux. If the distribution of the intrinsic SN Ia peak brightness is Gaussian in magnitude, the dominant signature of weak lensing is the shift of the peak of the observed magnitude distribution toward the faint end due to demagnification. This is as expected, since the magnitude scale stretches out the distribution at small flux, and compresses the distribution at large flux.

Current SN Ia data show both signatures of weak lensing of SNe Ia in the observed brightness distribution of SNe Ia, although at low statistical significance due to the small number of SNe Ia available for analysis (Wang, 2005).

With the hundreds of thousands of low to medium-redshift SNe Ia expected from future ground-based surveys, and the thousands of high-redshift SNe Ia from a space mission, it will be possible to rigorously study weak lensing effects on the SN Ia peak brightness distribution, and derive parameters that characterize $p_L(L_{int})$ and $p(\mu)$. Since $p(\mu)$ is a probe of cosmology in itself, it can be used to tighten constraints on cosmological parameters, and cross-check the dark energy constraints (Wang, 1999; Dodelson and Vallinotto, 2005; Cooray, Holz, and Huterer, 2006; Munshi and Valageas, 2006).

4.4.4
Other Systematic Uncertainties of SNe Ia

An optimally designed SN Ia survey should not suffer from other well known systematic uncertainties (not discussed here), such as the contamination by non Type Ia supernovae, and the Malmquist bias. Such a survey is feasible (Wang *et al.*, 2004; Crotts *et al.*, 2005; Phillips *et al.*, 2006), and could obtain over 10 000 SNe Ia from an ultra-deep survey of $\sim$20–30 (deg)^2 (Wang, 2000b), all with well-sampled light curves (in the restframe *J* band) and good-quality spectra (with several spectra per SN Ia sequenced in time).

There are also a couple of possible, though probably unlikely sources of systematic uncertainties, gray dust and SN Ia peak luminosity evolution.

Gray Dust
Gray dust, consisting of large dust grains, is difficult to detect by its reddening and could mimic the effect of dark energy (Aguirre, 1999). Gray dust can be constrained quantitatively by the cosmic far infrared background (Aguirre and Haiman, 2000), with no evidence found in favor of gray dust so far. Supernova flux correlation measurements can be used in combination with other lensing data to infer the level of dust extinction, and provide a viable method to eliminate possible gray dust contamination in SN Ia data (Zhang and Corasaniti, 2007). A broad wavelength coverage (for example, 0.8–4 µm) would allow us to tightly constrain the possibility of gray dust, since gray dust is not gray at wavelengths longer than around 1 µm.

Supernova Peak Luminosity Evolution

The evolution in SN Ia peak luminosity could arise due to progenitor population drift, since the most distant SNe Ia come from a stellar environment very different (a much younger universe) than that of the nearby SNe Ia. However, with sufficient statistics, we can subtype SNe Ia and compare SNe Ia at high redshift and low redshift that are similar in both light curves and spectra, thus overcoming the possible systematic effect due to progenitor population drift (Branch et al., 2001).

A sufficiently large sample of SNe Ia, say 10 000 at high z from a space mission, and 100 000 at low z from ground surveys, with well-sampled light curves and good quality spectra, will be critical in constraining SN Ia peak luminosity evolution. We can bin the SNe Ia by observational properties, such as color, light curve width, metallicity, spectral line ratios, and so on, and search for possible evolutionary effects. For example, in order to search for the possible drift in asymmetry with epoch, a very large number of SNe Ia per redshift bin would be needed to average over the aspect angle (Wheeler, 2009).

Figure 4.15 shows a sequence of simulated SN Ia spectra corresponding to SN Ia models with various metallicities in the C+O layer at 10 days after explosion (Lentz et al., 2000). The parameter ξ indicates the amount of metallicity compared to solar. Note the variability in the spectra (especially in the UV range) as a result of the change in metallicity. In principle, the UV portion of the redshifted SN Ia spectra

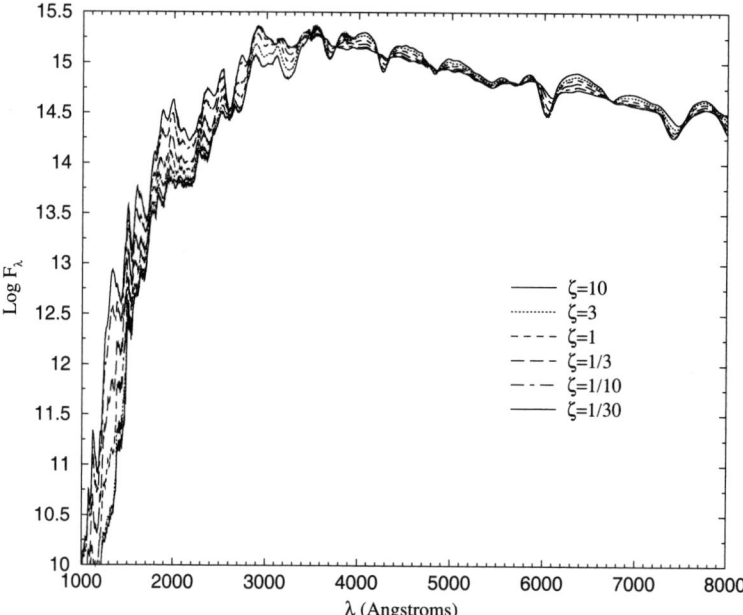

Figure 4.15 A sequence of simulated SN Ia spectra corresponding to SN Ia models with various metallicities in the C+O layer at 10 days after explosion (Lentz et al., 2000). The parameter ξ indicates the amount of metallicity compared to solar. Note the variability in the UV spectra as a result of the change in metallicity.

can be observed easily from the ground, and used to help sort SNe Ia into different metallicities. However, UV observations can lead to complications in calibration and data analysis, and dominate systematic errors (Kessler *et al.*, 2009).

The very large number of SNe Ia with well-sampled light curves and good-quality spectra will allow us to catch the most possible evolutionary effects through subtyping, as well as reducing the statistical scatter in each subtype. This in turn will allow us to improve the calibration of SNe Ia as cosmological standard candles.

4.5
Data Analysis Techniques

Here we discuss three key aspects of SN Ia data analysis: the fitting of SN Ia light curves to extract peak brightness, the flux-averaging analysis of SNe Ia to remove or reduce the weak lensing systematic effect, and the model-independent extraction of the cosmic expansion history $H(z)$ in uncorrelated redshift bins from SN Ia data for a flat universe.[5]

4.5.1
Light Curve Fitting

Ideally, SN Ia light curves should be fitted with a theoretical template based on SN Ia physics. The observable properties of SNe Ia should be determined by their final composition and some measure of mixing in the explosion. Woosley *et al.* (2007) explored the diversity of SN Ia light curves using a grid of 130 one-dimensional models. They found that the diversity of models far exceeds the actual diversity of SNe Ia found in nature. This suggests that our understanding of SN Ia explosions is still rather limited. Because of this, it is both practical and fruitful to use the empirical approach in fitting SN Ia light curves. There are two main empirical light curve-fitting techniques in use by observers, the multi-color light curve shape method (MLCS), and the spectral adaptive light curve template method (SALT).

The MLCS Method
The MLCS method was first developed by Phillips (1993), and Riess, Press, and Kirshner (1995). Jha, Riess, and Kirshner (2007) presented an updated version of MLCS, MLCS2k2. In this method, the observed light curves are first corrected for galactic extinction, K-correction, and time dilation. The corrected light curve, m_X, is fitted as follows:

$$m_X(t - t_0) = M_X^0 + \mu_0 + \zeta_X (\alpha_X + \beta_X/R_V) A_V^0 + P_X \Delta + Q_X \Delta^2, \quad (4.33)$$

where the bold-face denote quantities that depend on the SN restframe phase, t_0 is the epoch of maximum light in B, M_X^0 are the absolute magnitudes of the fidu-

---
5) This can be generalized to allow nonzero curvature when CMB data are included in the analysis, but then the $H(z)$ in different redshift bins become correlated.

cial SN Ia, μ_0 is the true distance modulus, R_V and A_V^0 are the host-galaxy extinction parameters, Δ is the luminosity/light-curve shape parameter, and $\boldsymbol{P}_X$ and $\boldsymbol{Q}_X$ are vectors describing the change in light curve shape as a quadratic function of Δ. Note that the usual prior of $A_V > 0$ in the MLCS method has been shown to be problematic in data analysis, and leads to results inconsistent with that from the SALT method (Kessler *et al.*, 2009). This indicates that $A_V > 0$ is too strong an assumption, given the unknown mixing of intrinsic SN Ia color and dust extinction.

The parameter ζ_X is defined as

$$\zeta_X \equiv \frac{A_X}{A_X^0}, \tag{4.34}$$

where A_X is the extinction in a given band X (it is a function of SN phase t). A_X^0 is defined as the extinction in passband X at maximum light in B. Thus, $\zeta_X(t = t_0) \equiv 1$ by definition. ζ_X can be calculated for a given set of filter bands and a set of time-sequenced SN Ia template spectra. ζ_X captures all the time-dependence in the extinction, and is insensitive to the total extinction and the extinction law R_V (Jha, Riess, and Kirshner, 2007).

The ratio A_X^0/A_V^0 can be fitted as follows (Cardelli, Clayton, and Mathis, 1989) (see Section 4.4.1):

$$\frac{A_X^0}{A_V^0} = \alpha_X + \frac{\beta_X}{R_V}, \tag{4.35}$$

where α_X and β_X are fitted coefficients.

There are five free parameters in the MLCS model that describe a given SN Ia: t_0, μ_0, Δ, A_V^0, and R_V. Using a training set, one can estimate initial values of the free parameters based on relative distances from the Hubble Law, and solve for the model vectors $\boldsymbol{M}_X^0$, $\boldsymbol{P}_X$, and $\boldsymbol{Q}_X$. Next, one can construct an empirical model covariance matrix S that incorporates the variance and covariance in the residuals of the training set data from the model (minus the variance and covariance in the training set data itself) (Jha, Riess, and Kirshner, 2007). The diagonal elements of the S matrix are derived from the variance about the model, while the off-diagonal elements are estimated from two-point correlations (in the same passband at different epochs, in different passbands at the same epoch and in different passbands at different epochs) (Riess *et al.*, 1998b).

The best-fit model parameters are found via χ^2 minimization, with

$$\chi^2 = \boldsymbol{r}^T C^{-1} \boldsymbol{r}, \tag{4.36}$$

where $\boldsymbol{r}$ is the vector of light curve residuals (in all bands) for a given set of model parameters, and $C = S + N$ (N is the covariance matrix of noise in the observed light curve).

The SALT Method

The SALT method was first developed by Guy *et al.* (2005) (SALT1 (Guy *et al.*, 2005)), and further refined by Guy *et al.* (2007) (SALT2 (Guy *et al.*, 2007)). This method explicitly addresses the problem of uncertainties due to the variability of the large features of SNe Ia spectra. Using a single SN Ia spectral sequence, for example, the most widely used spectral sequence provided by Nugent, Kim, and Perlmutter (2002), could result in significant systematic effects in the distance measurements. The main goal of the SALT approach is to provide the best "average" spectral sequence and the principal components responsible for the diversity of SNe Ia, so that the model can account for possible variations in SNe Ia spectra at any given phase.

In SALT1 (Guy et al., 2005), broadband corrections are applied to the spectral sequence of Nugent, Kim, and Perlmutter (2002) as a function of phase, wavelength and a stretch factor, so that the spectra integrated in response functions match the observed light-curves. SALT1 only uses multi-band light curves to train the model. The refined SALT method, SALT2 (Guy et al., 2007), includes both photometric and spectroscopic data to improve the model resolution in wavelength space, and is thus independent of any standard spectral sequence.

K-corrections are naturally built into the SALT model, thus they do not need to be done separately (this differs from the MLCS method). The flux normalization of each SN Ia is a free parameter of the SALT model; thus both nearby and distant SNe can be used in the training set without any priors on cosmology. In particular, the use of high-redshift SNe Ia enables the modeling of the restframe UV emission, which is critical to improving distance estimates for SNe Ia at $z > 0.8$.

The flux of a SN Ia is modeled by (Guy et al., 2007)

$$F(SN, p, \lambda) = x_0 \left[M_0(p, \lambda) + x_1 M_1(p, \lambda) + \ldots \right] \exp\left[c CL(\lambda) \right], \quad (4.37)$$

where p is the restframe time since the date of maximum luminosity in B-band (the phase), and λ the wavelength in the restframe of the SN. $M_0(p, \lambda)$ is the average spectral sequence whereas $M_i(p, \lambda)$ ($i = 1, 2, \ldots$) are additional components that describe the main variability of SNe Ia. $CL(\lambda)$ represents the average color correction law (no distinction between intrinsic or extinction by dust in the host galaxy – in contrast to the MLCS method). The optical depth is expressed using a color offset with respect to the average at the date of maximum light in the B-band, $c = (B - V)_{MAX} - \langle B - V \rangle$. This parametrization models the part of the color variation that is independent of phase, whereas the remaining color variation with phase is accounted for by the linear components. The parameter x_0 is the normalization of the SED sequence, and x_i ($i = 1, 2, \ldots$) are the intrinsic parameters of this SN (such as a stretch factor).

The parameters $\{x_i\}$ and c describe a given supernova. $\{M_i\}$ and CL are global properties of the model (rather than fitting analogous parameters to individual SN Ia as in the MLCS method), and are derived using a training set of SNe Ia in an iterative process that minimizes the χ^2 comparing the data with the model (Guy et al., 2007). The SALT code is publicly available.[6]

---

6) See http://supernovae.in2p3.fr/~guy/salt/

4.5.2
Flux-Averaging Analysis of SNe Ia

Since our universe is inhomogeneous in matter distribution, weak gravitational lensing by galaxies is one of the main systematic effects in the use of SNe Ia as cosmological standard candles. Flux averaging *justifies* the use of the distance-redshift relation for a smooth universe in the analysis of SN Ia data (Wang, 2000a). Flux averaging of SN Ia data is required to yield cosmological parameter constraints that are free of the bias induced by weak gravitational lensing (Wang, 2000a). To avoid missing the faint end (which is fortunately steep) of the magnification distribution of observed SNe Ia, only SNe Ia detected well above the threshold should be used in flux averaging. Here we present a consistent framework for flux-averaging analysis of SN Ia data, based on Wang (2000a) and Wang and Mukherjee (2004).

Why Flux Averaging?
The reason that flux averaging can remove/reduce gravitational lensing bias is that due to flux conservation, the average magnification of a sufficient number of standard candles at the same redshift is one.

The observed flux from a SN Ia can be written as

$$F(z) = F_{int}\mu\,,\quad F_{int} = F^{tr}(z|\mathbf{s}^{tr}) + \Delta F_{int}\,, \tag{4.38}$$

where $F^{tr}(z|\mathbf{s}^{tr})$ is the predicted flux due to the true cosmological model parametrized by the set of cosmological parameters $\{\mathbf{s}^{tr}\}$, ΔF_{int} is the uncertainty in SN Ia peak brightness due to intrinsic variations in SN Ia peak luminosity and observational uncertainties, and μ is the magnification due to gravitational lensing by intervening matter. Therefore

$$\Delta F^2 = \mu^2 \left(\Delta F_{int}\right)^2 + \left(F_{int}\right)^2 \left(\Delta\mu\right)^2\,. \tag{4.39}$$

Without flux averaging, we have

$$\chi^2_{N_{data}}(\mathbf{s}^{tr}) = \sum_i \frac{\left[F(z_i) - F^{tr}(z_i|\mathbf{s}^{tr})\right]^2}{\sigma^2_{F,i}}$$

$$= \sum_i \frac{\left[F^{tr}(z_i)(\mu_i - 1)\right]^2 + \mu_i^2 \left[\Delta F^{(i)}_{int}\right]^2}{\sigma^2_{F,i}}$$

$$+ 2\sum_i \frac{F^{tr}(z_i)\Delta F^{(i)}_{int}\mu_i(\mu_i - 1)}{\sigma^2_{F,i}}$$

$$\simeq \sum_i \frac{\left[F^{tr}(z_i)\right]^2 \xi^2_\mu}{\sigma^2_{F,i}} + N_{data}\,, \tag{4.40}$$

where

$$\xi^2_\mu \equiv \left\langle(\mu - \langle\mu\rangle)^2\right\rangle = \left\langle(\mu - 1)^2\right\rangle\,. \tag{4.41}$$

4.5 Data Analysis Techniques

Flux averaging leads to the flux in each redshift bin

$$\overline{F}(\overline{z}_{i_{\text{bin}}}) = F^{\text{tr}}(\overline{z}_{i_{\text{bin}}})\langle \mu \rangle_{i_{\text{bin}}} + \langle \mu \Delta F_{\text{int}} \rangle_{i_{\text{bin}}}. \tag{4.42}$$

For a sufficiently large number of SNe Ia in the ith bin, $\langle \mu \rangle_{i_{\text{bin}}} = 1$. Hence

$$\chi^2_{N_{\text{bin}}}(\mathbf{s}^{\text{tr}}) \simeq \sum_{i_{\text{bin}}}^{N_{\text{bin}}} \frac{\left[\langle \mu \Delta F_{\text{int}} \rangle_{i_{\text{bin}}}\right]^2}{\sigma^2_{F,i_{\text{bin}}}} \simeq \sum_{i_{\text{bin}}}^{N_{\text{bin}}} \frac{\left[\langle \Delta F_{\text{int}} \rangle_{i_{\text{bin}}}\right]^2}{\sigma^2_{F,i_{\text{bin}}}} \simeq N_{\text{bin}}. \tag{4.43}$$

Comparison of Eqs. (4.43) and (4.40) shows that flux averaging can remove/reduce the gravitational lensing effect, and leads to a smaller χ^2 per degree of freedom for the true model, compared to that from without flux averaging.

Flux Statistics Versus Magnitude Statistics

Normally distributed measurement errors are required if the χ^2 parameter estimate is to be a maximum likelihood estimator (Press *et al.*, 2007). Hence, it is important that we use the χ^2 statistics with an observable that has an error distribution closest to Gaussian.

It is usually assumed that the distribution of observed SN Ia peak brightness is Gaussian in *magnitudes*. Therefore, for a given set of cosmological parameters $\{\mathbf{s}\}$

$$\chi^2 = \sum_i \frac{\left[\mu_0(z_i) - \mu_0^p(z_i|\mathbf{s})\right]^2}{\sigma^2_{\mu_0}}, \tag{4.44}$$

where $\mu_0^p(z) = 5\log(d_L(z)/\text{Mpc}) + 25$, and $d_L(z) = (1+z)r(z)$ is the luminosity distance.

However, while we do not have a very clear understanding of how the intrinsic dispersions in SN Ia peak luminosity is distributed, the distribution of observational uncertainties in SN Ia peak brightness is Gaussian in *flux*, since CCDs have replaced photometric plates as detectors of photons.

Hence, we assume that the intrinsic dispersions in SN Ia peak brightness is Gaussian in *flux*, and not in magnitude. Thus,

$$\chi^2_{N_{\text{data}}}(\mathbf{s}) = \sum_i \frac{\left[F(z_i) - F^p(z_i|\mathbf{s})\right]^2}{\sigma^2_{F,i}}. \tag{4.45}$$

If the peak brightness of SNe Ia is given in magnitudes with symmetric error bars, $m_{\text{peak}} \pm \sigma_m$, we can obtain equivalent errors in flux as follows:

$$\sigma_F \equiv \frac{F(m_{\text{peak}} + \sigma_m) - F(m_{\text{peak}} - \sigma_m)}{2}. \tag{4.46}$$

We will refer to Eq. (4.44) as "magnitude statistics", and Eq. (4.45) as "flux statistics". A consistent framework for flux averaging is only straightforward in "flux statistics". If the dispersions in SN Ia peak brightness were Gaussian in magnitude, flux averaging would introduce a small bias (Wang, 2000a).

A Recipe for Flux Averaging

The fluxes of SNe Ia in a redshift bin should only be averaged *after* removing their redshift-dependence, which is a model-dependent process. For χ^2 statistics using MCMC or a grid of parameters, here are the steps in flux averaging:

1. Convert the distance modulus of SNe Ia into "fluxes",

$$F(z_j) \equiv 10^{-(\mu_0(z_j)-25)/2.5} = \left(\frac{d_L^{\text{data}}(z)}{\text{Mpc}}\right)^{-2}. \tag{4.47}$$

2. For a given set of cosmological parameters $\{s\}$, obtain "absolute luminosities", $\{\mathcal{L}(z_j)\}$, by removing the redshift-dependence of the "fluxes", that is,

$$\mathcal{L}(z_j) \equiv d_L^2(z_j|s)F(z_j). \tag{4.48}$$

3. Flux average the "absolute luminosities", $\{\mathcal{L}_j^i\}$, in each redshift bin i to obtain $\{\overline{\mathcal{L}}^i\}$:

$$\overline{\mathcal{L}}^i = \frac{1}{N}\sum_{j=1}^{N}\mathcal{L}_j^i(z_j^i), \quad \overline{z}_i = \frac{1}{N}\sum_{j=1}^{N}z_j^i. \tag{4.49}$$

4. Place $\overline{\mathcal{L}}^i$ at the mean redshift $\overline{z}_i$ of the ith redshift bin, now the binned flux is

$$\overline{F}(\overline{z}_i) = \overline{\mathcal{L}}^i/d_L^2(\overline{z}_i|s). \tag{4.50}$$

The 1σ error on each binned data point $\overline{F}^i$, σ_i^F, is taken to be the root mean square sum of the 1σ errors on the unbinned data points in the i-th redshift bin, $\{F_j^i\}$ ($j = 1, 2, \ldots, N$), multiplied by $1/\sqrt{N}$ (see Wang (2000b)).

5. For the flux-averaged data, $\{\overline{F}(\overline{z}_i)\}$, we find

$$\chi^2 = \sum_i \frac{\left[\overline{F}(\overline{z}_i) - F^p(\overline{z}_i|s)\right]^2}{\sigma_{F,i}^2}, \tag{4.51}$$

where $F^p(\overline{z}_i|s) = (d_L(z|s)/\text{Mpc})^{-2}$.

Marginalization Over H_0 in SN Ia Flux Statistics

Because of calibration uncertainties, SN Ia data need to be marginalized over H_0 if SN Ia data are combined with data that are sensitive to the value of H_0. For example, if we use the angular scale of the sound horizon at recombination l_a which depends on $\Omega_m h^2$, while the scaled Hubble parameter $E(z) = H(z)/H_0$ (which appears in the derivation of all distance-redshift relations) depends on Ω_m, a dependence on H_0 is implied.

The marginalization of SN Ia data over H_0 was derived in Wang and Garnavich (2001) for the usual magnitude statistics (assuming that the intrinsic dispersion in SN Ia peak brightness is Gaussian in magnitudes). Here we present the formalism for marginalizing SN Ia data over H_0 in the flux averaging of SN Ia data using flux statistics (see Eq. (4.45)). Public software for implementing SN Ia flux averaging with marginalization over H_0 (compatible with *cosmomc*) is available.[7]

In the χ^2 for SN Ia data after flux averaging, Eq. (4.51), the predicted SN Ia flux $F^p(z_i|s) = [d_L(z_i|s)/\text{Mpc}]^{-2} \propto h^2$. Assuming that the dimensionless Hubble parameter h is uniformly distributed in the range [0,1], it is straightforward to integrate over h in the probability distribution function to obtain

$$p(s|0 \leq h \leq 1) = e^{-\chi^2/2} = \frac{\int_0^1 dx e^{-g(x)}}{\int_0^1 dx e^{-g_0(x)}} \quad (4.52)$$

where

$$g(x) \equiv \sum_i \frac{\left[\overline{F}(\overline{z}_i) - x^2 F_*^p(\overline{z}_i|s)\right]^2}{2\sigma_{\overline{F},i}^2},$$

$$g_0(x) \equiv (x^2 - 1)^2 \sum_i \frac{\overline{F}(\overline{z}_i)^2}{2\sigma_{\overline{F},i}^2}, \quad (4.53)$$

where $F_*^p(\overline{z}_i|s) = F^p(\overline{z}_i|s, h = 1)$.

4.5.3
Uncorrelated Estimate of H(z)

We conclude this chapter by presenting a method that allows a robust and intuitive interpretation of SN Ia data, the uncorrelated measurement of the expansion history $H(z)$ in arbitrary redshift bins (Wang and Tegmark, 2005). This method is both easy to implement and easy to interpret.

We assume spatial flatness for simplicity. Thus, SNe Ia measure the luminosity distance $d_L(z) = (1 + z)r(z)$, where the comoving distance

$$r(z) = \int_0^z \frac{dz'}{H(z')} = H_*^{-1} \int_0^z \frac{dz'}{h(z')}, \quad (4.54)$$

where $c = 1$, $H_* = 100 \,\text{km}\,\text{s}^{-1}\,\text{Mpc}^{-1}$ and

$$h(z) \equiv H(z)/H_* = h(0) \left[\Omega_m (1 + z)^3 + \Omega_X X(z)\right]^{1/2}, \quad (4.55)$$

with $X(z) \equiv \rho_X(z)/\rho_X(0)$ denoting the dark energy density function. Determining if and (if so) how the dark energy density $X(z)$ depends on cosmic time is the main

---

7) http://www.nhn.ou.edu/~wang/SNcode/.

observational goal in the current quest to illuminate the nature of dark energy. Given a precise measurement of the matter density fraction Ω_m (from galaxy redshift surveys, for example), the dark energy density function $X(z)$ can be trivially determined from $H(z)$ via Eq. (4.55).

Assuming that the redshifts of SNe Ia are accurately measured, we can neglect redshift uncertainties, and simply treat the measured comoving distances $r(z_i)$ to the supernovae as the observables. In terms of μ_0, the distance modulus of SNe Ia, we have

$$\frac{r(z)}{1\,\mathrm{Mpc}} = \frac{1}{1+z} 10^{\mu_0/5-5}. \tag{4.56}$$

Let us write the comoving distance measured from the ith SN Ia as

$$r_i = r(z_i) + n_i, \tag{4.57}$$

where the noise vector satisfies $\langle n_i \rangle = 0$, $\langle n_i n_j \rangle = \sigma_i^2 \delta_{ij}$.

Transforming to Measurements of $H(z)^{-1}$

As a first step in our method, we sort the supernovae by increasing redshift $z_1 < z_2 < \ldots$, and define the quantities

$$x_i \equiv \frac{r_{i+1} - r_i}{z_{i+1} - z_i} = \frac{\int_{z_i}^{z_{i+1}} \frac{dz'}{H(z')} + n_{i+1} - n_i}{z_{i+1} - z_i} = \overline{f}_i + (n_{i+1} - n_i)/\Delta z_i, \tag{4.58}$$

where $\Delta z_i \equiv z_{i+1} - z_i$ and $\overline{f}_i$ is the average of $1/H(z)$ over the redshift range (z_i, z_{i+1}). Note that x_i gives an unbiased estimate of the average of $1/H(z)$ in the redshift bin, since $\langle x_i \rangle = \overline{f}_i$. Thus the quantities x_i are direct (but noisy) probes of the cosmic expansion history. Assembling the numbers x_i into a vector $\boldsymbol{x}$, its covariance matrix $\mathbf{N} \equiv \langle \boldsymbol{x}\boldsymbol{x}^t \rangle - \langle \boldsymbol{x} \rangle \langle \boldsymbol{x} \rangle^t$ is tridiagonal, satisfying $N_{ij} = 0$ except for the following cases:

$$N_{i,i-1} = -\frac{\sigma_i^2}{\Delta z_{i-1} \Delta z_i},$$

$$N_{i,i} = \frac{\sigma_i^2 + \sigma_{i+1}^2}{\Delta z_i^2},$$

$$N_{i,i+1} = -\frac{\sigma_{i+1}^2}{\Delta z_i \Delta z_{i+1}}. \tag{4.59}$$

The new data vector $\boldsymbol{x}$ clearly retains all the cosmological information from the original data set (the comoving distance measurements r_i), since the latter can be trivially recovered from $\boldsymbol{x}$ up to an overall constant offset by inverting Eq. (4.58). In summary, the transformed data vector $\boldsymbol{x}$ expresses the SN Ia information as a large number of unbiased but noisy measurements of the cosmic expansion history in very fine redshift bins, corresponding to the redshift separations between neighboring supernovae.

Averaging in Redshift Bins

The second step in our method is to average these noisy measurements x into minimum-variance measurements y_b of the expansion history in some given redshift bins. For instance, Figure 4.16b shows an example with seven bins, $b = 1, \ldots, 7$. Let the vector x^b denote the piece of the x-vector corresponding to the bth bin, and let N_b denote the corresponding covariance matrix. The weighted

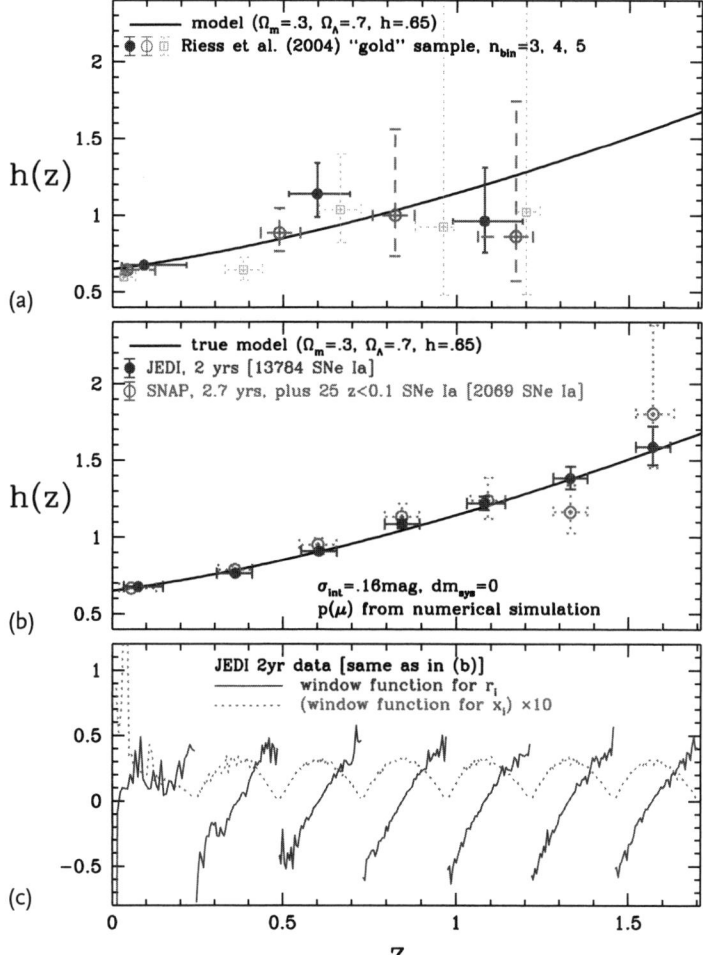

Figure 4.16 The cosmic expansion history (the dimensionless Hubble parameter $h(z)$) is measured in uncorrelated redshift bins from the Riess et al. "gold" sample (Riess et al., 2004) (a) and from simulated future data (b) for the NASA/JDEM mission concepts JEDI (solid points) and SNAP (dotted points) (Wang and Tegmark, 2005). The measured $h(z)^{-1}$ in a given redshift bin is simply the sum of the comoving supernova distances in that bin, weighted by the corresponding solid curve in (c), which roughly speaking subtracts more nearby supernovae from more distant ones.

average y_b can then be written

$$y_b = \mathbf{w}^b \cdot \mathbf{x}^b \tag{4.60}$$

for some weight vector $\mathbf{w}^b$ whose components sum to unity, that is, $\sum_i w_i^b = 1$ or equivalently $\mathbf{e} \cdot \mathbf{w}^b = 1$, where $\mathbf{e}$ is a vector containing all ones; $e_i = 1$. To find the best weight vector $\mathbf{w}^b$, we minimize the variance

$$\Delta y_b^2 \equiv \langle y_b^2 \rangle - \langle y_b \rangle^2 = \mathbf{w}^{b^t} \mathbf{N}_b \mathbf{w}^b \tag{4.61}$$

subject to the constraint that the weights add up to unity, that is, $\mathbf{e} \cdot \mathbf{w}^b = 1$. This constrained minimization problem is readily solved with the Lagrange multiplier method, giving

$$\mathbf{w}^b = \frac{\mathbf{N}_b^{-1} \mathbf{e}}{\mathbf{e}^t \mathbf{N}_b^{-1} \mathbf{e}}. \tag{4.62}$$

Substituting this back into Eq. (4.61) gives the size of the corresponding error bar:

$$\Delta y_b = \left(\mathbf{e}^t \mathbf{N}_b^{-1} \mathbf{e} \right)^{-1/2}. \tag{4.63}$$

Figure 4.16c shows the seven weight vectors $\mathbf{w}^b$ (dotted) corresponding to the seven measurements y_b in Figure 4.16b (solid points). We will also refer to the weight vectors as *window functions*, since they show the contributions to our measurements from different redshifts. Note that each window function vanishes outside its redshift bin, and that all seven of them share a characteristic bump shape roughly corresponding to an upside-down parabola vanishing at the bin endpoints. To illustrate the z-range that each measurement probes, we plot it at the median of the window function with horizontal bars ranging from the twentieth to the eightieth percentile.

Since the measurements y_b are linear combinations of the x_i which are in turn linear combinations of the r_i, we can also re-express our measurements directly as linear combinations of the original supernova comoving distances:

$$y_b = \sum_i \tilde{w}_i^b r_i, \tag{4.64}$$

where the new window functions

$$\tilde{w}_i^b \equiv \frac{w_{i-1}^b}{\Delta z_{i-1}} - \frac{w_i^b}{\Delta z_i} \tag{4.65}$$

would be essentially the negative derivative of the old window functions if all redshift intervals were the same. These new window functions are also plotted in Figure 4.16c (solid "sawtooth" curves), and are seen to be roughly linear (as expected for a parabola derivative), effectively subtracting supernovae at the near end of the bin from those at the far end.

A major advantage of this method is its transparency and simplicity. If one fits some parametrized model of $H(z)$ to the SN Ia data by maximizing a likelihood function, then the resulting parameter estimates will be some complicated (and generically nonlinear) functions of all the data points r_i. In contrast, the measurement y_5 in our fifth bin in Figure 4.16b is simply a linear combination of the comoving distance measurements for the supernovae in the fifth bin ($1.0 < z_i < 1.4$) as defined by the fifth window function in the bottom panel, so it is completely clear how each particular supernova affects the final result. In particular, the supernovae outside of this redshift range do not affect the measurement at all.

In creating uncorrelated redshift bins, those x_i straddling neighboring bins should be discarded. For say 7 bins there are only 6 such numbers, so this involves a rather negligible loss. The advantage is that it ensures that two measurements y_b and $y_{b'}$ have completely uncorrelated error bars if $b \neq b'$, since their window functions $\tilde{w}$ have no supernovae in common.

To minimize the bias in the $H(z)$-measurement due to weak lensing, we use flux averaging (see Section 4.5.2). Specifically, we compress the full supernova data set r_i into a smaller number of flux averaged supernovae assigned to the mean redshift in each of a large number of bins of width Δz. We use $\Delta z = 0.05$ for the current data and $\Delta z = 0.005$ for the simulated data. Note that since we have assumed a Gaussian distribution in the magnitudes of SNe Ia at peak brightness, flux averaging leads to a tiny bias of $-\sigma_{\text{int}}^2 \ln 10/5$ mag (Wang, 2000b). This tiny bias can be easily removed in the data analysis.

As a side effect, this averaging in narrow bins makes the denominators Δz_i roughly equal in Eq. (4.58), so that the only noticeable source of wiggles in the window functions in Figure 4.16c is Poisson noise, that is, that some of these narrow bins contain more supernovae than others. The method works without this averaging step as well. In that case, the window functions w wiggle substantially because of variations in the redshift spacing between supernovae, since very little weight is given to x_i if Δz_i happens to be tiny. However, we find that the supernova window functions $\tilde{w}$ remain rather smooth and well-behaved functions of redshift, as expected – two supernovae very close together with the same noise level σ_i automatically get the same weight. This means that our method effectively averages such similar redshift supernovae anyway, even if we do not do so by hand ahead of time. The difference between flux averaging and this automatic averaging is simply that we average their fluxes rather than their comoving distances – these two types of averaging are not equivalent since the flux is a nonlinear function of the comoving distance.

4.6
Forecast for Future SN Ia Surveys

The method for deriving uncorrelated estimates of $H(z)$ discussed in Section 4.5.3 can be applied to real data, as well as simulated future data. Figure 4.16 shows the results of applying this method to both real data (a) and simulated data (b), with the dimensionless expansion rate of the universe $h(z)$, $h(z) \equiv H(z)/H_*$, measured in between three and seven uncorrelated redshift bins.

Figure 4.16a uses the "gold" set of 157 SNe Ia published by Riess *et al.* (2004). The error bars are seen to be rather large, and consistent with a simple flat $\Omega_m = 0.3$ concordance model where the dark energy is a cosmological constant. As was shown in Tegmark (2002) using information theory, the relative error bars on the cosmic expansion history $H(z)$ scale as

$$\frac{\Delta H}{H} \propto \frac{\sigma}{N^{1/2} \Delta z^{3/2}}, \tag{4.66}$$

for N supernovae with noise σ. Here Δz is the width of the redshift bins used, so one pays a great price for narrower bins: halving the bin size requires eight times as many supernovae. The origin of this $(\Delta z)^{-3/2}$-scaling is intuitively clear: the noise averages down as $(\Delta z)^{-1/2}$, and there is an additional factor of $(\Delta z)^{-1}$ from effectively taking the derivative of the data to recover $H(z)^{-1}$ from the integral in Eq. (4.58)[8]. Figure 4.16c shows that the method effectively estimates this derivative by subtracting supernovae at the near end of the bin from those at the far end of the bin and dividing by Δz.

This means that for accurately measuring $H(z)$ and thereby the density history of the dark energy, numbers really do matter. For example, in order to measure dark energy density function to 10% accuracy in seven uncorrelated redshift bins as in Figure 4.16b (with $\Delta z = 0.243$), we need to have around 14 000 SNe Ia. We have simulated SN Ia data by placing supernovae at random redshifts, with the number of SNe Ia per 0.1 redshift interval given by a distribution. The intrinsic brightness of each SN Ia at peak brightness is drawn from a Gaussian distribution with a dispersion of $\sigma_{\text{int}} = 0.16$ mag. As the fiducial cosmological model, we used a flat universe with $\Omega_\Lambda = 0.7$.

We can compare two mission concepts for the NASA/DOE Joint Dark Energy Mission (JDEM): the Joint Efficient Dark-energy Investigation (JEDI) (Wang *et al.*, 2004; Crotts *et al.*, 2005; Cheng *et al.*, 2006) and the Supernova Acceleration Probe (SNAP) (Aldering *et al.*, 2004). For JEDI (solid points in Figure 4.16b), the number of SNe Ia per 0.1 redshift interval is obtained by fitting the measured SN Ia rate as function of redshift (Cappellaro, Evans, and Turatto, 1999; Hardin *et al.*, 2000; Dahlen *et al.*, 2004) to a model assuming a conservative delay time between star formation and SN Ia explosion of 3.5 Gyr. For SNAP (dotted points in Figure 4.16b), the number of SNe Ia per 0.1 redshift interval is taken from Figure 9 in Aldering *et al.* (2004).

We consider two kinds of SN Ia systematic uncertainties: weak lensing due to intervening matter and a systematic bias due to K-corrections. We include the weak lensing effect by assigning a magnification μ drawn from a probability distribution

---

8) Analogous estimates of the equation of state $w(z)$ have a painful $(\Delta z)^{-5/2}$-scaling, since they effectively involve taking the second derivative of the data (Tegmark, 2002).

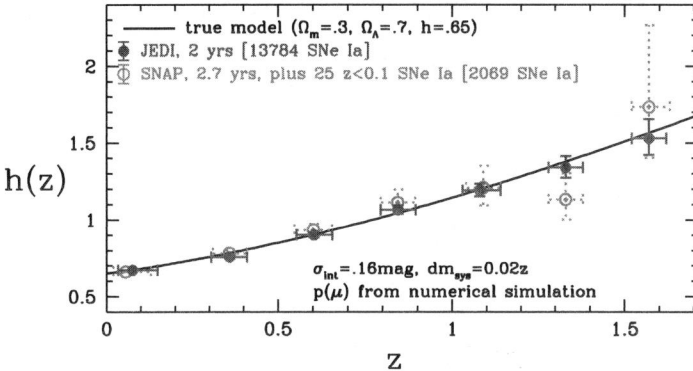

Figure 4.17 Same as Figure 4.16b, but with a systematic bias of $dm_{sys} = 0.02z$ from K-correction uncertainties added in addition to weak lensing noise computed from numerical simulations (Wang and Tegmark, 2005).

$p(\mu)$, extracted using an improved version of the universal probability distribution function (UPDF) method (see Section 4.4.3). The total uncertainty in each SN Ia data point is $\sqrt{\sigma_{int}^2 + \sigma_{lens}(z)^2}$, with $\sigma_{lens}(z)$ from Eq. (4.32). We consider a systematic bias of $\Delta m_{sys} = 0.02z$ due to K-corrections following Wang and Garnavich (2001).

We did not include the systematic bias due to K-corrections in Figure 4.16b, in order to compare the real data (Riess et al., 2004) and simulated data on an equal footing.

In Figure 4.17, we show the effect of adding the systematic bias due to K-corrections, in addition to the weak lensing noise. Comparing Figure 4.17 with Figure 4.16b, we see that the systematic bias does not have a significant effect on the uncorrelated estimates of $H(z)$. This is because our method effectively reduces a global systematic bias into a local bias with a much smaller amplitude.

Assuming no systematic bias and no lensing, the error on $h(z)$ scales as $(\Delta z)^{-3/2}$ (Wang and Tegmark, 2005). Contrasting current and future data with roughly the same redshift bin size, JEDI would shrink the error bars of $h(z)$ by more than an order of magnitude, so the potential improvement with a successful JDEM would be dramatic.

Note that since the measured quantity $h(z)^{-1}$ typically is not a straight line, the measured average of this curve over a redshift bin will generally lie either slightly above or below the curve at the bin center. Figure 4.17 shows that this bias is substantially smaller than the measurement uncertainties, since $h(z)$ and $h(z)^{-1}$ are rather well-approximated by straight lines over the narrow redshift bins that we have used.

A second caveat when interpreting Figure 4.16 and Figure 4.17 is that the absolute calibration of SNe Ia is not perfectly known – changing this simply corresponds to multiplying the function $h(z)$ by a constant, that is, to scaling the measured curve vertically.

4.7
Optimized Observations of SNe Ia

It is possible to optimize the use of SNe Ia as standard candles by obtaining their NIR light curves, and high-quality spectra. This provides guidance to the design of supernova surveys.

NIR Observations

There is strong physical motivation for the observations of SN Ia light curves in the restframe NIR bands. SNe Ia are better standard candles at NIR wavelengths compared to the optical wavelengths (Krisciunas, Phillips, and Suntzeff, 2004; Phillips et al., 2006; Wood-Vasey et al., 2008). Figure 4.18 shows the Hubble diagram of SNe Ia in the NIR, *without* the usual light curve width correction. The smaller intrinsic dispersion of SN Ia peak luminosity in the NIR can be explained by the theoretical modeling of SN Ia light curves using time-dependent multi-group radiative transfer calculations (Kasen, 2006). Figure 4.19 shows the dispersion in peak magnitude (measured at the first light curve maximum) as a function of wavelength band for SN Ia models with ^{56}Ni masses between 0.4 and 0.9 $M_\odot$ (Kasen, 2006).

The observations of NIR spectra have also been shown to provide important clues to the uniformity of normal SNe Ia (Marion et al., 2006, 2009), and can be used to improve SNe Ia as distance indicators.

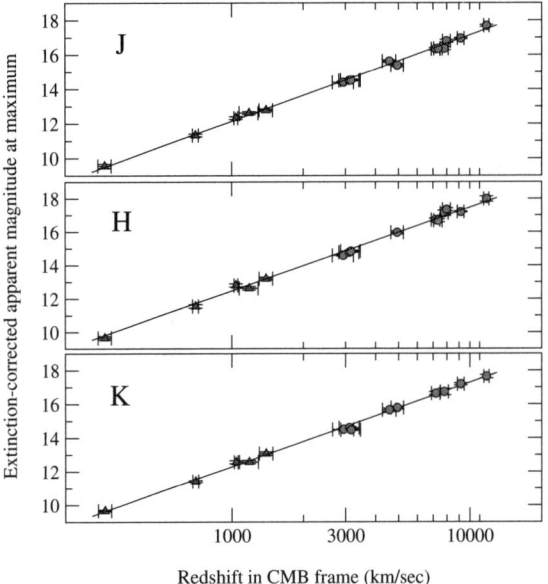

Figure 4.18 Hubble diagrams of SNe Ia in the NIR bands (Krisciunas, Phillips, and Suntzeff, 2004). Note that these SNe Ia have only been corrected for dust extinction; *no* corrections have been made for light curve width.

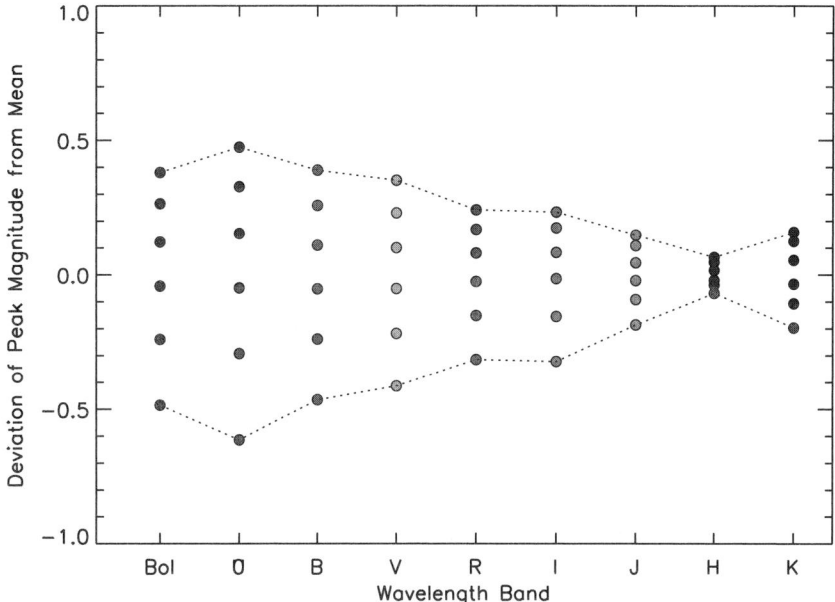

Figure 4.19 Dispersion in peak magnitude (measured at the first light curve maximum) as a function of wavelength band for SN Ia models with ^{56}Ni masses between 0.4 and 0.9 $M_\odot$ (Kasen, 2006).

An additional advantage of NIR observations of SNe Ia is that they are less affected by dust (see Section 4.4.1), thus enabling the separation of intrinsic color variation of SNe Ia from dust extinction. Krisciunas *et al.* (2007) showed that the J band is key to obtaining good estimates of dust extinction if R_V varies from object to object.

Spectral Luminosity Indicators in SNe Ia
The spectra of SNe Ia have been shown to provide calibration relations that decrease the scatter of SNe Ia in the Hubble diagram, and make SNe Ia better distance indicators.

The correlation between SN Ia spectroscopic features and luminosity has been found in the observational data. Garnavich *et al.* (2004) found that the 580/615 nm line depth ratio is a useful indicator of light curve decline rate, and therefore of luminosity (see Figure 4.20). Most recently, Bailey *et al.* (2009) used the Nearby Supernova Factory spectrophotometry of 58 SNe Ia to perform an unbiased search for flux ratios that correlate with SN Ia luminosity. They found that the 642/443 nm flux ratio is most strongly correlated with SN Ia absolute magnitudes.

The correlation of SN Ia spectroscopic features and luminosity can be understood by comparing theoretical modeling with observational data. Hachinger *et al.* (2008) found that the strength of the Si II λ5972 line may be a very promising spectroscopic luminosity indicator for SNe Ia. They showed that Si II λ6355 is sat-

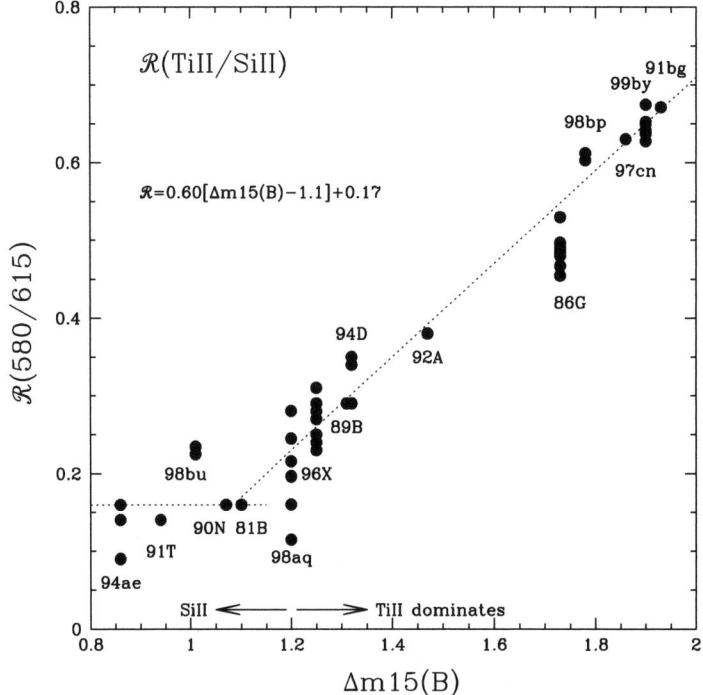

Figure 4.20 The 580/615 nm line depth ratio is a useful indicator of SN Ia decline rate, and thus of SN Ia intrinsic brightness (Garnavich et al., 2004).

urated, while Si II $\lambda 5972$ is stronger in less luminous SNe Ia, because of a rapidly increasing Si II/Si III ratio. Thus the correlation between Si II $\lambda 5972$ strength and luminosity is the effect of ionization balance.

Late-Time Light Curves

Another possibility for tightening the calibration of SNe Ia is to use the late-time light curves. There are theoretical reasons to expect the late-time light curves to be relevant in the peak luminosity of SNe Ia. For example, Milne, The, and Leising (2001) modeled the late light curves of SNe Ia, and predicted that the nonlocal and time-dependent energy deposition due to the transport of Comptonized electrons could produce 0.10 to 0.18 mag correction to the late time brightness of a SN Ia, depending on its position in the $\Delta m_{15}(B)$ sequence.

Wang et al. (2006) explored the nearly linear dependence of B magnitude on B–V color for a typical SN Ia for a post-maximum time window prescribed by the CMAGIC method (∼6 to ∼27 SN Ia restframe days after maximum light). They were able to use this to reduce the scatter of SNe Ia in a Hubble diagram.

Wang and Hall (2008) explored a new approach to calibrating SNe Ia using MCMC. MCMC has the unique advantage of being fast, efficient, and ideally suited for analyzing complex, multi-dimensional data. One can simultaneously model

(B_{max}, E, Δm_{15}, b_{lt}) for each of the SNe Ia in a sample of SNe Ia with late-time light curves, where the color term $E = B_{max} - V_{max}$, and b_{lt} is the late-time light curve slope.[9] They found tentative evidence for a correlation between the late-time light curve slope and the peak luminosity of SNe Ia; brighter SNe Ia seem to have shallower light curve slopes at $\gtrsim 60$ days from maximum light (Wang and Hall, 2008). If we correct B_{max} for color, light curve width, and late-time light curve slope, by subtracting

$$\Delta M^{cor} = \mathcal{R} E + \alpha (\Delta m_{15} - 1.1) + \alpha\prime |\Delta m_{15} - 1.1| + \beta b_{lt} \qquad (4.67)$$

where $\{\mathcal{R}, \alpha, \alpha\prime, \beta\}$ are the parameters determined using MCMC, the Hubble diagram dispersion is reduced to 0.12 mag (Wang and Hall, 2008).

---

[9] The slope at $\gtrsim 60$ days from maximum light, where the light curves are very close to straight lines.

5
Observational Method II:
Galaxy Redshift Surveys as Dark Energy Probe

Since the 1980s, galaxy redshift surveys have been used to map the large scale structure in the universe, and constrain cosmological parameters. Galaxy redshift surveys are powerful as dark energy probes, since they can allow us to measure the cosmic expansion history $H(z)$ through the measurement of baryon acoustic oscillations (BAO) in the galaxy distribution, and the growth history of cosmic large scale structure $f_g(z)$ through independent measurements of redshift-space distortions and the bias factor between the distribution of galaxies and that of matter (Wang, 2008b).

5.1
Baryon Acoustic Oscillations as Standard Ruler

The use of BAO as a cosmological standard ruler is a relatively new method for probing dark energy (Blake and Glazebrook, 2003; Seo and Eisenstein, 2003), but it has already yielded impressive observational results (Eisenstein *et al.*, 2005).

Measuring $H(z)$ and $D_A(z)$ from BAO
At the last scattering of CMB photons, the acoustic oscillations in the photon-baryon fluid became frozen, and imprinted their signatures on both the CMB (the acoustic peaks in the CMB angular power spectrum) and the matter distribution (the baryon acoustic oscillations in the galaxy power spectrum). Because baryons comprise only a small fraction of matter, and the matter power spectrum has evolved significantly since the last scattering of photons, BAO are much smaller in amplitude than the CMB acoustic peaks, and are washed out on small scales.

BAO in the observed galaxy power spectrum have the characteristic scale determined by the comoving sound horizon at the drag epoch (shortly after photon-decoupling), which is precisely measured by the CMB anisotropy data (Page *et al.*, 2003; Spergel *et al.*, 2007; Komatsu *et al.*, 2009). The observed BAO scales appear as slightly preferred redshift separation $s_\parallel$ and angular separation $s_\perp$:

$$s_\parallel \propto s H(z), \quad s_\perp \propto \frac{s}{D_A(z)}, \tag{5.1}$$

Dark Energy. Yun Wang
Copyright © 2010 WILEY-VCH Verlag GmbH & Co. KGaA, Weinheim
ISBN: 978-3-527-40941-9

where s is the sound horizon scale at the drag epoch, and the angular diameter distance $D_A(z) = r(z)/(1+z)$, with $r(z)$ denoting the comoving distance given by Eq. (2.24). Thus comparing the observed BAO scales with the expected values gives $H(z)$ in the radial direction, and $D_A(z)$ in the transverse direction.

Calibration of the BAO Scale

CMB data give us the comoving sound horizon at photon-decoupling epoch z_* (Eisenstein and Hu, 1998; Page *et al.*, 2003)

$$r_s(z_*) = \int_0^{t_*} \frac{c_s dt}{a} = H_0^{-1} \int_{z_*}^{\infty} dz \frac{c_s}{E(z)} = c H_0^{-1} \int_0^{a_*} \frac{da}{\sqrt{3(1+R_b)a^4 E^2(z)}}, \tag{5.2}$$

where a is the cosmic scale factor, $a_* = 1/(1+z_*)$, and

$$a^4 E^2(z) = \Omega_m(a + a_{eq}) + \Omega_k a^2 + \Omega_X X(z) a^4, \tag{5.3}$$

where the dark energy density function $X(z) \equiv \rho_X(z)/\rho_X(0)$, and the cosmic scale factor at the epoch of matter and radiation equality is given by

$$a_{eq} = \frac{\Omega_{rad}}{\Omega_m} = \frac{1}{1+z_{eq}}, \quad z_{eq} = 2.5 \times 10^4 \Omega_m h^2 \left(\frac{T_{CMB}}{2.7\,K}\right)^{-4}. \tag{5.4}$$

We have assumed three massless neutrino species, so that the radiation energy density today is (Kolb and Turner, 1990)

$$\rho_{rad}^0 = \frac{\pi^2}{30} g_*^0 T_{CMB}^4,$$

$$g_*^0 = 2 + \frac{7}{8} \times 2 \times 3 \times \left(\frac{4}{11}\right)^{4/3} \tag{5.5}$$

The sound speed c_s and the baryon/photon ratio R_b are given by

$$c_s^2 \equiv \frac{\delta p}{\delta \rho} \simeq \frac{c^2 \delta \rho_\gamma / 3}{\delta \rho_\gamma + \delta \rho_b} = \frac{c^2}{3\left(1 + \dot{\rho}_b/\dot{\rho}_\gamma\right)} = \frac{c^2}{3(1+R_b)} \tag{5.6}$$

$$R_b \equiv \frac{3\rho_b}{4\rho_\gamma} \equiv \overline{R_b} a, \quad \overline{R_b} = 31\,500 \Omega_b h^2 \left(\frac{T_{CMB}}{2.7\,K}\right)^{-4}. \tag{5.7}$$

We have used $\rho_\gamma \propto a^{-4}$ and $\rho_b \propto a^{-3}$.

Four-year COBE data give $T_{CMB} = 2.728 \pm 0.004$ K (95% C.L.) (Fixsen, 1996). The data from WMAP five-year observations give the redshift and the sound horizon at the photon-decoupling epoch

$$z_* = 1090.51 \pm 0.95, \quad r_s(z_*) = 146.8 \pm 1.8 \text{ Mpc}, \quad \text{at photon decoupling} \tag{5.8}$$

assuming $T_{CMB} = 2.725$ (Komatsu et al., 2009). The BAO scale measured in galaxy redshift surveys correspond to the sound horizon scale at the *drag epoch* (Hu and Sugiyama, 1996).

The drag epoch occurs when the photon pressure (or "Compton drag") can no longer prevent gravitational instability in the baryons. Thus there is no reason for the photon-decoupling epoch, z_*, to be the same as the drag epoch, z_d. The scattering in the photon/baryon fluid leads to an exchange of momentum, with momentum densities for photons and baryons given by (Hu and Sugiyama, 1996):

$$(\rho_\gamma + p_\gamma) V_\gamma = \frac{4}{3}\rho_\gamma V_\gamma \quad \text{for photons}$$

$$(\rho_b + p_b) V_b \simeq \rho_b V_b \quad \text{for baryons} , \quad (5.9)$$

where V_γ and V_b are the photon and baryon bulk velocities. As a consequence of momentum conservation, the rate of change of the baryon velocity due to Compton drag is scaled by a factor of R_b^{-1} compared with the photon case, which means that (Hu and Sugiyama, 1996)

$$\dot{\tau}_d = \frac{\dot{\tau}}{R_b} , \quad (5.10)$$

where τ_d and τ are the Compton optical depths for baryons and photons, respectively. Since the epoch of photon decoupling is defined by $\tau(z_*) = 1$, and the drag epoch is defined by $\tau_d(z_d) = 1$, $z_* = z_d$ only if $R_b = 1$. We live in a universe with a low baryon density, $R(z_*) < 1$ (see Eq. (5.7)), thus $\tau_d(z_d) = 1$ requires $z_d < z_*$, that is, the drag epoch occurs *after* photon decoupling (Hu and Sugiyama, 1996).

The redshift of the drag epoch z_d is well-approximated by (Eisenstein and Hu, 1998)

$$z_d = \frac{1291(\Omega_m h^2)^{0.251}}{1 + 0.659(\Omega_m h^2)^{0.828}} \left[1 + b_1(\Omega_b h^2)^{b_2}\right], \quad (5.11)$$

where

$$b_1 = 0.313(\Omega_m h^2)^{-0.419} \left[1 + 0.607(\Omega_m h^2)^{0.674}\right], \quad (5.12)$$

$$b_2 = 0.238(\Omega_m h^2)^{0.223} . \quad (5.13)$$

Using this fitting formula for z_d, Komatsu et al. (2009) found that from the WMAP five-year observations

$$s \equiv r_s(z_d) = 153.3 \pm 2.0 \text{ Mpc} , \quad z_d = 1020.5 \pm 1.6 \quad \text{at drag epoch} . \quad (5.14)$$

5.2
BAO Observational Results

The power of BAO as a standard ruler resides in the fact that the BAO scale can in principle be measured in both radial and transverse directions, with the radial measurement giving $H(z)$ directly, and the transverse measurement giving

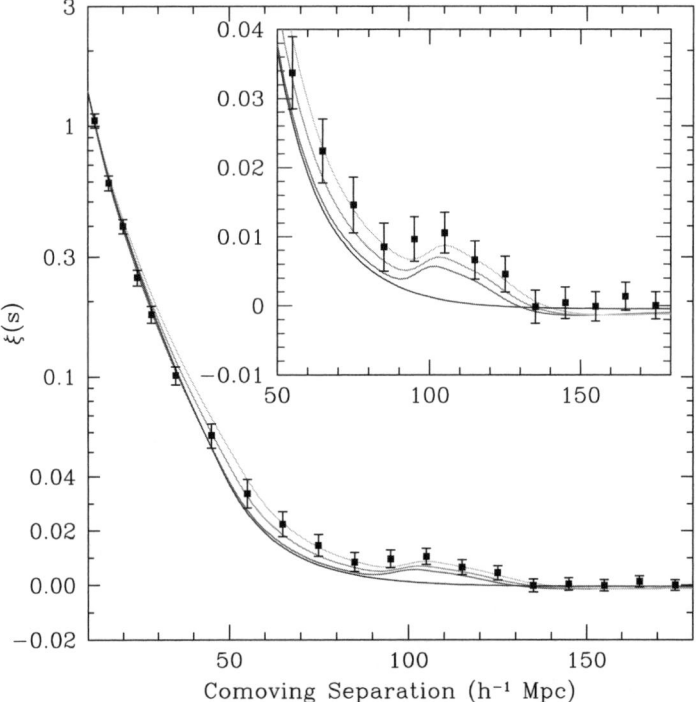

Figure 5.1 The large scale redshift-space correlation function of the SDSS LRG sample measured by Eisenstein et al. (2005). The error bars are from the diagonal elements of the mock catalog covariance matrix (the points are correlated). Note that the vertical axis mixes logarithmic and linear scalings. The inset shows an expanded view with a linear vertical axis. The models are $\Omega_m h^2 = 0.12$ (top), 0.13 (middle), and 0.14 (bottom), all with $\Omega_b h^2 = 0.024$ and $n = 0.98$ and with a mild nonlinear prescription folded in. The featureless smooth line shows a pure CDM model ($\Omega_m h^2 = 0.105$), which lacks the acoustic peak. The bump at $100\,h^{-1}$ Mpc scale is statistically significant.

$D_A(z)$. However, there are only a few published papers on measuring the BAO scale from the existing galaxy redshift survey data, and most of them extract a spherically averaged BAO scale (Eisenstein et al., 2005; Hutsi, 2006; Percival et al., 2007).

Eisenstein et al. (2005) and Hutsi (2006) found roughly consistent spherically averaged correlation functions using SDSS data, with about the same BAO scale. Figure 5.1 shows the galaxy correlation function $\xi(s)$ measured from the SDSS data by Eisenstein et al. (2005). This BAO scale measurement is usually quoted in the form of

$$A_{\rm BAO} \equiv \left[r^2(z_m) \frac{cz_m}{H(z_m)} \right]^{1/3} \frac{(\Omega_m H_0^2)^{1/2}}{cz_m} = 0.469 \left(\frac{n_S}{0.98} \right)^{-0.35} \pm 0.017 \tag{5.15}$$

where $z_m = 0.35$, and n_S denotes the power-law index of the primordial matter power spectrum. Note that A_{BAO} essentially measures the product of a volume-averaged distance

$$d_V \propto \left[cH^{-1}(z)D_A(z)^2\right]^{1/3}, \tag{5.16}$$

multiplied by the square root of the matter density ($\rho_m(z) \propto \Omega_m h^2$). The one-dimensional marginalized values are $\Omega_m h^2 = 0.130 \pm 0.010$, and $d_V(z_m) = 1370 \pm 64$ Mpc, assuming a fixed value of $\Omega_b h^2 = 0.024$ (Eisenstein et al., 2005). The product of $d_V(z_m)$ and $\Omega_m h^2$ is more tightly constrained than $d_V(z_m)$ or $\Omega_m h^2$ by the data, because the measured values of $d_V(z_m)$ and $\Omega_m h^2$ are correlated. Note that $A_{BAO} \propto d_V \cdot (\Omega_m h^2)^{1/2}$ is independent of the Hubble constant h, and its measured value is independent of a dark energy model (Eisenstein et al., 2005).

Clearly, the BAO constraint in Eq. (5.15) from Eisenstein et al. (2005) is not just a simple measurement of the BAO feature; it also relies on the constraints on $\Omega_m h^2$ from measuring the power spectrum turnover scale (related to matter-radiation equality). The latter makes the BAO constraint from Eisenstein et al. (2005) less

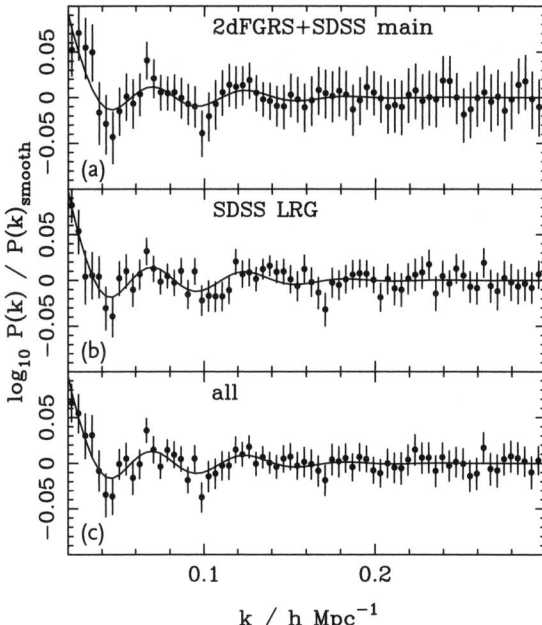

Figure 5.2 BAO in power spectra derived by Percival et al. (2007), MNRAS, Wiley-Blackwell, from (a) the combined SDSS and 2dF main galaxies, (b) the SDSS DR5 LRG sample, and (c) the combination of these two samples. The data are correlated and the errors are calculated from the diagonal terms in the covariance matrix. A standard ΛCDM distance-redshift relation was assumed to calculate the power spectra with $\Omega_m = 0.25$, $\Omega_\Lambda = 0.75$. The power spectra were then fitted with a cubic spline × BAO model, assuming the fiducial BAO model calculated using CAMB. The BAO component of the fit is shown by the solid line in each panel.

robust than it would be otherwise. A new analysis of the SDSS data to derive truly robust and detailed BAO constraints would be very useful for placing dark energy constraints (Dick, Knox, and Chu, 2006).

Percival *et al.* (2007) found that the power spectra from combined SDSS and 2dF data give spherically averaged BAO scales at $z = 0.2$ and $z = 0.35$ that are inconsistent with the prediction of the fiducial flat ΛCDM model at 2.4σ; this is in contradiction to the SN Ia data (which are consistent with the fiducial ΛCDM model at $z \lesssim 0.5$ at 1σ, see Riess *et al.* (2007)). Percival *et al.* (2007) found a similar discrepancy between SDSS main and SDSS LRG samples. Figure 5.2 shows BAO in power spectra calculated from (a) the combined SDSS and 2dF main galaxies, (b) the SDSS DR5 LRG sample, and (c) the combination of these two samples. The data are solid symbols with 1σ errors calculated from the diagonal terms in the covariance matrix.

The efforts to extract the BAO scale in both radial and transverse directions have led to contradicting results. Figure 5.3 shows the contour plots of the redshift-space two-point correlation function measured from a SDSS LRG sample (similar to DR3) by Okumura *et al.* (2008). The baryonic feature appears marginally as ridge structures around the scale $s = (s_\perp^2 + s_\parallel^2)^{1/2} \simeq 100 \, h^{-1}$ Mpc, and the

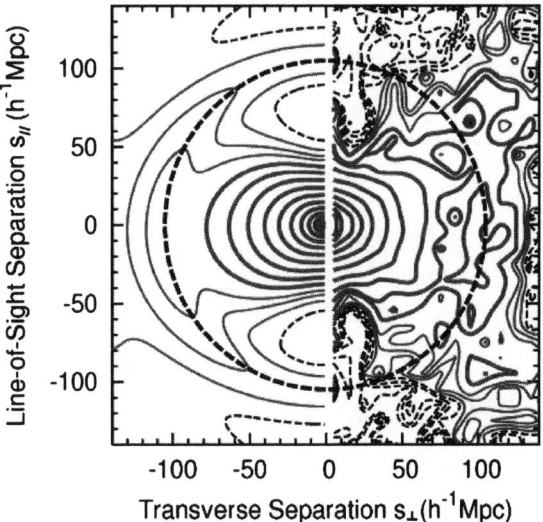

Figure 5.3 Contour plots of the redshift-space two-point correlation function measured from the SDSS LRG sample by Okumura *et al.* (2008). The right half of the figure shows their measurement, and the left half shows the corresponding analytical formula derived by Matsubara (2004) using a linear perturbation theory (Matsubara, 2004). The thin dashed lines show $\xi < -0.01$ increasing logarithmically with increments of 0.25, and $-0.01 \leq \xi < 0$ linearly with increments of 0.0025. The solid thin lines show $0 \leq \xi < 0.01$ increasing linearly with increments of 0.0025, and the solid thick lines $\xi \geq 0.01$ logarithmically with increments of 0.25. The baryonic feature appears marginally as ridge structures around the scale $s = (s_\perp^2 + s_\parallel^2)^{1/2} \simeq 100 \, h^{-1}$ Mpc, and the dashed circle traces the peaks of the baryon ridges.

dashed circle traces the peaks of the baryon ridges. Okumura et al. (2008) found that current galaxy redshift survey data are not adequate for extracting the BAO scale in both radial and transverse directions to measure $H(z)$ and $D_A(z)$. However, an independent analysis by Gaztanaga, Cabre, and Hui (2008) found that $H(z)$ can be measured quite accurately from the SDSS DR6 data, with $H(z = 0.24) = 79.7 \pm 2.1(\pm 1.0)\,\mathrm{km\,s^{-1}\,Mpc^{-1}}$ for $z = 0.15$–0.30, and $H(z = 0.43) = 86.5 \pm 2.5(\pm 1.0)\,\mathrm{km\,s^{-1}\,Mpc^{-1}}$ for $z = 0.40$–0.47. The difference between the results of Okumura et al. (2008) and Gaztanaga, Cabre, and Hui (2008) cannot be explained by the statistics of the data used (DR3 versus DR6).

Resolving the dramatic discrepancy between Okumura et al. (2008) and Gaztanaga, Cabre, and Hui (2008) in the analysis of the radial and transverse BAO scales is of critical importance to the understanding of BAO systematics, and the accurate forecasting of the capabilities of planned future galaxy redshift surveys. All current forecasts of future surveys assume that both radial and transverse BAO scales can be accurately extracted, and use either the Fisher matrix formalism (which gives the smallest possible errors) or methods based on numerical simulations that are not yet fully validated by application to real data.

5.3
BAO Systematic Effects

The systematic effects of BAO as a standard ruler are: bias between galaxy and matter distributions, nonlinear effects, and redshift-space distortions (Blake and Glazebrook, 2003; Seo and Eisenstein, 2003). Cosmological N-body simulations are required to quantify these effects (Angulo et al., 2005; Seo and Eisenstein, 2005; Springel et al., 2005; White, 2005; Jeong and Komatsu, 2006; Koehler, Schuecker, and Gebhardt, 2007; Angulo et al., 2008).

To be specific in our discussion on the systematic effects in the BAO scale measurement, we will use the results from Angulo et al. (2008) to illustrate. All the results from Angulo et al. (2008) shown here are from a numerical simulation covering a comoving cube volume of side $1340\,h^{-1}$ Mpc, in which dark matter is represented by more than 3 billion particles (1448^3), with the particle mass of $5.49 \times 10^{10}\,h^{-1}\,M_\odot$. This simulation corresponds to a comoving volume of $2.41\,h^{-3}\,\mathrm{Gpc}^3$, more than three times the volume of the catalog of SDSS LRGs used in the BAO detection by Eisenstein et al. (2005). It assumes a ΛCDM model with $\Omega_m = 0.25$, $\Omega_\Lambda = 0.75$, $\sigma_8 = 0.9$, and $h = H_0/(100\,\mathrm{km\,s^{-1}\,Mpc^{-1}}) = 0.73$.

In the current picture of structure formation in the universe, primordial matter density perturbations (which are responsible for the observed CMB anisotropy) seeded the cosmic large scale structure. Matter density fluctuations grew with time. Dense regions became denser, and galaxy-cluster and galaxy haloes formed first in such regions. Galaxy formation (in which baryons played a critical part) occurred in galaxy haloes.

Since we have to use galaxies to trace the matter density field, it is important for a numerical simulation to assign galaxies properly. Angulo et al. (2008) used a semi-analytic model to describe the key physical processes which are thought to determine the formation and evolution of galaxies; this approach mirrors the hybrid schemes introduced by Kauffmann, Nusser, and Steinmetz (1997) and Benson et al. (2000). This model makes an *ab initio* prediction of which dark matter haloes should contain galaxies by modeling the physics of the baryonic component of the universe (Baugh et al., 2005; Baugh, 2006). The specific model used by Angulo et al. (2008) reproduces the abundance of Lyman-break galaxies at $z = 3$ and $z = 4$, and the number counts of sub-mm detected galaxies (with a median redshift $z \sim 2$). It also gives a rough match to the abundance of luminous red galaxies (Almeida et al., 2008), and a reasonable match to the observed properties of local galaxies (Nagashima et al., 2005b,a; Almeida, Baugh, and Lacey, 2007).

5.3.1
Nonlinear Effects

On a very large scale, the growth of density perturbations is linear, and the different comoving wavelength scales are not coupled (see Section 2.3). When the amplitude of density perturbations on a given scale reaches order unity, nonlinear growth occurs, that is, the evolution of the different wavelength modes becomes increasingly coupled, leading to a departure from linear evolution. Thus nonlinear effects erase the BAO in the matter power spectrum on small scales, distort the matter power spectrum on quasi-linear scales, and degrade the BAO signal on linear scales. The characteristic comoving scale for nonlinearity increases with cosmic time, as density perturbations on larger and larger comoving scales grow to be of order unity in amplitude. Figure 5.4 shows the nonlinear growth of the matter power spectrum measured from a numerical simulation by Angulo et al. (2008). Nonlinear effects have to be removed or corrected for in the data analysis in order to obtain robust BAO scale measurements (see Jeong and Komatsu (2006); Koehler, Schuecker, and Gebhardt (2007); Smith, Scoccimarro, and Sheth (2007); Crocce and Scoccimarro (2008)).

The most troublesome consequence of nonlinear effects is the shift in the observed BAO scale in galaxy redshift survey data from the CMB-calibrated prediction. It is most intuitive to consider this effect in real space, where the nonlinear growth of density perturbations damps and shifts the BAO peak at $\sim 100\,h^{-1}$ Mpc, because the large scale bulk flows cause the differential motions of the galaxy pairs initially separated by the sound horizon scale at the drag epoch (Eisenstein et al., 2007).

Eisenstein and collaborators (Eisenstein et al., 2007; Seo et al., 2008) introduced a method to "reconstruct" the linear power spectrum from a nonlinearly evolved galaxy distribution in order to minimize the impact of nonlinear effects on the constraining power of BAO as a dark energy probe. They found that the shifts of the BAO peak can be predicted numerically, and can be substantially reduced (to less than 0.1% at $z = 0.3$–1.5) using a simple "density-field reconstruction" method (Eisenstein et al., 2007; Seo et al., 2008).

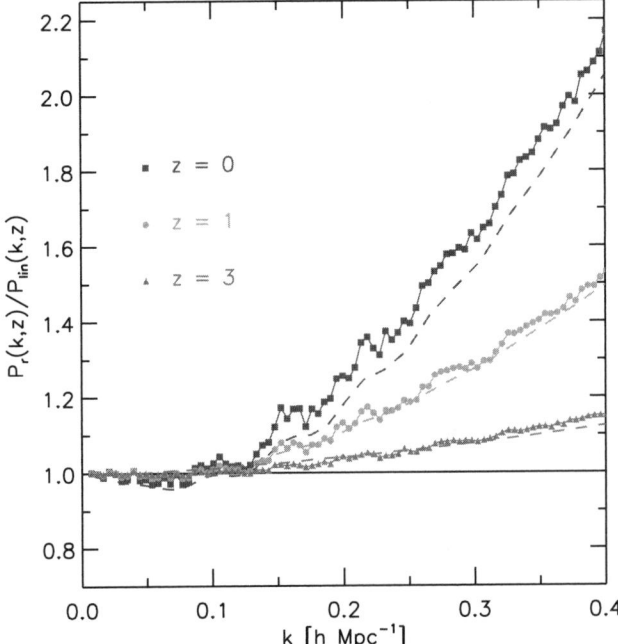

Figure 5.4 The nonlinear growth of the matter power spectrum (Angulo et al. (2008), MNRAS, Wiley-Blackwell). The power spectrum in real-space (measured at the redshift indicated by the key) is divided by the power spectrum at $z = 15$, after taking into account the change in the growth factor. Any deviation of the resulting ratio from unity indicates a departure from linear perturbation theory. The dashed lines show the same ratio as predicted using the ansatz of Smith et al. (2003) which transforms the linear power spectrum into the nonlinear power spectrum. The simulation corresponds to a comoving volume of $2.41 h^{-3}$ Gpc^3.

Padmanabhan, White, and Cohn (2009) reformulated this reconstruction method within the Lagrangian picture of structure formation, and found that this reconstruction does *not* reproduce the linear density field, at second order. They showed that it does reduce the damping of the BAO due to nonlinear structure formation. In particular, they showed that reconstruction reduces the mode-coupling term in the power spectrum, thus reducing the bias in the estimated BAO scale when the reconstructed power spectrum is used. Note that the reconstruction technique has only been demonstrated for dark matter, and not yet for haloes or galaxies.

5.3.2
Redshift-Space Distortions

Redshift-space distortions are the consequence of peculiar motions on the measurement of the power spectrum from a galaxy redshift survey. Peculiar motions produce different types of distortion to the power spectrum. On large scales, coherent bulk flows out of voids and into overdense regions lead to an enhancement in the density inferred in redshift-space, and hence to a boost in the recovered power. On small scales, the random motions of objects inside virialized dark matter haloes cause structures to appear elongated when viewed in redshift-space ("the finger of God" effect), leading to a damping of the power.

The enhancement of the power spectrum due to redshift-space distortions, under the assumption of linear perturbation theory for an observer situated at infinity (the plane parallel approximation), is given by (Kaiser, 1987):

$$\frac{P_s(k,\mu)}{P_r(k,\mu)} = \left(1 + \beta\mu^2\right)^2, \tag{5.17}$$

where $P_s(k,\mu)$ is the power spectrum in redshift-space, $P_r(k,\mu)$ is the power spectrum in real-space, and $\mu = \mathbf{k} \cdot \hat{\mathbf{r}}/k$, with $\hat{\mathbf{r}}$ denoting the unit vector along the line of sight. The redshift-space distortion parameter β is defined as

$$\beta(z) \equiv \frac{f_g(z)}{b(z)}, \tag{5.18}$$

where $f_g(z)$ denotes the growth rate, and $b(z)$ denotes the bias factor. Equation (5.17) can be derived using Eq. (2.41) and requiring that the number of galaxies is conserved when we go from real to redshift-space (Hamilton, 1998).

The enhancement of the spherically averaged power spectrum is

$$\frac{P_s(k)}{P_r(k)} = 1 + \frac{2}{3}\beta + \frac{1}{5}\beta^2, \tag{5.19}$$

which follows from integration over μ.

Equation (5.19) can be modified to include the damping effect due to the "the finger of God" effect (Angulo et al., 2008):

$$\frac{P_s(k)}{P_r(k)} = \frac{1 + \frac{2}{3}\beta + \frac{1}{5}\beta^2}{1 + k^2\sigma^2}, \tag{5.20}$$

where σ is a free parameter associated with the pairwise velocity dispersion (see Eq. (5.102)).

Figure 5.5 shows the ratio of the matter power spectrum measured in redshift-space, to the matter power spectrum measured in real-space (Angulo et al., 2008). Clearly, Eq. (5.20) provides a good description for redshift-space distortion to the matter power spectrum.

Since we cannot directly measure the matter power spectrum, we have to study the redshift-space distortion to the power spectrum of the type of object used as matter tracer in the galaxy redshift survey. The form of the redshift-space distortion to the power spectrum depends on the type of object under consideration.

In current theories of galaxy formation, dark matter haloes are hosts to galaxies. Angulo et al. (2008) found that Eq. (5.20) is a poor description of the redshift-space distortions to the dark matter halo power spectrum, but is a reasonable description of the redshift-space distortions to the galaxy power spectrum.

The small scale redshift-space distortions ("the finger of God" effect) can be removed from data in the BAO measurement using a nonlinear "finger-of-God" com-

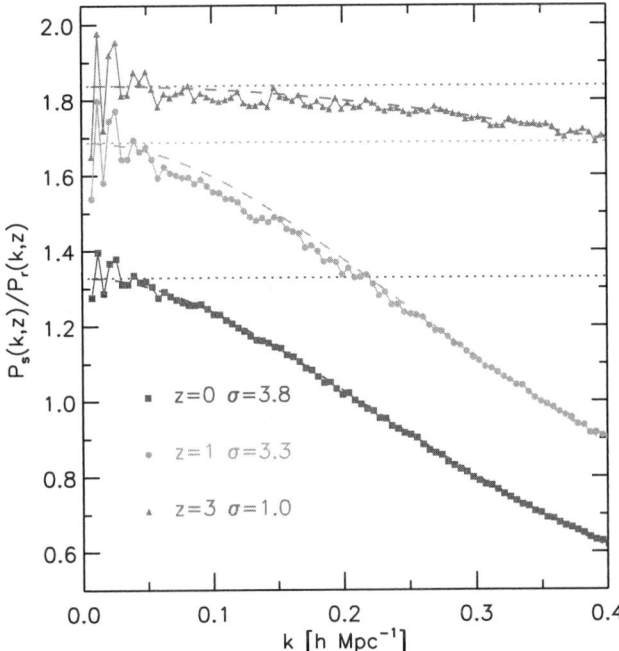

Figure 5.5 The ratio of the power spectrum measured for the dark matter in redshift-space, that is, including the impact of peculiar motions in the distance determination, to the power spectrum measured in real-space (Angulo et al. (2008), MNRAS, Wiley-Blackwell). The deviation from unity shows the redshift-space distortion to the nonlinear power spectrum. The results are shown for selected output redshifts, as indicated by the key. The horizontal dotted lines indicate the boost in the redshift-space power expected due to coherent flows, as predicted by Eq. (5.19). The dashed lines show a simple fit to the distortions (see Eq. (5.20)). The simulation corresponds to a comoving volume of $2.41 \, h^{-3} \, \text{Gpc}^3$.

pression step *before* the power spectrum analysis, in which a "friends-of-friends" algorithm is used to identify the clustering of matter (Tegmark et al., 2004). However, this may introduce a degree of arbitrariness in the results. We can use the version of Eq. (5.20) before spherical averaging and its counter part in the correlation function analysis to fully model both the large scale compression and the small scale "finger of God" effect due to redshift-space distortions.

Note that the redshift-space distortions on large scales do not modify the BAO, and can be used to measure the linear redshift-space distortion parameter β.

5.3.3
Scale-Dependent Bias

The bias factor between the tracer distribution measured by the galaxy redshift survey and the matter distribution depends on the tracer used. Angulo et al. (2008) showed that the clustering of haloes is *not* a shifted version of that of the dark

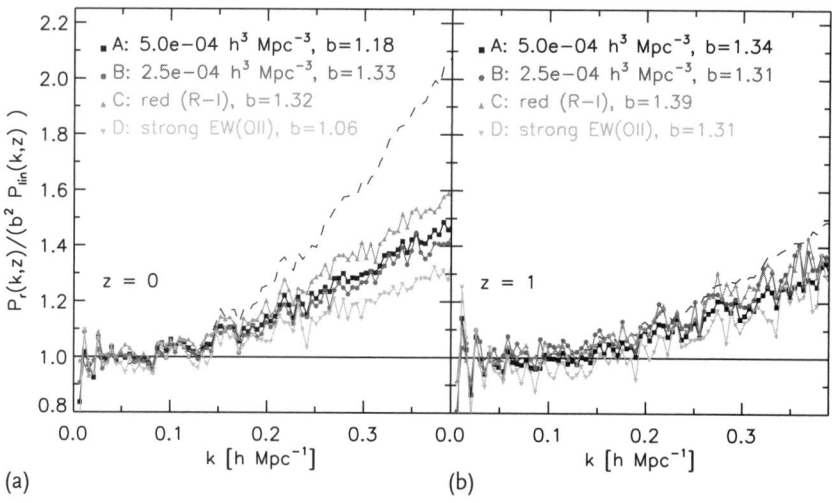

Figure 5.6 The power spectrum of different galaxy samples measured in real-space, divided by the square of an effective bias parameter and the appropriately scaled linear perturbation theory power spectrum (Angulo et al. (2008), MNRAS, Wiley-Blackwell). The sample definition and the value of the effective bias used are given by the key. The power spectrum of the dark matter spectrum in real-space, also divided by the linear perturbation theory spectrum, is shown by the dashed line. (a) shows the ratios at $z = 0$ and (b) at $z = 1$. The simulation corresponds to a comoving volume of $2.41 \, h^{-3} \, \text{Gpc}^3$.

matter, in contrast to current theoretical models. The bias between dark matter haloes and dark matter is scale-dependent, and the variation in the degree of scale-dependence with redshift is not monotonic. The scale-dependence of the bias for galaxies is less than that of dark matter haloes, but still significant (Angulo et al., 2008).

Figure 5.6 shows the real space power spectrum for four different samples of galaxies, divided by the square of an effective bias parameter and the appropriately scaled linear perturbation theory power spectrum. Samples A and B denote galaxies from an R-magnitude limited survey with a given space density. Sample C contains the reddest 50% of galaxies from sample A (selected using the $R-I$ color). Sample D contains the 50% of galaxies from sample A with the strongest emission lines, selected using the equivalent width of OII[3727]. The dashed line shows the real-space power spectrum of matter divided by the appropriate linear perturbation theory power spectrum.

Note that since nonlinear effects are independent of the galaxy sample, the differences in the power spectra of the four different galaxy samples in Figure 5.6 indicate that bias depends strongly on the galaxy sample, and that bias is scale-dependent. Therefore, scale-dependent bias must be properly modeled if quasi-linear scales are included in the analysis of BAO scales.

5.4
BAO Data Analysis Techniques

We will discuss two approaches to probing the BAO, using the galaxy power spectrum (Section 5.4.1) and using the galaxy two-point correlation function (Section 5.4.2). In both methods, the actual galaxy distribution is compared with a catalog of randomly distributed galaxies. These randomly distributed galaxies have the same redshift, magnitude, and mask constraints as the real data.

The BAO scales extracted from the two different analysis methods provide an important cross-check. We will discuss the potential of each method for mitigating systematic effects.

A way to test how well we can extract the BAO scale from real data is to apply the two analysis techniques to simulated data. To measure the power spectrum or correlation function of galaxies, one must convert the angular positions and redshifts of the galaxies into comoving spatial separations. This requires assuming a set of values of the cosmological parameters, including the dark energy parameters. The effect of a change in the value of dark energy parameters is to change the separations between pairs of galaxies, which leads to a change in the appearance of the galaxy power spectrum and correlation function. For small perturbations away from the true dark energy parameters, one can assume that the change in the measured galaxy power spectrum and correlation function can be represented by a rescaling of the wavenumber from k_{true} to k_{app} for the power spectrum, and a rescaling of the length scale from r_{true} to r_{app} for the correlation function. For simplicity, we will focus on spherically averaged data. The *scale parameter*, α, describes the change in the recovered BAO scale:

$$\alpha = \frac{k_{\text{app}}}{k_{\text{true}}} \quad \text{power spectrum} \; ; \quad \alpha = \frac{r_{\text{true}}}{r_{\text{app}}} \quad \text{correlation function} \, . \tag{5.21}$$

If the dark energy parameters are estimated correctly, then there is no shift in the BAO in the estimated power spectrum and $\alpha = 1$. In the case of a wide-angle, deep galaxy survey with spectroscopic redshifts, the scale parameter can be approximated by (Angulo et al., 2008):

$$\alpha \approx \left(\frac{D_A(z, X_{\text{assumed}})}{D_A(z, X_{\text{true}})} \right)^{-2/3} \left(\frac{H(z, X_{\text{assumed}})}{H(z, X_{\text{true}})} \right)^{1/3} , \tag{5.22}$$

where $X(z)$ denotes the dark energy density function (see Eq. (1.19)).

The accuracy and precision of the BAO scale measurement is reflected by that of the scale parameter α. This in turn depends on the modeling of the BAO in the data analysis. A common misconception is that the location of the BAO peaks in the galaxy power spectrum or two-point correlation function corresponds *exactly* to the sound horizon scale at the drag epoch. This misconception can lead to biased estimates of the BAO scale and hence biased estimates of cosmological and dark energy parameters. To accurately extract the BAO scale, the galaxy power spectrum and two-point correlation function must be modeled as completely as possible.

5.4.1
Using the Galaxy Power Spectrum to Probe BAO

The real space galaxy power spectrum is related to the matter power spectrum as follows:

$$P_g(k, z) = b(z) P_m(k, z), \tag{5.23}$$

where $b(z)$ is the bias factor between galaxy and matter distributions.

The Matter Power Spectrum

The matter power spectrum is defined as

$$P_m(k) \equiv |\delta_k|^2 \tag{5.24}$$

where δ_k is the Fourier transform of the matter density perturbation $\delta(r)$, defined as

$$\delta_k \equiv \int \delta(r) e^{i k \cdot r} d^3 r. \tag{5.25}$$

Therefore

$$\delta(r) \equiv \frac{\rho(r) - \bar{\rho}}{\bar{\rho}} = \frac{1}{(2\pi)^3} \int \delta_k e^{-i k \cdot r} d^3 k, \tag{5.26}$$

with $\rho(r)$ and $\bar{\rho}(r)$ denoting the matter density at position r and the mean matter density, respectively. Note that

$$\int e^{i k \cdot r} d^3 r = (2\pi)^3 \delta^D(k), \tag{5.27}$$

where δ^D denotes the Dirac delta function.

The theoretical matter power spectrum in the linear regime is given by

$$P(k)_{\text{lin}} = P_0 k A_s^2(k) T^2(k), \tag{5.28}$$

where P_0 is a normalization constant, $A_s^2(k)$ is the power spectrum of primordial matter density fluctuations, and $T(k)$ is the matter transfer function. The primordial power spectrum is determined by unknown inflationary physics in the very early universe, and can be measured directly from data in a model-independent manner (Wang, Spergel, and Strauss, 1999; Mukherjee and Wang, 2003). For simplicity, the primordial matter power spectrum is usually parametrized as a power-law:

$$k A_s^2(k) = k^{n_s}. \tag{5.29}$$

The matter transfer function $T(k)$ describes how the evolution of matter density perturbations depends on scale.

In the inflationary paradigm of the very early universe, density perturbations began as quantum fluctuations produced during inflation within the horizon for microphysics, the Hubble radius $H(t)^{-1}$ (where $H(t)$ is the Hubble parameter). The Hubble radius remained roughly constant during inflation, while the universe underwent extremely rapid expansion, stretching the physical scales of density perturbations ($\lambda_{\text{phys}} \propto a(t)\lambda$ for a comoving wavelength λ). Thus density perturbations crossed outside the microphysics horizon during inflation. After inflation, the universe was radiation dominated (with $a(t) \propto t^{1/2}$), then became matter dominated (with $a(t) \propto t^{2/3}$) after the matter-radiation equality epoch z_{eq} (see Eq. (5.4)). The Hubble radius grew faster than the cosmic scale factor $a(t)$ during both radiation and matter domination, since $H^{-1}(t) = [\dot{a}/a]^{-1} \propto t$. Thus density perturbations re-entered the microphysics horizon after inflation; those that exited the microphysics horizon last during inflation (the smallest scales) re-entered first. Since matter density perturbations could not grow until the universe became matter dominated (see Section 2.3), the growth of matter density perturbations is scale-dependent. This is encoded in the matter transfer function $T(k)$, and depends on the physics at matter-radiation equality and photon-decoupling. If dark energy perturbations are negligible, $T(k)$ only depends on the matter density $\rho_m \propto \Omega_m h^2$ and baryon density $\rho_b \propto \Omega_b h^2$, and on the dimensionless Hubble constant h through the choice of h/Mpc as the unit for k. It is most convenient and reliable to calculate $T(k)$, normalized such that $T(k \to 0) = 1$, using a public high precision CMB code such as *CMBFAST* (Seljak and Zaldarriaga, 1996) or *CAMB* (Lewis, Challinor, and Lasenby, 2000).

The galaxy power spectrum can be measured from data using the FKP method (Feldman, Kaiser, and Peacock, 1994). This method uses galaxy catalogs obtained from galaxy surveys and a much larger synthetic galaxy catalog with the same angular and radial selection functions.

Basic Idea Behind the FKP Method for Estimating $P_g(k)$

It is the locations of galaxies, and not the smooth matter density field $\rho(r)$ that is observed. The basic idea behind the FKP method is to take the Fourier transform of the distribution of real galaxies, minus the transform of a synthetic catalog with the same angular and radial selection function as the real galaxies but otherwise without structure. It also incorporates a weight function $w(r)$ which is adjusted to optimize the performance of the power-spectrum estimator. It defines a weighted galaxy fluctuation field, with a convenient normalization, to be

$$F(r) \equiv \frac{w(r)\left[n_g(r) - \alpha_s n_s(r)\right]}{\left[\int d^3 r\, \bar{n}^2(r) w^2(r)\right]^{1/2}}, \qquad (5.30)$$

where $\bar{n}(r)$ is the expected mean space density of galaxies given the angular and luminosity selection criteria, and

$$n_g(r) = \sum_i \delta^D(r - r_i^g), \quad n_s(r) = \sum_i \delta^D(r - r_i^s) \qquad (5.31)$$

with r_i denoting the location of the i-th galaxy from the real (with superscript "g") or synthetic (with superscript "s") catalog. The synthetic catalog has a number density that is $1/\alpha_s$ times that of the real catalog. The synthetic catalog is created assuming that galaxies form a Poisson sample of the density field, $\rho/\bar{\rho}$ (Peebles, 1980).

Denoting the Fourier transform of $F(r)$ as $F(k)$, it can be shown that

$$\langle |F(k)|^2 \rangle = \int \frac{d^3k'}{(2\pi)^3} P_g(k') |G(k-k')|^2 + (1+\alpha_s) \frac{\int d^3r\, \bar{n}(r) w^2(r)}{\int d^3r\, \bar{n}^2(r) w^2(r)} \quad (5.32)$$

where the window function is

$$G(k) \equiv \frac{\int d^3r\, \bar{n}(r) w(r) e^{ik\cdot r}}{\left[\int d^3r\, \bar{n}^2(r) w^2(r)\right]^{1/2}}. \quad (5.33)$$

For a typical galaxy redshift survey, $G(k)$ is a compact function with width $\sim 1/D$, where D characterizes the depth of the survey. Assuming that we have a "fair sample" of the matter density distribution, then

$$\langle |F(k)|^2 \rangle \simeq P_g(k) + P_{\text{shot}}, \quad (5.34)$$

where the constant shot noise component

$$P_{\text{shot}} \equiv \frac{(1+\alpha_s) \int d^3r\, \bar{n}(r) w^2(r)}{\int d^3r\, \bar{n}^2(r) w^2(r)}. \quad (5.35)$$

The FKP estimator of $P_g(k)$ is thus

$$\hat{P}_g(k) = |F(k)|^2 - P_{\text{shot}}, \quad (5.36)$$

with the final estimator of $P_g(k)$ given by averaging $\hat{P}_g(k)$ over a shell in k-space:

$$\hat{P}_g(k) \equiv \frac{1}{V_k} \int_{V_k} d^3k'\, \hat{P}_g(k'), \quad (5.37)$$

where V_k is the volume of the shell.

Practical Implementation of the FKP Method for Estimating $P_g(k)$

The radially averaged power spectrum from the FKP estimator is

$$\hat{P}_g(k) = \frac{1}{N_k} \sum_{k<|k|<k+\delta k} \left[|F(k)|^2 - S(0) \right] \quad (5.38)$$

where N_k is the number of modes in the shell, and $F(k)$ and $S(k)$ are given by

$$F(k) = \int d^3r\, w(r) [n_g(r) - \alpha_s n_s(r)] e^{ik\cdot r} \to \sum_g w(r_g) e^{ik\cdot r_g} - \alpha_s \sum_s w(r_s) e^{ik\cdot r_s},$$

$$S(k) = (1+\alpha_s) \int d^3r\, \bar{n}(r)\, w^2(r)\, e^{ik\cdot r} \to \alpha_s(1+\alpha_s) \sum_s w^2(r_s) e^{ik\cdot r_s}.$$

$$(5.39)$$

The variance of the estimated $P(k)$, for any shell thickness, is

$$\sigma_P^2(k) = \frac{2}{N_k^2} \sum_{k'} \sum_{k''} |P_g Q(k' - k'') + S(k' - k'')|^2 \qquad (5.40)$$

where k and k' are constrained to lie in the shell, and

$$Q(k) = \int d^3r \, \bar{n}^2(r) w^2(r) e^{ik \cdot r} \to \alpha_s \sum_s \bar{n}(r_s) w^2(r_s) e^{ik \cdot r_s} . \qquad (5.41)$$

The weight function $w(r)$ is chosen such that it minimizes the variance $\sigma_P^2(k)$. This leads to

$$w(r) = \frac{1}{1 + \bar{n}(r) P_g(k)} . \qquad (5.42)$$

Note that the weight function depends on the assumed value for $P_g(k)$. The optimal estimator results from allowing a range of $P_g(k)$ and then selecting an optimal value for $P_g(k)$. For convenience, we can adjust the normalization of the weight function so that

$$\int d^3r \, \bar{n}^2(r) w^2(r) \to \alpha_s \sum_s \bar{n}(r_s) w^2(r_s) = 1 . \qquad (5.43)$$

If the shell intercepts a sufficiently large number of coherent volumes, then the fractional error in the estimated $P_g(k)$ is reasonably small. Then the fluctuations in the power will become Gaussian distributed, and the likelihood for any particular theory represented by $P_{g,\text{th}}(k)$ is

$$L[P_{g,\text{th}}(k)] = p[P_i | P_{g,\text{th}}(k)]$$
$$= \frac{1}{(2\pi)^{N/2} |C|} \exp\left\{-\frac{C_{ij}^{-1}}{2} \left[\hat{P}_{g,i} - P_{g,\text{th}}(k_i)\right]\left[\hat{P}_{g,j} - P_{g,\text{th}}(k_j)\right]\right\}, \qquad (5.44)$$

where $\hat{P}_{g,i}$ is the vector of estimates, and the correlation matrix for the binned estimates of $\hat{P}_g$ is

$$C_{ij} \equiv \left\langle \delta \hat{P}_g(k_i) \delta \hat{P}_g(k_j) \right\rangle = \frac{2}{N_k N_{k'}} \sum_k \sum_{k'} |P_g Q(k - k') + S(k - k')|^2 , \qquad (5.45)$$

where k and k' lie in the shells around k_i and k_j, respectively. Note that C_{ij} depends on $P_g(k)$.

The original FKP technique is a direct Fourier method; one first chooses a k grid with sufficient grid size and spacing and then obtains the Fourier transform by performing direct summation at each grid point (see Eqs. (5.38) and (5.39)), instead of using fast Fourier transform (FFT). However, it is possible to modify this method so that one can use FFT (for example, see Cole et al. (2005)). Unlike in the direct Fourier method, one has to assign galaxies to a linearly spaced grid using an interpolation method such as cloud in cell, nearest grid point, or triangular shaped cloud assignment scheme (Hockney and Eastwood, 1988). This induces gridding noise and one needs to correct the resultant power spectrum for this. However, the direct Fourier method is much slower than the FFT method. In both methods, the final step involves obtaining power in thin spherical shells in k space to get the power spectrum. This resultant power spectrum is convolved with the window function of the survey as these surveys are volume limited. Therefore, one needs to deconvolve the obtained power spectrum with the window function of the survey. In practice, it is much more convenient to convolve the theoretical power spectrum with the survey window function, and compare it with the measured power spectrum (without deconvolution).

The power spectra for combined 2 dF and SDSS data shown in Figure 5.2 were estimated by Percival et al. (2007) using a modified version of the FKP method, such that FFT is used instead of direct summation at each grid point. Note that the BAO signature can be seen clearly. However, there is a clear difference between the BAO scale present in the combined 2dF and SDSS main data and the BAO scale apparent in the SDSS LRG data.

Mitigation of Systematic Effects in BAO Scale Extraction from $P_g(k)$
Simulated data must be used to study how the BAO scale extraction from the measured galaxy power spectrum is affected by systematic effects. The BAO scale can be extracted by fitting the measured galaxy power spectrum to the linear perturbation theory power spectrum with appropriate modifications to allow for our ignorance on dark energy parameters and to model nonlinear effects. A simple method to model power spectrum data consists of the following steps (Angulo et al., 2008; Percival et al., 2007):

1. Construct a smooth reference spectrum $P_{g,\text{ref}}$ from the measured galaxy power spectrum. $P_{g,\text{ref}}$ results from a coarse rebinning of the measured power spectrum that erases any oscillatory features such as BAO. For example, one can use a cubic spline fit over the wavenumber range $0.0046 < (k/h\,\text{Mpc}^{-1}) < 1.2$, using the measured spectrum smoothed over 25 bins in wavenumber (Angulo et al., 2008). The spline is constrained to pass through the data points.
2. Compute the ratio, $R(k)$, of the measured galaxy power spectrum, $P_g(k)$, to the reference power spectrum, $P_{g,\text{ref}}(k)$:

$$R(k) = \frac{P_g(k)}{P_{g,\text{ref}}(k)}. \tag{5.46}$$

3. Generate a linear perturbation theory matter power spectrum, $P^L(k)$, using a high precision CMB code (such as *CAMB* (Lewis, Challinor, and Lasenby, 2000) or *CMBFAST* (Seljak and Zaldarriaga, 1996)). The set of cosmological

parameters assumed is the same as that of the simulated data if one is testing the accuracy of BAO scale extraction only. For real data, the set of cosmological parameters should be varied in a maximum likelihood analysis. Next, define a smooth reference spectrum for $P^L(k)$, P^L_{ref}, in the same manner as described in Step 1, using the same wavenumber bins. Finally, compute the ratio, P^L/P^L_{ref}.

4. Modify the linear theory ratio, P^L/P^L_{ref}, as follows:

$$R_L(k) = \left[\frac{P^L(\alpha k)}{P^L_{ref}(\alpha k)} - 1\right] \times W(k, k_{nl}) + 1, \quad (5.47)$$

where the scale parameter α mimics a change in dark energy parameters (see Eq. (5.21)), and the Gaussian filter $W(k)$ describes the damping of the oscillations beyond some characteristic wavenumber:

$$W(k) = \exp\left(-\frac{k^2}{2k^2_{nl}}\right), \quad (5.48)$$

with k_{nl} as a free parameter. Thus there are two free parameters, α and k_{nl}.

5. Compute the likelihood for a grid of models, each specified by values of (k_{nl}, α). The likelihood is given by (assuming Gaussian errors):

$$-2\ln L = \chi^2 = \sum_i \left(\frac{R^i - R^i_L}{\sigma^i/P^i_{ref}}\right)^2 \quad (5.49)$$

where the summation is over wavenumber and σ^i is the error on the power spectrum estimated in the i-th bin.

6. Derive confidence limits on α and k_{nl} in a likelihood analysis.

Figures 5.7–5.9 show the results from Angulo et al. (2008), from a high resolution simulation corresponding to a comoving volume of 2.41 h^{-3} Gpc^3 (Angulo et al., 2008), and an ensemble of 50 low-resolution simulations (each with the same comoving volume but less resolution) to estimate the cosmic variance of the high resolution simulation. The dark matter haloes have mass in excess of $5.4 \times 10^{12}\, h^{-1}\, M_\odot$. The galaxies form an R-magnitude limited sample with a space density of $\bar{n} = 5 \times 10^{-4}\, h^3\, \text{Mpc}^{-3}$.

Figure 5.7 shows the values obtained for α from the power spectrum at various redshifts of the dark matter (triangles), dark matter haloes (circles) and galaxies (squares) (Angulo et al., 2008). There is a trend for the best-fitting value to deviate away from unity with decreasing redshift, although the result at $z = 0$ is still within 1σ of $\alpha = 1$ for dark matter and dark matter haloes. Figure 5.8 shows the best-fitting value of the damping scale k_{nl} as a function of redshift, for the same tracers of the matter density distribution as in Figure 5.7, in real-space (a) and redshift-space (b).

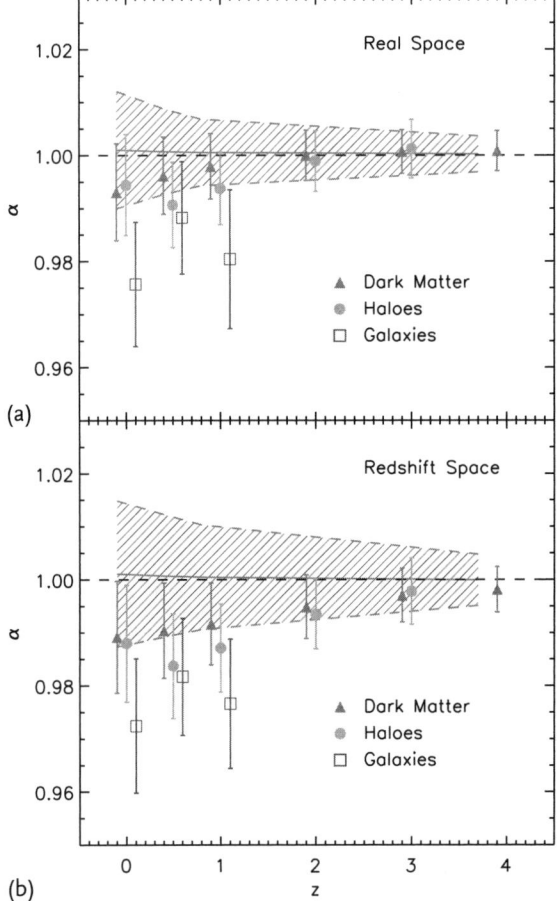

Figure 5.7 The best-fitting value of the scale factor α as a function of redshift, for different tracers of the matter density distribution, in real-space (a) and redshift-space (b) (Angulo et al. (2008), MNRAS, Wiley-Blackwell). The symbols show results from the high resolution simulation corresponding to a comoving volume of $2.41\, h^{-3}\, Gpc^3$: dark matter (triangles), dark matter haloes with mass in excess of $5.4 \times 10^{12}\, h^{-1}\, M_\odot$ (circles) and galaxies from an R-magnitude limited sample with a space density of $\bar{n} = 5 \times 10^{-4}\, h^3\, Mpc^{-3}$ (squares). The error bars show the 1σ range on α, calculated from $\Delta\chi^2$. The hatched region shows the central 68% range of the results obtained using the dark matter in an ensemble of low resolution simulations. Recall that $\alpha = 1$ corresponds to an unbiased measurement of the BAO scale (hence of dark energy parameters).

Figure 5.9 shows the recovered value of the scale parameter α for various galaxy samples (Angulo et al., 2008). Note that the accuracy and precision of the estimated α depends on the galaxy sample. For example, using a catalog of R-magnitude-limited galaxies with space density of $5 \times 10^{-4}\, h^3\, Mpc^{-3}$ (sample A) or red galaxies (sample C), one could measure the BAO scale more accurately (smaller bias in

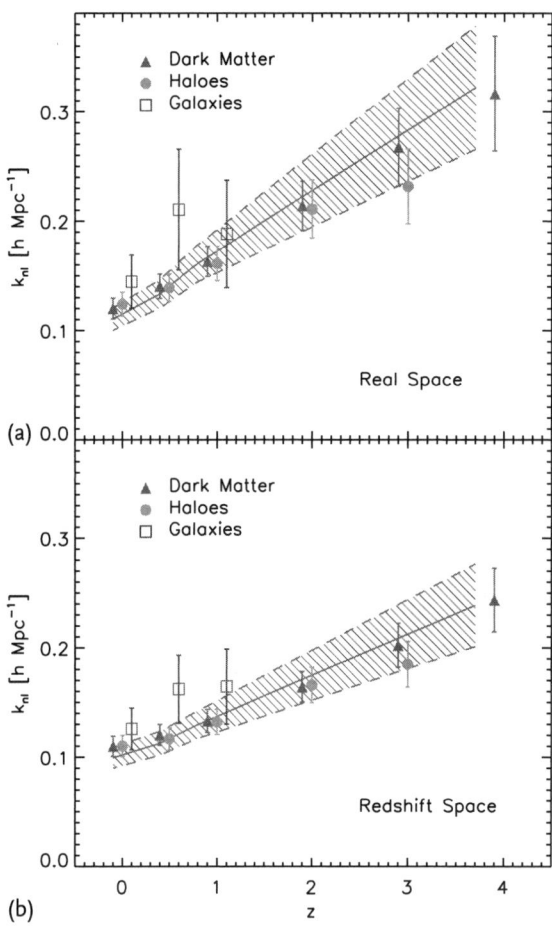

Figure 5.8 The best-fitting value of the damping scale k_{nl} as a function of redshift, for different tracers of the matter density distribution, in real-space (a) and redshift-space (b) ((Angulo et al., 2008), MNRAS, Wiley-Blackwell). The symbols show results from the high resolution corresponding to a comoving volume of $2.41\,h^{-3}\,Gpc^3$: dark matter (triangles), dark matter haloes with mass in excess of $5.4 \times 10^{12}\,h^{-1}\,M_\odot$ (circles) and galaxies (squares). The error bars show the 1σ range on k_{nl}. The hatched region shows the central 68% range of the results obtained using the dark matter in an ensemble of low-resolution simulations.

α) and more precisely (smaller dispersion in α) than using a catalog of galaxies chosen by the strength of their emission lines (sample D).

Note that the size of the systematic shift of the estimated α away from $\alpha = 1$ for the galaxy samples is comparable to the random measurement errors for the simulation (Angulo et al., 2008). It will require a larger simulation volume to reduce the size of random errors, and to ascertain whether such shifts reflect genuine limits of the method discussed here.

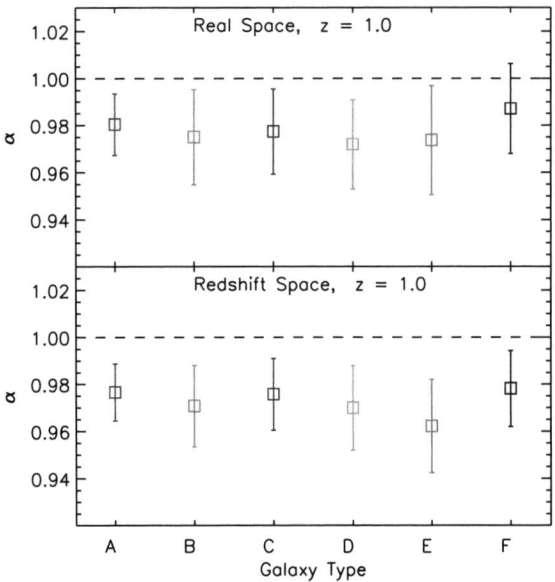

Figure 5.9 The recovered value of the scale parameter α for various galaxy samples (Angulo et al. (2008), MNRAS, Wiley-Blackwell). Sample A is R-magnitude-limited to reach a space density of $5 \times 10^{-4}\,h^3\,\mathrm{Mpc}^{-3}$. Sample B is magnitude-limited to reach half the space density of sample A. Sample C contains the reddest 50% of galaxies from sample A, using the R–I color. Sample D contains the 50% of galaxies from sample A with the strongest emission lines, using the equivalent width of OII[3727]. Sample E contains the bluest 50% of galaxies from sample A, using the R–I color. Sample F contains the 50% of galaxies from sample A with the weakest emission lines, using the equivalent width of OII[3727]. The simulation corresponds to a comoving volume of $2.41\,h^{-3}\,\mathrm{Gpc}^3$. Recall that $\alpha = 1$ corresponds to an unbiased measurement of the BAO scale (hence of dark energy parameters).

A more accurate model for the power spectrum is given by the "dewiggled" power spectrum (Tegmark et al., 2006; Eisenstein et al., 2006):

$$P_{\mathrm{dw}}(k) = P_{\mathrm{lin}}(k)G(k) + P_{\mathrm{nw}}(k)\left[1 - G(k)\right], \tag{5.50}$$

where $P_{\mathrm{lin}}(k)$ is the linear theory power spectrum and $P_{\mathrm{nw}}(k)$ is a smooth, linear theory, cold dark matter only power spectrum, with the same shape as $P_{\mathrm{lin}}(k)$ but without any baryonic oscillations (i.e., $P_{\mathrm{ref}}^{\mathrm{L}}$). The weight function $G(k)$ is given by

$$G(k) \equiv \exp\left[-(k/\sqrt{2}k_\star)^2\right], \tag{5.51}$$

describing the transition from large scales ($k \ll k_\star$), where $P_{\mathrm{dw}}(k)$ follows linear theory, to small scales ($k \gg k_\star$) where the acoustic oscillations are completely damped by nonlinear effects.

Equation (5.50) provides a phenomenological description of the modification of the BAO by nonlinear effects found in numerical simulations. Importantly, it can be justified using the renormalized perturbation theory (RPT) developed by Crocce and Scoccimarro (2006, 2008). According to RPT, the first term on the right hand side of Eq. (5.50) describes the growth of a single mode, quantified by the propagator function $G(k)$. In the high-k limit the propagator is given by the Gaussian form in Eq. (5.51) with $k_\star$ given by (Crocce and Scoccimarro, 2006; Matsubara, 2008)

$$k_\star = \left[\frac{1}{3\pi^2}\int dk\, P_{\text{lin}}(k)\right]^{-1/2}. \tag{5.52}$$

The second term on the right hand side of Eq. (5.50) can be interpreted as the power generated by the coupling of Fourier modes on small scales, $P_{\text{mc}}(k)$. The term $P_{\text{mc}}(k)$ is negligible on large scales (small k), but dominates the total power on small scales (high k). For the scales relevant to the BAO analysis ($k \sim k_\star$), P_{mc} has a similar amplitude to $P_{\text{nw}}(k)[1 - G(k)]$.

The limitation of Eq. (5.50) can be explained by RPT as well (Crocce and Scoccimarro, 2008). According to RPT, the propagator $G(k)$ only behaves as a Gaussian in the high-k limit. In addition, the term P_{mc} shows acoustic oscillations, although of a much smaller amplitude than $P(k)$, while $P_{\text{nw}}(k)[1 - G(k)]$ is a smooth function.

Equation (5.50) can be improved by modifying $P_{\text{nw}}(k)$ to model the change in the overall shape of the power spectrum due to nonlinear evolution:

$$P_{\text{dw}}^{\text{nl}}(k) = \left(\frac{1 + Qk^2}{1 + Ak + Bk^2}\right) P_{\text{dw}}(k) = f(k) P_{\text{dw}}(k). \tag{5.53}$$

The factor $f(k)$ could also be used to model a scale-dependent bias factor. This model for nonlinear evolution is based on the Q-model of Cole et al. (2005), modified by the addition of a new parameter, B, in order to improve its accuracy at high k. Fixing $B = Q/10$ gives the approximate behavior of the nonlinear power spectrum at large k (Sanchez, Baugh, and Angulo, 2008).

Figure 5.10 shows a comparison of the real-space dark matter power spectrum averaged over the simulation ensemble (open points) with the linear theory power spectrum (dot-dashed line), the "dewiggled" power spectrum from Eq. (5.50) (solid line), and its nonlinear version from Eq. (5.53) (dashed line) computed with $Q = 13$ and $A = 1.5$ (Sanchez, Baugh, and Angulo, 2008). The improved modeling of the power spectrum still leads to biased estimate of α similar to that shown in Figure 5.7, but with $\alpha > 1$ (Sanchez, Baugh, and Angulo, 2008).

The biased estimates of α from the measured power spectra (see Figures 5.7 and 5.9) are a consequence of the mode-coupling shifts due to nonlinear effects (Crocce and Scoccimarro, 2008). This may indicate a limit to the accuracy with which the BAO scale can be extracted from power spectrum data. In Fourier space, systematic effects such as redshift-space distortions and scale-dependent bias are important, and have to be minimized by dividing the measured power spectrum by a smooth reference power spectrum. This division by the smooth reference power spectrum leads to information loss that degrades the BAO scale accuracy and precision (Sanchez, Baugh, and Angulo, 2008). However, larger volume simulations will be needed to quantify the limit of accuracy of the $P(k)$ method of BAO analysis.

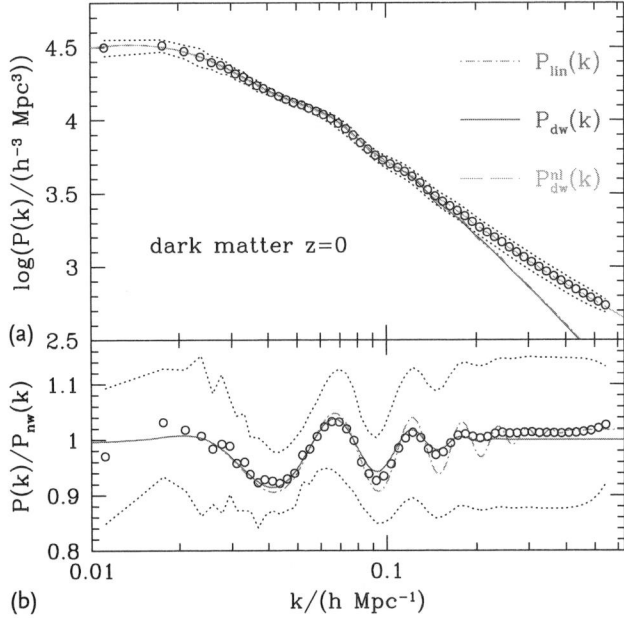

Figure 5.10 Results from the numerical simulations by Sanchez, Baugh, and Angulo (2008), MNRAS, Wiley-Blackwell. (a) A comparison of the real-space dark matter power spectrum averaged over the simulation ensemble (open points) with the linear theory power spectrum (dot-dashed line), the "dewiggled" power spectrum from Eq. (5.50) (solid line), and its nonlinear version from Eq. (5.53) (dashed line) computed with $Q = 13$ and $A = 1.5$. The dotted lines indicate the variance on $P(k)$ estimated from the ensemble. (b) The ratio of these power spectra to $P_{nw}(k)$.

5.4.2
Using Two-Point Correlation Functions to Probe BAO

To extract the BAO scales, we can also compute the two-point correlation function of galaxies in comoving coordinates. For spherically-averaged data, the BAO scale corresponds to a peak around the scale of the sound horizon at the drag epoch. Eisenstein *et al.* (2005) first demonstrated this with real data. Hutsi (2006) found similar results.

Definition of the Two-Point Correlation Function $\xi(r)$
The two-point correlation function ξ is defined as

$$\xi(r) \equiv \langle \delta(x+r)\delta(x) \rangle . \tag{5.54}$$

Thus the power spectrum and $\xi(r)$ are related by

$$P(k) \equiv |\delta_k|^2 = \int \xi(r) e^{ik \cdot r} d^3 r , \tag{5.55}$$

where δ_k is the Fourier transform of the matter density perturbation $\delta(r)$ (see Eq. (5.25)).

Measurement of the Two-Point Correlation Function $\xi(r)$

The two-point galaxy correlation function, ξ, can be measured by comparing the actual galaxy distribution to a catalog of randomly distributed galaxies. These randomly distributed galaxies have the same redshift, magnitude, and mask constraints as the real data. The pairs of galaxies are counted in bins of separation along the line of sight, π_s, and transverse to the line of sight, r_p, to estimate $\xi(r_p, \pi_s)$. In converting from redshift to distance, a fiducial model must be assumed, usually a flat universe model dominated by a cosmological constant, with $\Omega_m = 0.3$, $\Omega_\Lambda = 0.7$. Hence it is important to iterate the final results by changing the fiducial model to the best-fit model derived from the data.

Each galaxy and random galaxy can be given a weighting factor to account for both selection effects and to optimize the statistics. For example, to minimize the variance on the estimated $\xi(s)$ when the survey selection function $n(z_i)$ varies significantly, one can introduce the so-called "minimum-variance weighting" (Davis and Huchra, 1982; Davis and Peebles, 1983):

$$w_i = \frac{1}{1 + 4\pi n(z_i) J_3(s)}, \tag{5.56}$$

where the separation $s \equiv \sqrt{\pi_s^2 + r_p^2}$, $n(z)$ is the galaxy density distribution, and

$$J_3(s) = \int_0^s \xi(s') s'^2 ds'. \tag{5.57}$$

Hawkins et al. (2003) used $n(z)$ from the random catalog to ensure that the weights vary smoothly with redshift, and they found that results are insensitive to the precise form of J_3. Each galaxy pair (i, j) is given a weight $w_f w_i w_j$ (with w_f correcting for galaxies not observed due to effects such as fiber collisions), while each galaxy-random and random-random pair is given a weight $w_i w_j$.

An often-used minimum-variance estimator of ξ is that of Landy and Szalay (1993):

$$\xi(r_p, \pi_s) = \frac{DD - 2DR + RR}{RR}, \tag{5.58}$$

where r_p and π_s denote the transverse and line-of-sight separations in redshift space, respectively. DD is the normalized sum of weights of galaxy-galaxy pairs with separation (r_p, π_s), RR is the normalized sum of weights of random-random pairs with the same separation in the random catalog and DR is the normalized sum of weights of galaxy-random pairs with the same separation. DR is calculated by overlaying the real galaxy catalog and the simulated random galaxy catalog. DD, RR, and DR are normalized through dividing by the total number of pairs in each. Spherically averaging $\xi(r_p, \pi_s)$ at constant $s = \sqrt{\pi_s^2 + r_p^2}$ gives the redshift-space correlation function $\xi(s)$. Both Eisenstein et al. (2005) and Okumura et al. (2008) used the Landy & Szalay estimator in Eq. (5.58) to analyze SDSS LRG data.

If the rms scatter on $P(k)$, $\sigma_P^2(k)$, is computed (see Section 5.4.1), the covariance of the two-point correlation function can be calculated using (Cohn, 2006; Smith, Scoccimarro, and Sheth, 2008):

$$C_\xi(r, r') \equiv \langle [\xi(r) - \bar{\xi}(r)][\xi(r') - \bar{\xi}(r')] \rangle$$

$$= \int \frac{dk\, k^2}{2\pi^2} j_0(kr) j_0(kr') \sigma_P^2(k), \qquad (5.59)$$

where $\xi(r)$ and $\bar{\xi}(r)$ are the correlation function and its mean, respectively.

The BAO scale shown in Figure 5.1 is measured from the spherically-averaged redshift-space correlation function from SDSS LRG sample by Eisenstein *et al.* (2005). There is no verifiable detection of the radial and transverse BAO scales from current data (Okumura *et al.*, 2008; Gaztanaga, Cabre, and Hui, 2008). This may be an indication of systematic uncertainties.

Mitigation of Systematic Effects in BAO Scale Extraction from $\xi(r)$

Simulated data must be used to study how the BAO scale extraction from the measured galaxy two-point correlation function is affected by systematic effects. Using 50 low-resolution N-body simulations (each with a comoving volume of $2.41\,h^{-3}\,\text{Gpc}^3$ and with the dark matter followed using 448^3 particles), Sanchez, Baugh, and Angulo (2008) found that the BAO signature in the two-point correlation function is less affected by scale-dependent effects than that in the power spectrum.

The two-point correlation function can be obtained by taking the Fourier transform of Eq. (5.50):

$$\xi_{dw}(r) = \xi_{lin}(r) \otimes \tilde{G}(r) + \xi_{nw}(r) \otimes (1 - \tilde{G}(r)), \qquad (5.60)$$

where the symbol $\otimes$ denotes a convolution, and $\tilde{G}(r)$ is the Fourier transform of $G(k)$. The first term contains the information about the acoustic oscillations; it represents the convolution of the linear theory correlation function with a Gaussian kernel. This convolution implies that in the correlation function, the damping of the higher harmonic oscillations causes the acoustic peak to broaden and shift to smaller scales (Smith, Scoccimarro, and Sheth, 2008; Crocce and Scoccimarro, 2008).

Figure 5.11 compares the mean $z = 0$ real-space correlation function of the dark matter measured from an ensemble of simulations (open points) with the following models for the correlation function: (i) the linear theory correlation function $\xi_{lin}(r)$ (solid line), (ii) a nonlinear correlation function $\xi^{nl}(r)$ computed using halofit, without any damping of the acoustic oscillations (short-dashed line), (iii) the dewiggled linear theory correlation function $\xi_{dw}(r)$, computed as described by Eq. (5.60) (long-dashed line), and (iv) a dewiggled correlation function nonlinearized using halofit $\xi_{dw}^{nl}(r)$ (dot-dashed line) (Sanchez, Baugh, and Angulo, 2008). The error bars indicate the variance between the correlation functions measured from the different realizations in the simulation ensemble.

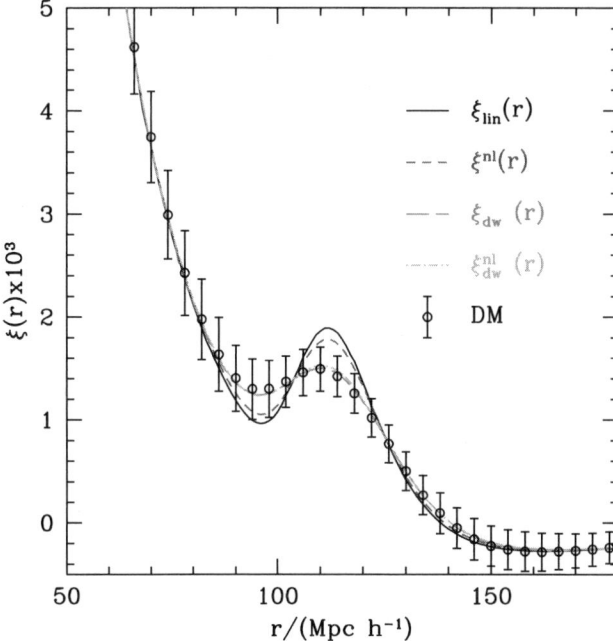

Figure 5.11 Comparison by Sanchez, Baugh, and Angulo (2008), MNRAS, Wiley-Blackwell, of the $z = 0$ real-space dark matter two-point correlation function averaged over the ensemble of simulations (open points) with: (i) the linear theory correlation function $\xi_{\text{lin}}(r)$ (solid line), (ii) an estimate of the nonlinear correlation function $\xi^{\text{nl}}(r)$ computed using halofit without damping of the acoustic oscillations (dashed line), (iii) the dewiggled linear theory correlation function $\xi_{\text{dw}}(r)$ defined by Eq. (5.60) (long-dashed line), and (iv) a dewiggled correlation function after being nonlinearized using halofit $\xi_{\text{dw}}^{\text{nl}}(r)$ (dot-dashed line). The error bars indicate the rms scatter between the different realizations in the ensemble of simulations.

Figure 5.11 shows that the acoustic peak in the two-point correlation function at redshift $z = 0$ shows strong deviations from the predictions of linear theory. Clearly, the linear theory dewiggled correlation function from Eq. (5.60) gives a very good description of the results of numerical simulations; this indicates that the damping of the oscillations is the most important effect to include in the modeling of the real space correlation function on large scales. The incorporation of the full change in the shape of $P(k)$ due to nonlinear evolution produces very little difference in the shape of the acoustic peak in the correlation function, but this effect might be important on intermediate scales ($r \simeq 70\,h^{-1}$ Mpc).

The scale parameter from Eq. (5.21) corresponds to an equivalent shift from scale r_{true} to $r_{\text{app}} = r_{\text{true}}/\alpha$ in the two-point correlation function. Deviation from $\alpha = 1$ indicates a biased estimate of the BAO scale, and the uncertainty on α indicates the precision of the BAO scale measurement. Sanchez, Baugh, and Angulo (2008) used Eq. (5.60) to analyze an ensemble of 50 low-resolution N-body simulations (each with a comoving volume of $2.41\,h^{-3}$ Gpc^3). The estimated scale parameter is slightly biased: $\alpha = 0.996 \pm 0.006$ at $z = 0$, $\alpha = 0.998 \pm 0.004$ at $z = 0.5$ and

$\alpha = 0.997 \pm 0.003$ at $z = 1$. The constraints on α become tighter with increasing redshift. This is because the higher harmonic oscillations are less damped as redshift increases (which accompanies an increase in the range of wavenumbers over which density perturbations are linear), thus the position of the BAO peak can be more precisely determined. The deviation from $\alpha = 1$ indicates the limitation of Eq. (5.60) in describing the full shape of the correlation function.

The model of Eq. (5.60) can be improved by utilizing the BAO information contained in $\xi_{mc}(r)$, the correlation function generated by the coupling of Fourier modes on small scales. According to renormalized perturbation theory (RPT), the main contribution to $\xi_{mc}(r)$ on the scale of BAO is of the form (Crocce and Scoccimarro, 2008)

$$\xi_{mc}(r) \propto \xi'_{lin} \xi^{(1)}_{lin}(r) , \qquad (5.61)$$

where ξ'_{lin} is the derivative of the linear theory correlation function and

$$\xi^{(1)}_{lin}(r) \equiv \hat{r} \cdot \nabla^{-1} \xi_{lin}(r) = 4\pi \int P_{lin}(k) j_1(kr) k dk . \qquad (5.62)$$

Thus an improved model of the correlation function is given by (Crocce and Scoccimarro, 2008)

$$\xi_{nl}(r) = \xi_{lin}(r) \otimes \tilde{G}(r) + A_{mc} \xi'_{lin} \xi^{(1)}_{lin}(r) , \qquad (5.63)$$

where A_{mc} is a free parameter. The second term in Eq. (5.63) can describe the shape of the residuals of the measured correlation function with respect to $\xi_{lin} \otimes \tilde{G}$ close to the BAO peak. At smaller scales, where the approximation is not so accurate, the model underestimates the correlation function (Sanchez, Baugh, and Angulo, 2008). Applying Eq. (5.63) to an ensemble of 50 low-resolution N-body simulations (each with a comoving volume of 2.41 h^{-3} Gpc^3), Sanchez, Baugh, and Angulo (2008) found that the scale parameter is measured more accurately: $\alpha = 1.003 \pm 0.008$ at $z = 0$, $\alpha = 1.002 \pm 0.005$ at $z = 0.5$ and $\alpha = 1.000 \pm 0.003$ at $z = 1$. This indicates that the implementation of a full calculation of ξ_{mc} using RPT over the full range of scales included in the analysis can lead to unbiased estimate of the BAO scale. Note also that reducing the bias in the estimate of the BAO scale generally leads to an increase in its statistical scatter.

In current analysis techniques explored, the correlation function analysis leads to more accurate and precise estimate of the BAO scale than the power spectrum method (Sanchez, Baugh, and Angulo, 2008). The main reason for this is that the correlation function is less affected by scale-dependent effects than the power spectrum. Thus in a correlation function analysis, the entire correlation function can be modeled (including the large scale shape), while in the power spectrum analysis, the information on amplitude and large scale shape is discarded in order to remove sensitivity to systematic effects such as nonlinear effects and redshift-space distortions (Sanchez, Baugh, and Angulo, 2008).

5.5
Future Prospects for BAO Measurements

Given real or simulated galaxy redshift survey data, one would need to extract the radial and transverse BAO scales from the data in order to estimate $H(z)$ and $D_A(z)$, *before* measuring the dark energy parameters, as illustrated by Sections 5.4.1 and 5.4.2. Robust forecast can only come from a Monte-Carlo-based approach that begins with extracting the radial and transverse BAO scales from realistically simulated galaxy catalogs.

Most of the BAO forecasts have been done using the Fisher matrix formalism, which gives the smallest possible statistical uncertainties (see Section 2.4.2). The Fisher matrix method allows an estimate of expected measurement uncertainties on $H(z)$ and $D_A(z)$ from a future galaxy redshift survey based on the assumed survey parameters, *without* analyzing simulated galaxy catalogs to extract the BAO scales. While the Fisher matrix forecasts are likely too optimistic, they are easy and straightforward to make, thus provide the most convenient way to estimate the expected constraints on dark energy from future galaxy redshift surveys. Here we discuss the Fisher matrix forecast methodology in detail.

In the limit where the length scale corresponding to the survey volume is much larger than the scale of any features in $P_g(k)$, we can assume that the likelihood function for the band powers of a galaxy redshift survey is Gaussian, and given by Eq. (5.44) with a measurement error in $\ln P(k)$ that is proportional to $[V_{\text{eff}}(k)]^{-1/2}$, with the effective volume of the survey defined as (see Eq. (5.45)):

$$V_{\text{eff}}(k,\mu) \equiv \int dr^3 \left[\frac{n(r)P_g(k,\mu)}{n(r)P_g(k,\mu)+1}\right]^2 = \left[\frac{nP_g(k,\mu)}{nP_g(k,\mu)+1}\right]^2 V_{\text{survey}}, \quad (5.64)$$

where the comoving number density n is assumed to only depend on the redshift (and constant in each redshift slice) for simplicity in the last part of the equation.

In order to propagate the measurement error in $\ln P_g(k)$ into measurement errors for the parameters p_i, we can use Eq. (2.129), and that

$$\mathcal{L}[P(k)] = \mathcal{L}(p_i), \quad (5.65)$$

with $\mathcal{L}[P(k)]$ given by Eq. (5.44). Ignoring the subdominant normalization factors, this gives an approximated Fisher matrix (Tegmark, 1997)

$$F_{ij} = \int_{k_{\min}}^{k_{\max}} \frac{\partial \ln P_g(k)}{\partial p_i} \frac{\partial \ln P_g(k)}{\partial p_j} V_{eff}(k) \frac{dk^3}{2(2\pi)^3}, \quad (5.66)$$

where p_i are the parameters to be estimated from data, and the derivatives are evaluated at parameter values of the fiducial model. Note that the Fisher matrix F_{ij} is the inverse of the covariance matrix of the parameters p_i if the p_i are Gaussian distributed.

"Wiggles Only" Method

In order to arrive at robust BAO forecasts, we may use the information contained in the BAO peaks only, and discard the information contained in the broad shape of $P_g(k)$ (Blake and Glazebrook, 2003; Seo and Eisenstein, 2007). The measurement of the BAO peaks gives measurements of $s/D_A(z)$ and $sH(z)$ (see Eq. (5.1)).

Note that Eq. (5.66) can be rewritten as

$$F_{ij} = V_{\text{survey}} \int_{-1}^{1} d\mu \int_{k_{\min}}^{k_{\max}} \frac{\partial P_g(k,\mu)}{\partial p_i} \frac{\partial P_g(k,\mu)}{\partial p_j} \cdot \left[\frac{1}{P_g(k,\mu) + n^{-1}}\right]^2 \frac{2\pi k^2 dk}{2(2\pi)^3}, \tag{5.67}$$

where $\mu = \hat{k} \cdot \hat{r}$, with $\hat{r}$ denoting the unit vector along the line of sight.

Seo and Eisenstein (2007) obtained simple fitting formulae for estimated errors in $s/D_A(z)$ and $sH(z)$ by approximating Eq. (5.67) with

$$F_{ij} \simeq V_{\text{survey}} \int_{-1}^{1} d\mu \int_{k_{\min}}^{k_{\max}} \frac{\partial P_b(k,\mu|z)}{\partial p_i} \frac{\partial P_b(k,\mu|z)}{\partial p_j}$$
$$\cdot \left[\frac{1}{P_g^{\text{lin}}(k,\mu|z) + n^{-1}}\right]^2 \frac{2\pi k^2 dk}{2(2\pi)^3}, \tag{5.68}$$

where $P_b(k,\mu|z)$ is the power spectrum that contains baryonic features. The linear galaxy power spectrum

$$P_g^{\text{lin}}(k,\mu|z) = P_{g,r}^{\text{lin}}(k|z) R(\mu) \tag{5.69}$$

$$P_{g,r}^{\text{lin}}(k|z) = [b(z)]^2 \left[\frac{G(z)}{G(0)}\right]^2 P_m^{\text{lin}}(k|z=0) \tag{5.70}$$

where $P_{g,r}^{\text{lin}}(k|z)$ is the linear galaxy power spectrum in real space, $b(z)$ is the bias factor, $G(z)$ is the growth factor, and $P_m^{\text{lin}}(k|z=0)$ is the present day linear matter power spectrum. $R(\mu)$ is the linear redshift-space distortion factor given by (see Eq. (5.17)) (Kaiser, 1987)

$$R(\mu) = \left(1 + \beta\mu^2\right)^2. \tag{5.71}$$

The power spectrum that contains baryonic features, $P_b(k,\mu)$, is given by (Seo and Eisenstein, 2007)

$$P_b(k,\mu|z) = \sqrt{8\pi^2} A_0 P_g^{\text{lin}}(k_{0.2},\mu|z) \frac{\sin(x)}{x} \cdot \exp\left[-(k\Sigma_s)^{1.4} - \frac{k^2 \Sigma_{\text{nl}}^2}{2}\right], \tag{5.72}$$

where we define

$$k_{0.2} \equiv 0.2\, h\, \text{Mpc}^{-1} \tag{5.73}$$

$$x \equiv \left(k_\perp^2 s_\perp^2 + k_\parallel^2 s_\parallel^2\right)^{1/2} \tag{5.74}$$

$$k_\parallel = \mathbf{k} \cdot \hat{\mathbf{r}} = k\mu \tag{5.75}$$

$$k_\perp = \sqrt{k^2 - k_\parallel^2} = k\sqrt{1-\mu^2}. \tag{5.76}$$

The nonlinear damping scale

$$\Sigma_{nl}^2 = (1-\mu^2)\Sigma_\perp^2 + \mu^2 \Sigma_\parallel^2$$

$$\Sigma_\parallel = \Sigma_\perp (1 + f_g)$$

$$\Sigma_\perp = 12.4\, h^{-1}\, \text{Mpc} \left(\frac{\sigma_8}{0.9}\right) \cdot 0.758 \frac{G(z)}{G(0)} p_{NL}$$

$$= 8.355\, h^{-1}\, \text{Mpc} \left(\frac{\sigma_8}{0.8}\right) \cdot \frac{G(z)}{G(0)} p_{NL}, \tag{5.77}$$

where the growth rate $f_g = d\ln G(z)/d\ln a$. The parameter p_{NL} indicates the remaining level of nonlinearity in the data; with $p_{NL} = 0.5$ (50% nonlinearity) as the best case, and $p_{NL} = 1$ (100% nonlinearity) as the worst case (Seo and Eisenstein, 2007). For a fiducial model based on WMAP3 results (Spergel et al., 2007) ($\Omega_m = 0.24$, $h = 0.73$, $\Omega_\Lambda = 0.76$, $\Omega_k = 0$, $\Omega_b h^2 = 0.0223$, $\tau = 0.09$, $n_s = 0.95$, $T/S = 0$), $A_0 = 0.5817$, $P_{0.2} = 2710\sigma_{8,g}^2$, and the Silk damping scale $\Sigma_s = 8.38\, h^{-1}\, \text{Mpc}$ (Seo and Eisenstein, 2007).

Defining

$$p_1 = \ln s_\perp^{-1} = \ln(D_A/s), \tag{5.78}$$

$$p_2 = \ln s_\parallel = \ln(sH), \tag{5.79}$$

substituting Eq. (5.72) into Eq. (5.68), and making the approximation of $\cos^2 x \sim 1/2$, we find

$$F_{ij} \simeq V_{\text{survey}} A_0^2 \int_0^1 d\mu\, f_i(\mu) f_j(\mu) \int_0^{k_{\max}} dk\, k^2$$

$$\cdot \left[\frac{P_m^{\text{lin}}(k|z=0)}{P_m^{\text{lin}}(k_{0.2}|z=0)} + \frac{1}{n P_g^{\text{lin}}(k_{0.2},\mu|z) e^{-k^2\mu^2\sigma_r^2}} \right]^{-2}$$

$$\cdot \exp\left[-2(k\Sigma_s)^{1.4} - k^2 \Sigma_{nl}^2\right], \tag{5.80}$$

where $P_g^{\text{lin}}(k_{0.2},\mu|z)$ is given by Eq. (5.69) with $k = k_{0.2}$, and $k_{\max} = 0.5\, h\, \text{Mpc}^{-1}$ (Seo and Eisenstein, 2007). Note that we have added the damping factor, $e^{-k^2\mu^2\sigma_r^2}$, due to redshift uncertainties, with

$$\sigma_r = \frac{\partial r}{\partial z} \sigma_z \tag{5.81}$$

where r is the comoving distance from Eq. (2.24). The functions $f_i(\mu)$ are given by

$$f_1(\mu) = \partial \ln x / \partial \ln p_1 = \mu^2 - 1 \tag{5.82}$$

$$f_2(\mu) = \partial \ln x / \partial \ln p_2 = \mu^2 . \tag{5.83}$$

The square roots of diagonal elements of the inverse of the Fisher matrix of Eq. (5.80) give the estimated smallest possible measurement errors on $s_\perp^{-1}$ and $s_\parallel$. The estimated errors are *independent* of cosmological priors, thus scale with (area)$^{-1/2}$, for a fixed survey depth.

Full P(k) Method
Since the full $P_g(k)$ is measured from a galaxy redshift survey, it is also useful to make forecasts of dark energy constraints using the full $P_g(k)$. The observed galaxy power spectrum can be reconstructed using a particular reference cosmology, including the effects of bias and redshift-space distortions (Seo and Eisenstein, 2003):

$$P_g^{\text{obs}}\left(k_\perp^{\text{ref}}, k_\parallel^{\text{ref}}\right) = \frac{[D_A(z)^{\text{ref}}]^2 H(z)}{[D_A(z)]^2 H(z)^{\text{ref}}} b^2 (1 + \beta \mu^2)^2 \\ \cdot \left[\frac{G(z)}{G(0)}\right]^2 P_m(k)_{z=0} + P_{\text{shot}} , \tag{5.84}$$

where $\mu = \mathbf{k} \cdot \hat{\mathbf{r}}/k$, with $\hat{\mathbf{r}}$ denoting the unit vector along the line of sight; $\mathbf{k}$ is the wavevector with $|\mathbf{k}| = k$. Hence $\mu^2 = k_\parallel^2/k^2 = k_\parallel^2/(k_\perp^2 + k_\parallel^2)$. The values in the reference cosmology are denoted by the subscript "ref", while those in the true cosmology have no subscript. Note that

$$k_\perp^{\text{ref}} = k_\perp D_A(z)/D_A(z)^{\text{ref}} , \quad k_\parallel^{\text{ref}} = k_\parallel H(z)^{\text{ref}}/H(z) . \tag{5.85}$$

Equation (5.84) characterizes the dependence of the observed galaxy power spectrum on $H(z)$ and $D_A(z)$ due to BAO, as well as the sensitivity of a galaxy redshift survey to the linear redshift-space distortion parameter β (see Eq. (5.18)).

The observed galaxy power spectrum in a given redshift shell centered at redshift z_i can be described by a set of parameters, $\{H(z_i), D_A(z_i), \overline{G(z_i)}, \beta(z_i), P_{\text{shot}}^i, n_S, \omega_m, \omega_b\}$, where n_S is the power-law index of the primordial matter power spectrum, $\omega_m = \Omega_m h^2$, and $\omega_b = \Omega_b h^2$ (h is the dimensionless Hubble constant). Note that $P(k)$ does *not* depend on h if k is in units of Mpc^{-1}, since the matter transfer function $T(k)$ only depends on ω_m and ω_b (Eisenstein and Hu, 1998),[10] if the dark energy dependence of $T(k)$ can be neglected. Note also that $T(k)$ is normalized such that $T(k \to 0) = 1$. Since $G(z)$, b, and the power spectrum normalization P_0 are completely degenerate in Eq. (5.84), they can be combined into a single parameter, $\overline{G(z_i)} \equiv b(z)G(z)P_0^{1/2}/G(0)$.

---

10) Massive neutrinos can suppress the galaxy power spectrum amplitudes by $\gtrsim 4\%$ on BAO scales (Eisenstein and Hu, 1999).

The square roots of diagonal elements of the inverse of the full Fisher matrix of Eq. (5.66) give the estimated smallest possible measurement errors on the assumed parameters. The parameters of interest are $\{H(z_i), D_A(z_i), \beta(z_i)\}$, all other parameters are marginalized over. Note that the estimated errors we obtain are *independent* of cosmological priors since no priors are explicitly imposed, and thus scale with (area)$^{-1/2}$ for a fixed survey depth. Priors on ω_m, ω_b, Ω_k, and n_S will be required to obtain the errors on dark energy parameters if only BAO data are considered.

In order to compare the "wiggles only" method and the full $P(k)$ method for BAO forecast, we must include the nonlinear effects in the same way in both methods. We can include nonlinear effects in the full power spectrum calculation by modifying the derivatives of $P_g(k)$ with respect to the parameters p_i as follows (Seo and Eisenstein, 2007):

$$\frac{\partial P_g(k,\mu|z)}{\partial p_i} = \frac{\partial P_g^{\text{lin}}(k,\mu|z)}{\partial p_i} \cdot \exp\left(-\frac{1}{2}k^2 \Sigma_{nl}^2\right). \tag{5.86}$$

The damping is applied to derivatives of $P_g(k)$, rather than $P_g(k)$, to ensure that no information is extracted from the damping itself (Seo and Eisenstein, 2007). Equation (5.67) becomes

$$F_{ij} = V_{\text{survey}} \int_{-1}^{1} d\mu \int_{k_{\min}}^{k_{\max}} \frac{\partial \ln P_g^{\text{lin}}(k,\mu)}{\partial p_i} \frac{\partial \ln P_g^{\text{lin}}(k,\mu)}{\partial p_j}$$
$$\cdot \left[\frac{n P_g^{\text{lin}}(k,\mu)}{n P_g^{\text{lin}}(k,\mu) + 1}\right]^2 e^{-k^2 \Sigma_{nl}^2} \frac{2\pi k^2 dk}{2(2\pi)^3}. \tag{5.87}$$

Under the same assumptions, the full $P_g(k)$ method can boost the figure of merit (FoM) for constraining dark energy by a factor of $\sim$3–4, compared to the "wiggles only" method (see, e.g., Wang (2009)), if no other data or priors are added, and redshift-space distortions are marginalized over in the $P(k)$ method (Seo and Eisenstein, 2003; Wang, 2006). The two methods give very similar constraints on the BAO scales $s/D_A(z)$ and $s\,H(z)$ (Seo and Eisenstein, 2007; Wang, 2009); the difference comes in the use of additional information from the broad shape of $P_g(k)$ in the full $P(k)$ method.

Galaxy Number Density

For a given galaxy redshift survey, the galaxy number density $n(z)$ and bias function $b(z)$ should be modeled using available data and supplemented by cosmological N-body simulations that include galaxies (Angulo et al., 2008). Since $n(z)$ and $b(z)$ depend on survey parameters such as the flux limit and the target selection method and efficiency, a more generic galaxy number density given by assuming $n P_g^r(k_{0.2}|z) = 3$ is often used in Fisher matrix forecasts (Seo and Eisenstein, 2007), where $P_g^r(k_{0.2}|z)$ is the real-space power spectrum of galaxies at $k = 0.2\,h\,\text{Mpc}^{-1}$

and redshift z. Note that this assumption means

$$n P_g^r(k_{0.2}|z) = P_m(k_{0.2}|0) n(z) b^2(z) \left[\frac{G(z)}{G(0)} \right]^2 = 3, \tag{5.88}$$

where $G(z)$ is the growth factor, and $b(z)$ is the bias factor. Assuming a fiducial cosmological model that fits all current observational data, $G(z)$ decreases by about a factor of 2 from $z = 0$ to $z = 2$, while $b(z)$ may increase with z somewhat and is dependent on the type of galaxies sampled by the survey.

For an ambitious yet feasible galaxy redshift survey of $H\alpha$ emission line galaxies, using a fully empirical $n(z)$ derived from current observational data (Geach et al., 2009), and a bias factor $b(z)$ derived from cosmological N-body simulations calibrated with current observational data (Orsi et al., 2009), $n P_g^r(k_{0.2}|z) > 3$ near the median redshift ($z_m \sim 1$), while $n P_g^r(k_{0.2}|z) < 3$ at $z \sim 2$, assuming a realistic efficiency for galaxy spectroscopy (Geach et al., 2009). This is as expected. The observed galaxy number density $n(z)$ from a flux-limited survey peaks at the median redshift, and decreases sharply in the high z tail, as the number of galaxies fainter than the flux limit increases. The increase in the bias factor $b(z)$ is not fast enough to compensate for the decrease in both $G(z)$ and $n(z)$ to satisfy $n P_g^r(k_{0.2}|z) \geq 3$ at $z \sim 2$. Therefore, assuming $n P_g^r(k_{0.2}|z) = 3$ is likely too optimistic, while assuming $n(z)[b(z)/b(0)]^2 P_g^r(k_{0.2}|z = 0) = 3$ could be a conservative alternative in Fisher matrix forecasts for a generic galaxy redshift survey (Wang, 2009).

5.6
Probing the Cosmic Growth Rate Using Redshift-Space Distortions

A galaxy redshift survey can allow us to measure both $H(z)$ and $f_g(z)$ (Guzzo et al., 2008; Wang, 2008b). The measurement of $f_g(z)$ can be obtained through independent measurements of redshift-space distortion parameter $\beta = f_g(z)/b$ (Kaiser, 1987) and the bias parameter $b(z)$ (which describes how light traces mass) (Guzzo et al., 2008).

5.6.1
Measuring Redshift-Space Distortion Parameter β

The parameter β can be measured directly from galaxy redshift survey data by studying the observed two-point redshift-space correlation function (Hawkins et al., 2003; Tegmark et al., 2006; da Angela et al., 2006; Ross et al., 2007). Hamilton (1998) reviewed various techniques for measuring β.

Peculiar velocities of galaxies lead to systematic differences between redshift-space and real-space measurements, and the effects are a combination of large scale coherent flows induced by the gravity of large scale structure, and a small scale random velocity of each galaxy (see Section 5.3.2). The large scale flows compress the contours of $\xi(r_p, \pi_s)$ along the π_s direction (along the line of sight), with the degree of compression determined by β. Kaiser (1987) showed that the coherent infall velocities lead to the following relation between the redshift-space power spectrum $P_s(k)$ and the real-space power spectrum $P_r(k)$ (see Eq. (5.17) (Kaiser, 1987)):

$$P_s(k, \mu) = \left(1 + \beta \mu^2\right)^2 P_r(k, \mu) , \qquad (5.89)$$

where $\mu = \mathbf{k} \cdot \hat{\mathbf{r}}/k$, with $\hat{\mathbf{r}}$ denoting the unit vector along the line of sight; $\mathbf{k}$ is the wavevector with $|\mathbf{k}| = k$. The small scale random motion of galaxies leads to a smearing in the radial direction (the "finger of God" effect).

Linear Regime

In the linear regime, the ratio of the spherically-averaged two point correlation function in redshift-space and real-space is given by

$$\frac{\xi(s)}{\xi(r)} = 1 + \frac{2\beta}{3} + \frac{\beta^2}{5} . \qquad (5.90)$$

Recall that the redshift-space correlation function $\xi(s)$ can be obtained by spherically averaging the measured redshift-space correlation function $\xi(r_p, \pi_s)$ (see Eq. (5.58)) at constant $s = \sqrt{\pi_s^2 + r_p^2}$ (see Section 5.4.2).

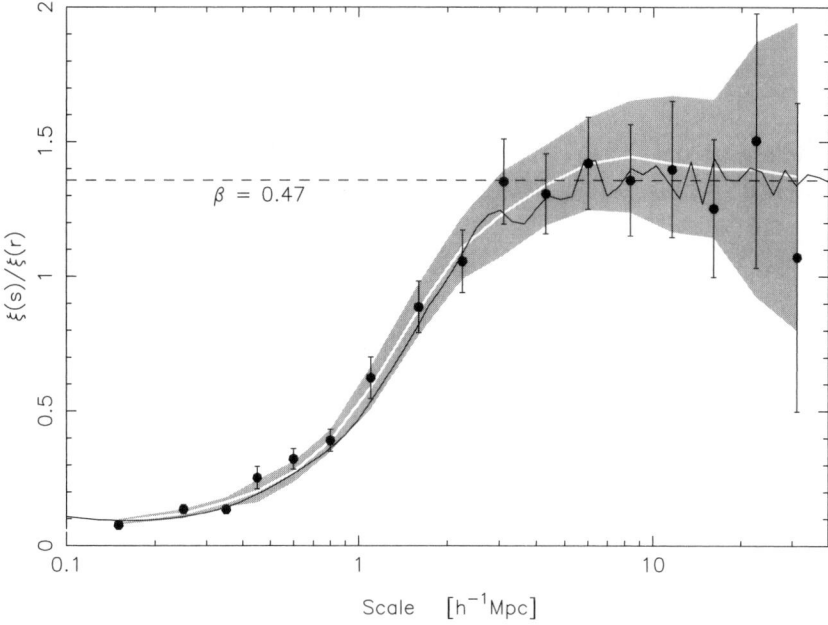

Figure 5.12 The ratio of $\xi(s)$ to $\xi(r)$ for the 2dFGRS data (solid points), and the Hubble volume simulation (solid line) (Hawkins et al. (2003), MNRAS, Wiley-Blackwell). The mean of the mock catalog results is also shown (white line), with the rms errors shaded. The error bars on the 2dF data are from the rms spread in mock catalog results.

Since the two-point correlation function is defined by the joint probability of finding galaxies centered within the volume elements dV_1 and dV_2 at a given separation, the projected correlation function (integrated along the line-of-sight) should be the same in real and redshift-space – both give the two-point angular correlation function. Thus the real-space correlation function $\xi(r)$ can be estimated by inverting the projected redshift-space correlation function $\Xi(r_p)$ (i.e., the angular correlation function) (Davis and Peebles, 1983):

$$\xi(r) = -\frac{1}{\pi} \int_r^\infty dr_p \frac{\Xi'(r_p)}{\left(r_p^2 - r^2\right)^{1/2}} \tag{5.91}$$

where

$$\Xi(r_p) = 2 \int_0^\infty d\pi_s \, \xi(r_p, \pi_s) \,. \tag{5.92}$$

Figure 5.12 shows the ratio of $\xi(s)$ to $\xi(r)$ for the 2dFGRS data (Peacock et al., 2001), obtained by Hawkins et al. (2003).

Nonlinear Regime

In the nonlinear regime, the parameter β can be measured by fitting the measured $\xi(r_p, \pi_s)$ (see Section 5.4.2) to a phenomenological model (Peebles, 1980)

$$\xi(r_p, \pi_s) = \int_{-\infty}^\infty dv \, f(v) \tilde{\xi}(r_p, \pi_s - v/H_0) \,, \tag{5.93}$$

where $\tilde{\xi}(r_p, \pi_s)$ is the *linear* redshift-space correlation function. Hamilton (1992) derived the model for $\tilde{\xi}(r_p, \pi_s)$ by translating Eq. (5.89) from Fourier space into real space:

$$\tilde{\xi}(r_p, \pi_s) = \xi_0(s) P_0(\mu) + \xi_2(s) P_2(\mu) + \xi_4(s) P_4(\mu) \,, \tag{5.94}$$

where $P_l(\mu)$ are Legendre polynomials, $\mu = \cos\theta$, with θ denoting the angle between the position vector r and π_s, and

$$\xi_0(s) = \left(1 + \frac{2\beta}{3} + \frac{\beta^2}{5}\right) \xi(r) \,, \tag{5.95}$$

$$\xi_2(s) = \left(\frac{4\beta}{3} + \frac{4\beta^2}{7}\right) \left[\xi(r) - \overline{\xi}(r)\right], \tag{5.96}$$

$$\xi_4(s) = \frac{8\beta^2}{35} \left[\xi(r) + \frac{5}{2}\overline{\xi}(r) - \frac{7}{2}\overline{\overline{\xi}}(r)\right], \tag{5.97}$$

where

$$\overline{\xi}(r) = \frac{3}{r^3} \int_0^r dr' \xi(r') r'^2 ,\qquad(5.98)$$

$$\overline{\overline{\xi}}(r) = \frac{5}{r^5} \int_0^r dr' \xi(r') r'^4 .\qquad(5.99)$$

The small scale random motions can be modeled by

$$f(v) = \frac{1}{\sigma_p \sqrt{2}} \exp\left(\frac{-\sqrt{2}|v|}{\sigma_p}\right) \qquad(5.100)$$

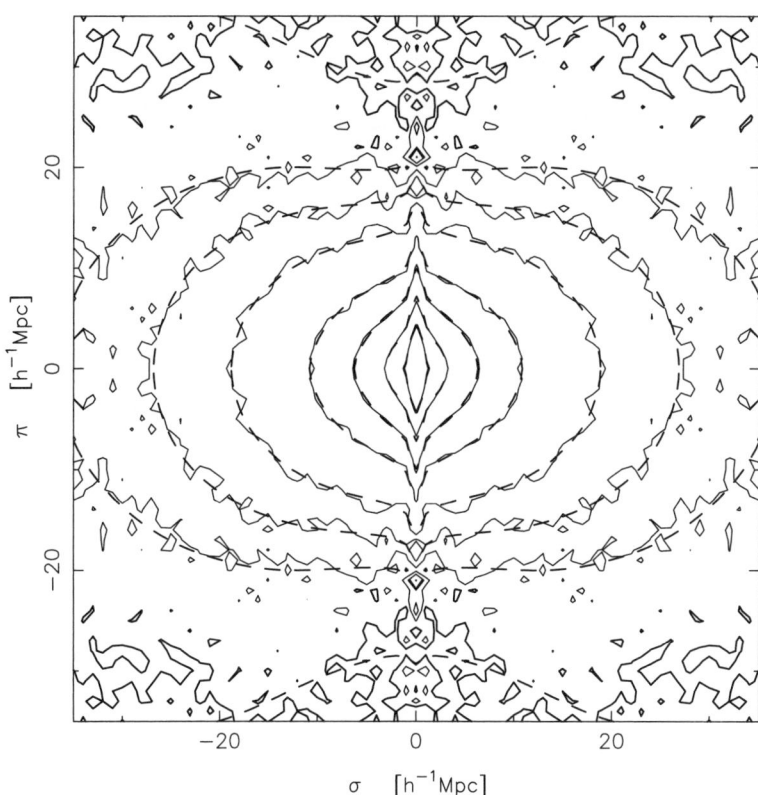

Figure 5.13 Contours of two-point correlation function $\xi(r_p, \pi_s)$ for the 2dFGRS data (solid lines) and the best-fitting model using the Hubble volume $\xi(r)$ fitted to scales $8 < s < 30\, h^{-1}$ Mpc (dashed lines) (Hawkins et al. (2003), MNRAS, Wiley-Blackwell). Contour levels are at $\xi = 4.0, 2.0, 1.0, 0.5, 0.2, 0.1, 0.05,$ and 0.0 (thick line). Note that the labels $\sigma = r_p$, and $\pi = \pi_s$.

where σ_p is the pairwise peculiar velocity dispersion. Convolution in real space becomes multiplication in Fourier space, so Eq. (5.93) becomes

$$P_s(k) = \hat{f}(k_\parallel) P_s^{\text{lin}}(k) = \hat{f}(k_\parallel) \left(1 + \beta\mu^2\right)^2 P_r(k), \qquad (5.101)$$

where $\hat{f}(k_\parallel)$ is the Fourier transform of $f(v)$ (Hamilton, 1998):

$$\hat{f}(k_\parallel) = \int_{-\infty}^{\infty} dv\, f(v) e^{ik_\parallel v} = \frac{1}{1 + \frac{1}{2}(\sigma_p k_\parallel)^2}, \qquad (5.102)$$

where $k_\parallel = k\mu$.

Figure 5.13 shows the 2dFGRS two-point galaxy correlation function $\xi(r_p, \pi_s)$, with the redshift-space distortions very clearly indicated. This yielded a measurement of $\beta = 0.49 \pm 0.09$ in a multi-parameter fit to $\xi(r_p, \pi_s)$ (Hawkins et al., 2003). The error bars are determined using the rms spread of results in mock catalogs. These catalogs are generated from Hubble Volume simulations in the fiducial cosmology.

Guzzo et al. (2008) showed how the estimators of β can be tested for both statistical and systematic errors.

5.6.2
Measuring the Bias Factor

In order to measure the growth rate $f_g(z) = b(z)\beta(z)$, we need to measure the bias factor $b(z)$, in addition to the linear redshift-space distortion factor $\beta(z)$. If we know that bias is linear, that is, $\delta_g(x) = b\delta_m(x)$, then $b(z) \simeq \sigma_{8,g}/\sigma_{8,m}$ (the ratio of σ_8 for galaxies and matter). Thus $f_g(z)\sigma_{8,m} \simeq \beta(z)\sigma_{8,g}$. The measurement of $\beta(z)$ and $\sigma_{8,g}$ thus provides a measurement of $f_g(z)\sigma_{8,m}$, which can be used directly to test gravity (Percival and White, 2009; White, Song, and Percival, 2009).

However, it is important to directly measure the bias factor, including its scale-dependence (which is one of the main systematic uncertainties in the BAO scale measurement, see Section 5.3.3). This can be done through the comparison of the measured probability distribution function of galaxy fluctuations with theoretical expectations (Sigad, Branchini, and Dekel, 2000; Marinoni et al., 2005). Here we focus on another method that utilizes the galaxy bispectrum.

We can assume that the galaxy density perturbation δ_g is related to the matter density perturbation $\delta(x)$ as follows (Fry and Gaztanaga, 1993):

$$\delta_g = f[\delta(x)] \simeq b_1\delta(x) + b_2\delta^2(x)/2$$
$$\simeq b_1\delta^{(1)}(x) + b_1\delta^{(2)}(x) + \frac{1}{2}b_2\left[\delta^{(1)}(x)\right]^2. \qquad (5.103)$$

Thus to second order (Matarrese, Verde, and Heavens, 1997)

$$\langle \delta_{g1}\delta_{g2}\delta_{g3}\rangle = b_1^3 \left\langle \delta_1^{(1)}\delta_2^{(1)}\delta_3^{(2)}\right\rangle + \text{cyc.} + \frac{b_1^2 b_2}{2}\left\langle \delta_1^{(1)}\delta_2^{(1)}\left[\delta_3^{(1)}\right]^2\right\rangle + \text{cyc.} \qquad (5.104)$$

where "cyc." refers to the permutations $\{231\}$ and $\{312\}$.

The galaxy bispectrum is defined by

$$\langle \delta_{gk_1} \delta_{gk_2} \delta_{gk_3} \rangle \equiv (2\pi)^3 \, B(k_1, k_2, k_3) \delta^D(k_1 + k_2 + k_3) \,. \tag{5.105}$$

Using the expression for $\delta_k^{(2)}$ from Catelan et al. (1995) and Eq. (5.104), we find

$$B(k_1, k_2, k_3) = P_g(k_1) P_g(k_2) \left[\frac{2J(k_1, k_2)}{b_1} + \frac{b_2}{b_1^2} \right] + \text{cyc.} \,, \tag{5.106}$$

where we have used

$$\langle \delta_{gk_1} \delta_{gk_2} \rangle = (2\pi)^3 P_g(k_1) \delta^D(k_1 + k_2) \,, \tag{5.107}$$

with δ_D denoting the Dirac delta function. Equation (5.107) follows from Eqs. (5.54), (5.55), (5.25) and (5.27). J is a function that depends on the shape of the triangle formed by (k_1, k_2, k_3) in k space, but only depends very weakly on cosmology (Matarrese, Verde, and Heavens, 1997):

$$J(k_1, k_2, \Omega_m) = 1 - B(\Omega_m) + \frac{k_1 \cdot k_2}{2 k_1 k_2} \left(\frac{k_1}{k_2} + \frac{k_2}{k_1} \right) + B(\Omega_m) \left(\frac{k_1 \cdot k_2}{k_1 k_2} \right)^2 \,, \tag{5.108}$$

where $B(\Omega_m) \simeq 2/7$ (assuming no coupling of dark energy to matter), and is insensitive to Ω_m (Bouchet et al., 1992; Catelan et al., 1995).

Verde et al. (2002) applied the galaxy bispectrum method for measuring b_i to the 2dFGRS data. Independent measurements of $\beta(z)$ and $b_i(z)$ have only been published for the 2dFGRS and VVDS data (Verde et al., 2002; Hawkins et al., 2003; Marinoni et al., 2005; Guzzo et al., 2008).

The large scale infall (parametrized by the redshift-space distortion parameter β, see Eq. (5.89)) and small scale smearing (parametrized by the pairwise velocity σ_p) lead to the power spectrum in redshift-space (see Eq. (5.101)):

$$P_s(k) = P(k) \frac{(1 + \beta \mu^2)^2}{1 + k^2 \mu^2 \sigma_p^2 / 2} \,. \tag{5.109}$$

Note that σ_p is implicitly divided by H_0. The bispectrum is modified similarly (Verde et al., 2002):

$$B_s(k_1, k_2, k_3) = (B_{12} + B_{23} + B_{31}) \left[\left(1 + \frac{a_V^2 k_1^2 \mu_1^2 \sigma_p^2}{2}\right) \left(1 + \frac{a_V^2 k_2^2 \mu_2^2 \sigma_p^2}{2}\right) \right.$$
$$\left. \times \left(1 + \frac{a_V^2 k_3^2 \mu_3^2 \sigma_p^2}{2}\right) \right]^{-1/2} \,, \tag{5.110}$$

where $k_3 = -k_1 - k_2$, and $\mu_i = r \cdot k_i/(rk_i)$. The adjustable parameter α_V depends on the shape of the triangle, and must be calibrated from simulations (Verde et al., 2002). Also

$$B_{12} = \left(1 + \beta\mu_1^2\right)\left(1 + \beta\mu_2^2\right)\left[\frac{\text{Ker}(k_1, k_2)}{b_1} + \frac{b_2}{b_1^2}\right] P_g(k_1) P_g(k_2). \quad (5.111)$$

The kernel function Ker is Jb_1, modified for redshift-space (Verde et al., 1998; Heavens, Matarrese, and Verde, 1998):

$$\text{Ker}(k_1, k_2) = J(k_1, k_2)b_1 + \mu^2 \beta b_1 K^{(2)}(k_1, k_2) + \mu_1^2 \mu_2^2 \beta^2 b_1^2$$
$$+ \frac{b_1^2 \beta}{2}(\mu_1^2 + \mu_2^2) + \frac{b_1^2 \beta}{2} \mu_1 \mu_2 \left(\frac{k_1}{k_2} + \frac{k_2}{k_1}\right)$$
$$+ \frac{b_1^2 \beta^2}{2} \mu_1 \mu_2 \left(\mu_2^2 \frac{k_2}{k_1} + \mu_1^2 \frac{k_1}{k_2}\right), \quad (5.112)$$

with $\mu = -\mu_3$, and (Catelan and Moscardini, 1994)

$$K^{(2)}(k_1, k_2) = \frac{3}{7} + \frac{k_1 \cdot k_2}{2k_1 k_2}\left(\frac{k_1}{k_2} + \frac{k_2}{k_1}\right) + \frac{4}{7}\left(\frac{k_1 \cdot k_2}{k_1 k_2}\right)^2. \quad (5.113)$$

Equations (5.110)–(5.112) show how the bispectrum can allow us to measure the bias parameters.

The bispectrum depends on β, σ_p, and P_g, in addition to the bias parameters b_1 and b_2. The bispectrum and power spectrum data come from Fourier transforming the galaxy number density distribution $n(r)$ through (see the discussion of the FKP method in Section 5.4.1)

$$F(r) \equiv \lambda w(r)\left[n(r) - \alpha_s n_s(r)\right], \quad (5.114)$$

where λ is a constant to be determined, and $n_s(r)$ is the number density of a random catalog with the same selection function as the real catalog, but with $1/\alpha_s$ times ($\alpha_s \leq 0.2$) as many particles. The weight $w(r) = 1/[1 + P_0 \bar{n}(r)]$ has been chosen to minimize the variance of higher-order correlation functions (Scoccimarro, 2000), where $\bar{n}(r)$ is the average number density of galaxies at position r, and P_0 is the power spectrum to be estimated. Since the results are not sensitive to P_0, it can be chosen to be a constant, for example, $P_0 = 5000\,\text{h}^{-3}\,\text{Mpc}^3$, to enable the use of a fast Fourier transform. If we set $\lambda = I_{22}^{-1/2}$, where (Matarrese, Verde, and Heavens, 1997)

$$I_{ij} \equiv \int d^3 r w^i(r) \bar{n}^j(r) \quad (5.115)$$

then the power spectrum may be estimated from

$$\langle |F_k|^2 \rangle = P_g(k) + \frac{I_{21}}{I_{22}}(1 + \alpha), \quad (5.116)$$

and Eq. (5.109) can be used to remove the redshift-space distortions. The bispectrum may be estimated from

$$\langle F_{k_1} F_{k_2} F_{k_3} \rangle = \frac{I_{33}}{I_{22}^{3/2}} \left\{ B_g(k_1, k_2, k_3) + \frac{I_{32}}{I_{33}} \left[P_g(k_1) + P_g(k_2) + P_g(k_3) \right] \right.$$
$$\left. + (1 - \alpha^2) \frac{I_{31}}{I_{33}} \right\}.$$
(5.117)

It is assumed implicitly that the power spectrum is roughly constant over the width of the survey window function in k-space.

The real parts of $F_{k_1} F_{k_2} F_{k_3}$ are taken as data, for triangles in k space ($k_1 + k_2 + k_3 = 0$). Each triangle yields an estimate of a linear combination of $c_1 \equiv 1/b_1$ and $c_2 \equiv b_2/b_1^2$ (see Eqs. (5.110) and (5.112)). Triangles of different shapes (i.e., different Ker[k_1, k_2]) must be used to lift the degeneracy between nonlinear gravity and nonlinear bias (Verde et al., 2002).

Verde et al. (2002) found that $b_1 = 1.04 \pm 0.11$ and $b_2 = -0.054 \pm 0.08$ for the 2dF galaxy redshift survey. Their results were marginalized over β and σ_p. Figure 5.14 shows their bispectrum measurement from 2dFGRS (Verde et al., 2002).

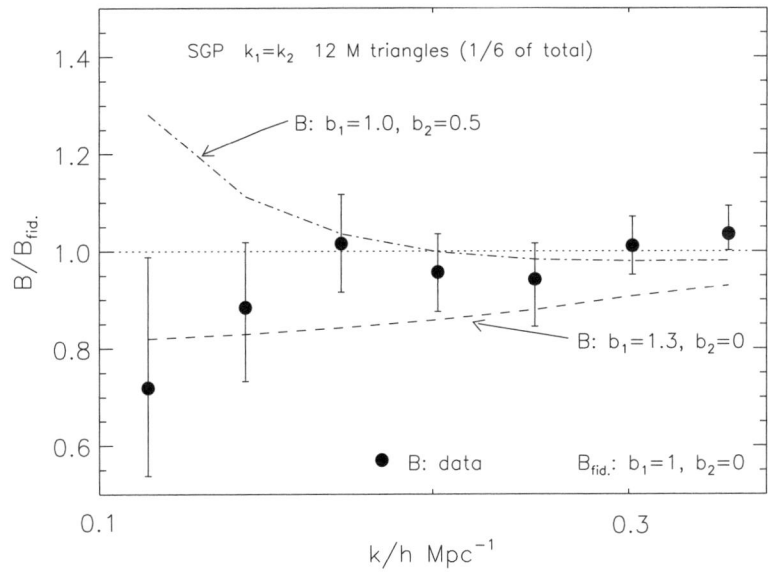

Figure 5.14 Ratio of the average measured bispectrum from 2dF galaxy redshift survey and the average perturbation theory predictions, relative to the bispectrum for a fiducial unbiased model B_{fid} (Verde et al. (2002), MNRAS, Wiley-Blackwell). The dashed line corresponds to $b_1 = 1.3, b_2 = 0$, and the dot-dashed line corresponds to $b_1 = 1.0, b_2 = 0.5$. The error bars are obtained via Monte Carlo from 16 mock catalogs, and are placed centrally on the mean of the estimates from the mock catalogs. This illustrates the level of bias in the estimator. The figure shows that there is no evidence of scale-dependent bias from the 2dF data.

5.6.3
Using $f_g(z)$ and $H(z)$ to Test Gravity

The cause for the observed cosmic acceleration could be an unknown energy component in the universe (i.e., dark energy), or a modification of general relativity (i.e., modified gravity). These two possibilities can be differentiated, since given the *same* cosmic expansion history $H(z)$, the modified gravity model is likely to predict a growth rate of cosmic large scale structure $f_g^{MG}(z)$ that differs from the prediction of general relativity $f_g^H(z)$ (see Section 1.2). The growth rate associated with dark energy, $f_g^H(z)$, depends only on $H(z)$ if dark energy is not coupled to dark matter, and dark energy perturbations are negligible (which is true except on very large scales) (Ma et al., 1999). The growth rate associated with modified gravity, $f_g^{MG}(z)$, depends on the details of how general relativity is modified.

A suitably designed galaxy redshift survey would allow the measurement of the cosmic expansion history $H(z)$ from BAO (see Section 5.1), and the growth rate of cosmic large scale structure

$$f_g(z) = \beta(z)b(z) \qquad (5.118)$$

from the independent measurements of the linear redshift-space distortion parameter β (see Section 5.6.1), and the bias factor between the galaxy and matter distributions $b(z)$ (see Section 5.6.2). The measurement of both $H(z)$ and $f_g(z)$ allows us to differentiate between dark energy and modified gravity.

Figure 5.15 shows the errors on $H(z)$ and $f_g(z) = \beta(z)b(z)$ for a dark energy model that gives the same $H(z)$ as a DGP gravity model with the same Ω_m^0, for a NIR galaxy redshift survey covering 11 931 (deg)^2, and the redshift range $0.5 < z < 2$ (assuming a conservative nonlinear cut equivalent to $p_{NL} = 0.6$), compared with current data (Wang, 2008b). We have neglected the very weak dependence of the transfer function on dark energy at very large scales in this model (Ma et al., 1999), and added an uncertainty in $\ln b$ (extrapolated from the 2dFGRS measurement) in quadrature to the estimated error on β.

Figure 5.15b shows the $f_g(z)$ for a modified gravity model (the DGP gravity model) with $\Omega_m^0 = 0.25$ (solid line), as well as a dark energy model that gives the same $H(z)$ for the same Ω_m^0 (dashed line). The cosmological constant model from Figure 5.15a is also shown (dotted line). Clearly, current data cannot differentiate between dark energy and modified gravity. A very wide and deep galaxy redshift survey provides measurement of $f_g(z)$ accurate to a few percent (see Figure 5.15b); this will allow an unambiguous distinction between dark energy models and modified gravity models that give identical $H(z)$ (see the solid and dashed lines in Figure 5.15b). A survey covering $\sim$14 000 (deg)^2 would rule out the DGP gravity model that gives the same $H(z)$ and Ω_m^0 at >99% confidence level (Wang, 2008b).

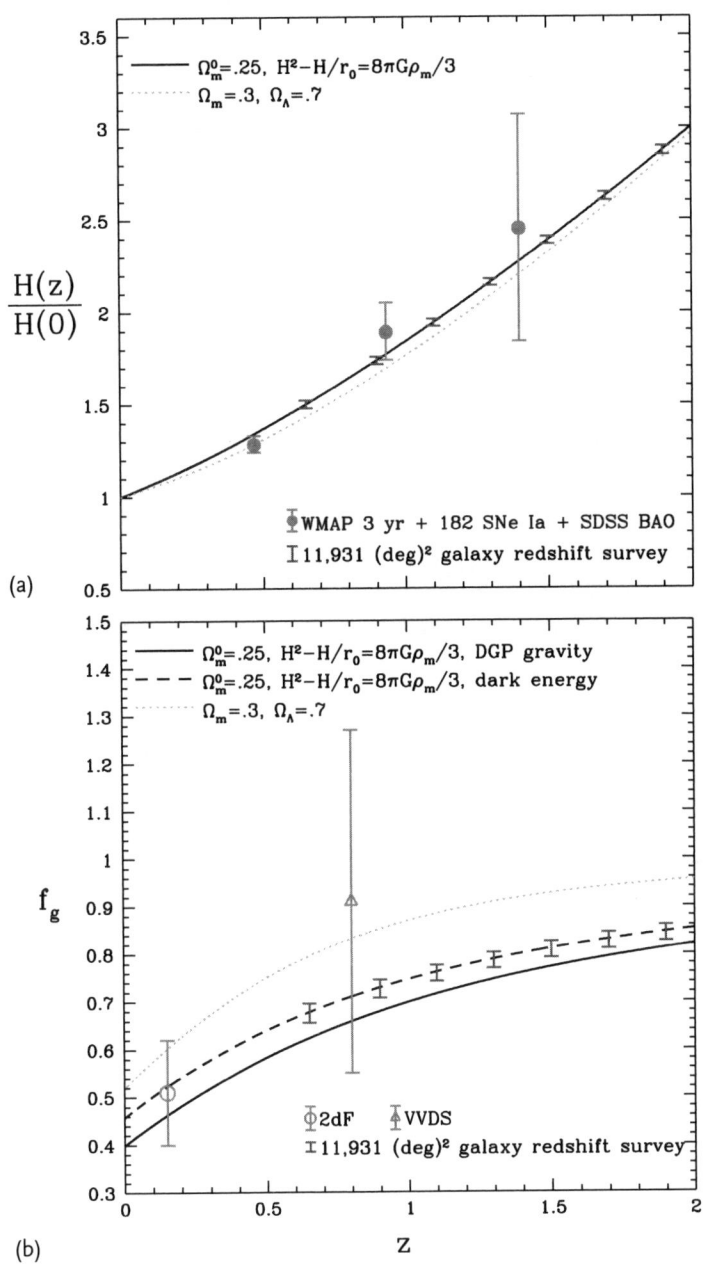

Figure 5.15 Current and expected future measurements of the cosmic expansion history $H(z)$ (a) and the growth rate of cosmic large scale structure $f_g(z)$ (b) (Wang, 2008b). The future data correspond to a NIR galaxy redshift survey covering $>10\,000$ (deg)^2 and $0.5 < z < 2$. If the $H(z)$ data (a) are fit by both a DGP gravity model and an equivalent dark energy model that predicts the same expansion history, a survey area of $\sim 14\,000$ (deg)^2 is required to rule out the DGP gravity model at $>99\%$ confidence level.

5.7
The Alcock–Paczynski Test

Alcock and Paczynski (1979) noted that if an astrophysical structure is spherically symmetric, then its measured radial and transverse dimensions can be used to constrain the cosmological model. A galaxy redshift survey enables the Alcock–Paczynski test to be carried out.

Features in the galaxy power spectrum, such as the BAO, should have the same length scale in the radial and transverse directions. The radial length scale is measured using

$$H(z)\Delta r_\| = c\Delta z , \qquad (5.119)$$

where Δz is the redshift interval spanned by $\Delta r_\|$. The transverse length scale is measured using

$$\Delta r_\perp = D_A(z)\Delta \theta , \qquad (5.120)$$

where $\Delta \theta$ is the angle subtended by $\Delta r_\perp$.

Thus the Alcock–Paczynski test of requiring that $\Delta r_\perp = \Delta r_\|$ leads to

$$\frac{H(z)D_A(z)}{c} = \frac{\Delta z}{\Delta \theta} . \qquad (5.121)$$

This provides a cross-check of the measured $H(z)$ and $D_A(z)$ derived from the BAO scale measurements.

Redshift-space distortions (a source of cosmological information themselves) introduce a systematic uncertainty in the Alcock–Paczynski test. Unless properly modeled and removed, redshift-space distortions can alter the measured length scale in the radial direction from galaxy redshift surveys, and bias the Alcock–Paczynski test.

6
Observational Method III: Weak Lensing as Dark Energy Probe

6.1
Weak Gravitational Lensing

Basic Concepts in Gravitational Lensing

The deflection of light by matter concentration is a prediction of general relativity that has been tested to high precision by observations. Most of the discussion here on the basic concepts in gravitational lensing follows that of Schneider, Ehlers, and Falco (1992).

The deflection angle due to gravitational lensing by a point mass lens M is

$$\hat{\alpha} = \frac{4GM}{c^2 \xi} = \frac{2R_S}{\xi}, \tag{6.1}$$

where $R_S = 2GM/c^2$ is the Schwarzschild radius, and the impact parameter $\xi \gg R_S$.

For a lens with an extended mass distribution, we need to integrate over the lens plane:

$$\hat{\alpha} = \int d^2 \xi' \frac{4G\Sigma(\xi')}{c^2} \frac{\xi - \xi'}{|\xi - \xi'|^2}, \tag{6.2}$$

where $\Sigma(\xi)$ is the lens mass density projected onto the lens plane.

The lens equation relates the geometric positions of the source and the lens with the deflection angle $\hat{\alpha}$ (Schneider, Ehlers, and Falco, 1992):

$$\eta = \frac{D_s}{D_d} \xi - D_{ds} \hat{\alpha}(\xi), \tag{6.3}$$

where D_d, D_s, and D_{ds} are the angular diameter distances from the observer to the lens, from the observer to the source, and from the lens to the source, respectively. The source position η in the source plane and the impact vector in the lens plane ξ are measured from the optical axis, the line connecting the observer and the lens. Figure 6.1 illustrates a typical gravitational lens system (Bartelmann and Schneider, 2001).

Dark Energy. Yun Wang
Copyright © 2010 WILEY-VCH Verlag GmbH & Co. KGaA, Weinheim
ISBN: 978-3-527-40941-9

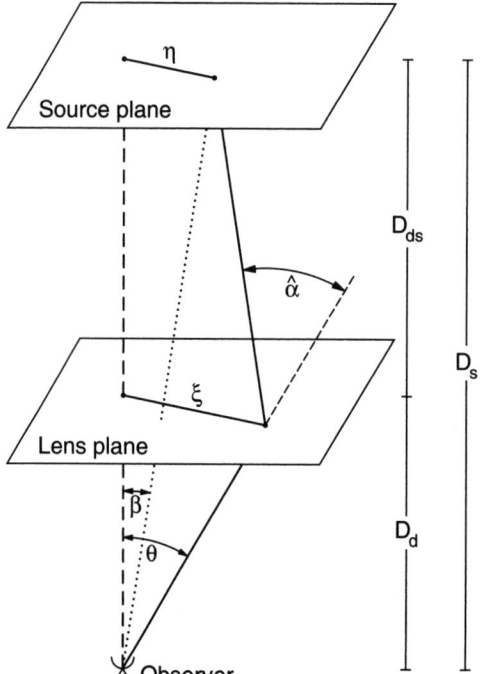

Figure 6.1 Illustration of a typical gravitational lens system (Bartelmann and Schneider, 2001).

We can define angular variables:

$$\beta \equiv \frac{\eta}{D_s}, \quad \theta \equiv \frac{\xi}{D_d}, \quad \alpha \equiv \frac{D_{ds}}{D_s}\hat{\alpha}. \tag{6.4}$$

The lens equation becomes

$$\beta = \theta - \alpha. \tag{6.5}$$

It is useful to understand gravitational lensing in terms of general relativity. The presence of a gravitational lens changes the time it takes for light from a source to reach the observer. This *time delay* has two sources: the *geometric time delay* caused by the increase in the path length traveled by the photons, and the *potential time delay*, the extra travel time of light caused by the presence of the gravitational potential. *Fermat's principle* states that images of the source must form where the total time delay is stationary (i.e., where light from the source interferes constructively) (Schneider, Ehlers, and Falco, 1992).

The total time delay is proportional to the Fermat potential (Schneider, Ehlers, and Falco, 1992)

$$\phi(\theta,\beta) = \frac{1}{2}(\theta-\beta)^2 - \Psi(\theta), \tag{6.6}$$

where Ψ is the deflection potential. The Fermat potential ϕ satisfies

$$\nabla \phi(\boldsymbol{\theta}, \boldsymbol{\beta}) = 0 \tag{6.7}$$

due to Fermat's principle. Equations (6.6) and (6.7) give the lens equation, Eq. (6.5), with

$$\boldsymbol{\alpha} = \nabla \Psi, \tag{6.8}$$

which explains why Ψ is called the deflection potential.

It is useful to note that for a source located on the optical axis, that is $\boldsymbol{\beta} = 0$, the image of the source forms an *Einstein ring* centered around the optical axis, with a radius (known as the *angular Einstein radius*)

$$\theta_E = \alpha = \frac{D_{ds}}{D_s} \hat{\alpha}. \tag{6.9}$$

For a point mass lens, the deflection angle $\hat{\alpha}$ is given by Eq. (6.1). The above equation gives

$$\theta_E^2 = \frac{4GM}{c^2} \frac{D_{ds}}{D_d D_s}. \tag{6.10}$$

We can define a critical surface mass density as follows

$$\Sigma_{cr} \equiv \frac{M}{\pi (D_d \theta_E)^2} = \frac{c^2 D_s}{4\pi G D_d D_{ds}}, \tag{6.11}$$

which is *independent* of the lens mass M.

Now we can define a dimensionless surface mass density for a lens with an arbitrary mass distribution:

$$\kappa \equiv \frac{\Sigma(D_d \boldsymbol{\theta})}{\Sigma_{cr}}, \tag{6.12}$$

where $\Sigma(D_d \boldsymbol{\theta})$ is the mass density of the lens projected onto the lens plane.

Using the definitions from Eqs. (6.4) and (6.12), Eq. (6.2) can be written as

$$\boldsymbol{\alpha}(\boldsymbol{\theta}) = \frac{1}{\pi} \int d^2 \boldsymbol{\theta}' \kappa(\boldsymbol{\theta}') \frac{\boldsymbol{\theta} - \boldsymbol{\theta}'}{|\boldsymbol{\theta} - \boldsymbol{\theta}'|^2}. \tag{6.13}$$

Thus the deflection potential (defined by Eq. (6.8)) is given by

$$\Psi(\boldsymbol{\theta}) = \frac{1}{\pi} \int d^2 \boldsymbol{\theta}' \kappa(\boldsymbol{\theta}') \ln|\boldsymbol{\theta} - \boldsymbol{\theta}'|. \tag{6.14}$$

Inverting the above gives

$$\kappa(\boldsymbol{\theta}) = \frac{1}{2} \nabla^2 \Psi(\boldsymbol{\theta}). \tag{6.15}$$

Definition of Weak Lensing

The above discussion on the basic concepts in gravitational lensing provides a foundation for understanding the basic framework of weak lensing. The notation and discussion of cosmological weak lensing in this chapter follows that of Hoekstra and Jain (2008).

For small sources such as galaxies, weak gravitational lensing maps the source's surface brightness distribution as follows:

$$f^{\text{obs}}(\theta_i) = f^s(\mathcal{A}_{ij}\theta_j), \tag{6.16}$$

where $\mathcal{A}$ is the Jacobian matrix of the mapping:

$$\mathcal{A} = \frac{\partial(\delta\theta_i)}{\partial\theta_j} = \delta_{ij} - \Psi_{,ij} = \begin{pmatrix} 1-\kappa-\gamma_1 & -\gamma_2 \\ -\gamma_2 & 1-\kappa+\gamma_1 \end{pmatrix} \tag{6.17}$$

$$= (1-\kappa)\begin{pmatrix} 1-g_1 & -g_2 \\ -g_2 & 1+g_1 \end{pmatrix}, \tag{6.18}$$

where Ψ is the 2D lensing potential (the same as the deflection potential defined in Eq. (6.8)), $\Psi_{,ij} \equiv \partial^2\Psi/\partial\theta_i\partial\theta_j$, κ is the lensing convergence, γ is the shear, and g is the reduced shear defined as

$$g(\boldsymbol{\theta}) \equiv \frac{\gamma(\boldsymbol{\theta})}{1-\kappa(\boldsymbol{\theta})}. \tag{6.19}$$

Equation (6.18) clearly shows that the observable of weak lensing shape measurements is the reduced shear, and *not* the shear.

The lensing convergence is given by

$$\kappa = \frac{1}{2}\nabla^2\Psi(\boldsymbol{\theta}) = \int d\chi\, W(\chi)\delta(\chi,\chi\boldsymbol{\theta}), \tag{6.20}$$

where the Laplacian operator is defined as $\nabla^2 \equiv \partial^2/\partial\boldsymbol{\theta}^2$ (under the flat sky approximation), and χ is the comoving distance. Clearly, the lensing convergence κ is the same as the dimensionless surface mass density for the lens defined in Eq. (6.12) (see Eq. (6.15)).

The complex shear $\gamma \equiv \gamma_1 + i\gamma_2 = \gamma\exp(2i\alpha)$, where α is the orientation angle of the shear. The components of the shear field are related to the lensing potential by

$$\gamma_1 = \frac{1}{2}(\Psi_{,11} - \Psi_{,22}), \quad \gamma_2 = \Psi_{,12}. \tag{6.21}$$

Note that we focus on the second derivatives of the lensing potential Ψ, since Ψ itself cannot be measured (adding a constant to Ψ does not change the physical lensing pattern), and $\Psi_{,i}$ is just a deflection angle (which cannot be measured without knowing the "intrinsic" positions of galaxies).

For a flat universe, $d\chi = dz/H(z)$. The lensing window function is

$$W(\chi) = \frac{3}{2}\Omega_m H_0^2 a^{-1}(\chi)\chi \int d\chi_s\, n_s(\chi_s)\frac{\chi_s - \chi}{\chi_s}, \tag{6.22}$$

where $n_s(\chi_s)$ denotes the number density of source galaxies. Note that for a single source at redshift z_s, $n_s(\chi_s)$ is a Dirac delta function; we can readily reproduce Eq. (6.22) by comparing Eq. (6.20) with Eq. (6.12).

The solid angles subtended by the source and the image are related through

$$\omega = \mu(\boldsymbol{\theta})\omega^{(s)}, \tag{6.23}$$

where the magnification $\mu(\boldsymbol{\theta})$ is given by

$$\mu = \frac{1}{\det \mathcal{A}} = \frac{1}{(1-\kappa)^2 - |\gamma|^2}. \tag{6.24}$$

Note that we have assumed that the galaxies are at cosmological distances, so that their observed angular sizes are small enough for the lensing mapping Jacobian $\mathcal{A}$ to be constant over the entire galaxy.

The weak lensing regime is defined to satisfy $\kappa \ll 1$. Therefore, the convergence κ gives the magnification of an image, and the shear $g \simeq \gamma$ gives the ellipticity induced on an initially circular image. If the galaxies were randomly oriented in the absence of lensing, the matter density field can be reconstructed from the measured ellipticities of the sources.

Galaxy Ellipticity

The vast majority of galaxies from a weak lensing observation are faint, and do not have regular shapes. Thus we must define galaxy shapes to account for the irregularity of galaxy images, and to adapt properly to observational data (Bartelmann and Schneider, 2001).

The surface brightness of a galaxy image is a function of the angular position $\boldsymbol{\theta}$, $I(\boldsymbol{\theta})$. In order to measure $I(\boldsymbol{\theta})$ at large angular separations from the center of the image, we must assume that the galaxy image is isolated. The center of the image, $\bar{\boldsymbol{\theta}}$, is defined by

$$\bar{\boldsymbol{\theta}} = \frac{\int d^2\theta \, q_\mathrm{I}[I(\boldsymbol{\theta})]\boldsymbol{\theta}}{\int d^2\theta \, q_\mathrm{I}[I(\boldsymbol{\theta})]}, \tag{6.25}$$

where $q_\mathrm{I}(I)$ is a suitably chosen weight function. Note that the weight function $q_\mathrm{I}(I)$ usually depends on $\bar{\boldsymbol{\theta}}$; for example, it is often chosen to be a Gaussian centered at $\bar{\boldsymbol{\theta}}$ with width σ (Hirata, 2009). Once we have chosen $q_\mathrm{I}(I)$, we can define the tensor of the second brightness moments (Blandford et al., 1991),

$$Q_{ij} = \frac{\int d^2\theta \, q_\mathrm{I}[I(\boldsymbol{\theta})](\theta_i - \bar{\theta}_i)(\theta_j - \bar{\theta}_j)}{\int d^2\theta \, q_\mathrm{I}[I(\boldsymbol{\theta})]}, \quad i,j \in \{1,2\}. \tag{6.26}$$

Note that $q_\mathrm{I}(I)$ should be chosen such that the integrals converge in Eqs. (6.25) and (6.26).

It is useful to define a *complex ellipticity* given by

$$\epsilon \equiv \frac{Q_{11} - Q_{22} + 2i Q_{12}}{Q_{11} + Q_{22} + 2(Q_{11} Q_{22} - Q_{12}^2)^{1/2}}. \tag{6.27}$$

This definition of ellipticity has the advantage that the expectation value of the nth moment, $\langle \epsilon^n \rangle_{\epsilon_s}$, is independent of the source ellipticity distribution (Seitz and Schneider, 1997).

For an ellipse with major and minor axes a and b,

$$\epsilon = \left(\frac{1 - b/a}{1 + b/a} \right) \exp(2i\alpha), \tag{6.28}$$

where α is the position angle of the major axis.

The transformation between source and image ellipticity in terms of ϵ is given by (Seitz and Schneider, 1997)

$$\epsilon^{(s)} = \begin{cases} \dfrac{\epsilon - g}{1 - g^* \epsilon} & \text{for } |g| \leq 1 \\ \dfrac{1 - g\epsilon^*}{\epsilon^* - g^*} & \text{for } |g| > 1 \end{cases} \tag{6.29}$$

where g is the reduced shear given by Eq. (6.19), and g^* is its complex conjugate. The inverse transformation is given by interchanging ϵ and $\epsilon^{(s)}$ and replacing g by $-g$ in Eq. (6.29). Thus, assuming that there are no observational distortions, the observed ellipticity e^{obs} is related to its unlensed value e^{int} as follows:

$$\epsilon^{obs} = \frac{\epsilon^{int} + g}{1 + g^* \epsilon^{int}}. \tag{6.30}$$

Note that there are definitions in the literature that differ from Eqs. (6.28)–(6.30), such that ellipticity is related to two times the shear (see e.g., Dodelson (2003)).

Note that the transformation of image ellipticities depends only on the reduced shear, and not on the shear and the surface mass density separately (see Eq. (6.29)). This is as expected, since Eq. (6.18) shows that the pre-factor $(1-\kappa)$ only affects the size of an image, and not its shape. Thus the reduced shear g or functions of g are the only observables in the measurements of image ellipticities.

Correlation Functions

The weak lensing signal is encoded in angular correlation functions measured from the galaxy shape catalogs. These are projections of the three-dimensional correlation functions using Limber's approximation (Limber, 1953), that is assuming that galaxies are at very large distances, and that radial separations of galaxies are much smaller than the mean distance to the galaxies on the length scales of interest. See Bartelmann and Schneider (2001) for a derivation of the weak lensing angular correlation functions from their three-dimensional counterparts.

The two-point correlation function of the shear, for galaxies in the i-th and j-th redshift bins, is defined as

$$\xi_{\gamma_i \gamma_j}(\theta) \equiv \langle \gamma_i(\boldsymbol{\theta}_1) \cdot \gamma_j^*(\boldsymbol{\theta}_2) \rangle, \tag{6.31}$$

where $\theta \equiv |\boldsymbol{\theta}_1 - \boldsymbol{\theta}_2|$. The shears are measured with respect to a coordinate system defined by the separation vector between two galaxies (Dodelson, 2003). It can be shown that the two-point correlation function of the convergence is identical to that of the shear. The shear power spectrum at angular wavenumber l is the Fourier transform of $\xi_{\gamma_i \gamma_j}$, and is related to the matter power spectrum $P_m(k, z)$ as follows:

$$C_{\gamma_i \gamma_j}(l) = \int_0^\infty dz \frac{W_i(z) W_j(z)}{\chi(z)^2 H(z)} P_m\left(\frac{l}{\chi(z)}, z\right), \qquad (6.32)$$

where the window functions $W_i(z)$ and $W_j(z)$ are given by Eq. (6.22).

The galaxy-convergence power spectrum is

$$C_{g_i \kappa_j}(l) = \int_0^\infty dz \frac{W_{gi}(z) W_j(z)}{\chi(z) H(z)} P_g\left(\frac{l}{\chi(z)}, z\right), \qquad (6.33)$$

where W_{gi} is the normalized redshift distribution of the foreground galaxies (the lens), and $P_g(k, z)$ is the 3D galaxy power spectrum. The galaxy-convergence power spectrum can be measured, since it is related to the tangential shear, the tangential component of the background ellipticities azimuthally averaged over circular annuli centered on the foreground (lens) galaxy (see e.g., Hoekstra, Yee, and Gladders (2002); Sheldon et al. (2004)).

A third power spectrum that can be measured from the same galaxy catalogs is the galaxy-galaxy power spectrum, $C_{g_i g_j}(l)$. The three power spectra, $C_{\gamma_i \gamma_j}(l)$, $C_{g_i \kappa_j}(l)$, and $C_{g_i g_j}(l)$, can be measured for multiple photometric redshift bins (divided using redshifts estimated from the multi-band photometry), and contain all the two-point statistical information on galaxy clustering and lensing that can be extracted from multi-band imaging data of galaxies. The shear-shear power spectra, $C_{\gamma_i \gamma_j}(l)$, are sensitive to the cosmic expansion history $H(z)$ through the geometric factors $W_i(z)$ and $W_j(z)$, and the growth factor of cosmic large scale structure $G(z) = \delta_m(\mathbf{r}|z)/\delta_m(\mathbf{r}|0)$ through the matter power spectrum $P_m(k, z)$. An advantage of weak lensing as a dark energy probe is that the bias factor between the galaxy and matter distributions can be determined using the combination of $C_{g_i g_j}(l)$ and $C_{g_i \kappa_j}(l)$. This strengthens the cosmological constraints derived from galaxy clustering (see Chapter 5) by providing an independent measurement of the bias factor.

Higher-order statistics, such as the shear three-point correlation function, can also be measured in principle, and could provide constraints complementary to the two-point shear correlation functions.

E and B Decomposition

Gravitational lensing by a point mass lens leads to a tangential distortion pattern centered on the lens, with the deflection angle vector, $\boldsymbol{\alpha} = \nabla \Psi$ (Eq. (6.8)), pointing away from the lens in the lens plane (see Figure 6.1 and Eq. (6.5)). Thus the deflection field is curl-free (i.e., $\nabla \times \boldsymbol{\alpha} = 0$), and this remains true for a linear superposition of point masses, that is, a general mass field. In general, the shear γ at a point can be written as a symmetric and traceless tensor (with only two degrees of freedom) (Stebbins, 1996), and decomposed into a gradient or "E" component that results from a curl-free deflection field, and a curl or pseudo-scalar "B" component that results from a divergence-free deflection field (Crittenden et al., 2002). To first order, the "B" component should be zero if only weak lensing is at work. However, weak lensing in the presence of source redshift clustering can produce a weak "B" component (Schneider, Van Waerbeke, and Mellier, 2002).

A detection of nonzero B-modes usually indicates a nongravitational contribution to the shear field, which is likely a systematic contamination to the lensing signal. The E and B components of the shear correlation function can be written as follows (Crittenden et al., 2002; Pen, Van Waerbeke, and Mellier, 2002):

$$\xi_{E,B}(\theta) = \frac{\xi_+(\theta) \pm \xi'(\theta)}{2}, \tag{6.34}$$

where ξ' is given by (Schneider, Van Waerbeke, and Mellier, 2002)

$$\xi'(\theta) = \xi_-(\theta) + \int_\theta^\infty \frac{d\vartheta}{\vartheta} \xi_-(\theta) \left[4 - 12 \left(\frac{\theta}{\vartheta} \right)^2 \right] \tag{6.35}$$

and

$$\xi_\pm(\theta) \equiv \xi_{tt}(\theta) \pm \xi_{\times\times}(\theta) = \frac{1}{2\pi} \int_0^\infty d\ell\, \ell\, C_{\kappa\kappa}(\ell) J_{0,4}(\ell\theta), \tag{6.36}$$

where $\xi_{tt}(\theta)$ and $\xi_{\times\times}(\theta)$ are the tangential and rotated ellipticity correlation functions, θ is the angular separation between galaxy pairs, and $J_{0,4}$ are Bessel functions of the first kind. The convergence power spectrum $C_{\kappa\kappa}(l) = C_{\gamma\gamma}(l)$ (see Eq. (6.32)).

6.2
Weak Lensing Observational Results

The first conclusive detection of cosmic shear was made in 2000 (Wittman et al., 2000; van Waerbeke et al., 2000; Bacon, Refregier, and Ellis, 2000; Kaiser, Wilson, and Luppino, 2000). The Canada-France-Hawaii Telescope Legacy Survey (CFHTLS) (http://www.cfht.hawaii.edu/Science/CFHTLS/) is the largest weak lensing survey carried out to date. By its completion, the CFHTLS Wide weak lensing survey will have imaged 140 (deg)² in the five filters defined by the SDSS to a depth of $i' = 24.5$.

CFHTLS is a five-year project, with deep and wide observations carried out using the MEGACAM instrument mounted at the prime focus of the telescope. The MEGACAM camera consists of an array of 9×4 CCDs (2048×4612 pixels each), with the pixel size at MEGAPRIME focus of $0.186''$, and a field of view of $1° \times 1°$ (Boulade et al., 2003).

The 2008 results from the CFHTLS are shown in Figures 6.2–6.4 (Fu et al., 2008). Figure 6.2 shows the E-mode and B-mode two-point shear correlation function (defined by Eq. (6.34)), Figure 6.3 shows the variance of the aperture-mass (in circular apertures), measured from the combined 57 pointings of the CFHTLS (Fu

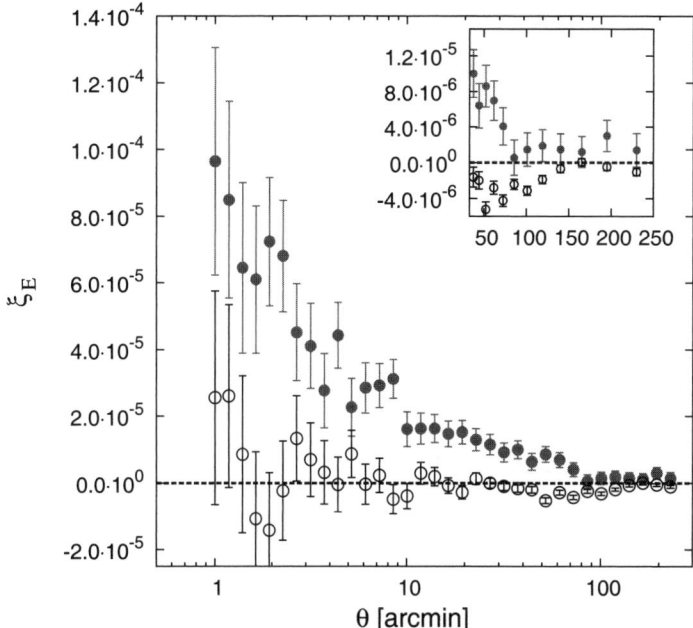

Figure 6.2 Two-point shear correlation function from the combined 57 pointings of the CFHTLS (Fu et al., 2008). The error bars of the E-mode include statistical noise added in quadrature to the non-Gaussian cosmic variance. Only statistical uncertainty contributes to the error budget for the B-mode. Filled points show the E-mode, open points the B-mode. The enlargement in the panel shows the signal in the angular range 35′–230′.

et al., 2008). The variance of the aperture-mass (in circular apertures) is defined as (Schneider, Van Waerbeke, and Mellier, 2002)

$$\langle M_{\rm ap}^2 \rangle_{\rm E,B}(\theta) = \int_0^{2\theta} \frac{d\vartheta\,\vartheta}{2\theta^2} \left[T_+\left(\frac{\vartheta}{\theta}\right) \xi_+(\vartheta) \pm T_-\left(\frac{\vartheta}{\theta}\right) \xi_-(\vartheta) \right] \quad (6.37)$$

where the filter functions $T_{+/-}$ are defined by

$$T_+(x) = \frac{6(2-15x^2)}{5}\left[1 - \frac{2}{\pi}\arcsin(x/2)\right] + \frac{x\sqrt{4-x^2}}{100\pi}$$
$$\times \left(120 + 2320x^2 - 754x^4 + 132x^6 - 9x^8\right) H(2-x)\,;$$

$$T_-(x) = \frac{192}{35\pi} x^3 \left(1 - \frac{x^2}{4}\right)^{7/2} H(2-x)\,, \quad (6.38)$$

where H denotes the Heaviside step function.

For the CFHTLS data, the aperture-mass statistics, $\langle M_{\rm ap}^2 \rangle$, gives more stringent constraints on cosmological parameters than the shear correlation function, ξ_E (Fu

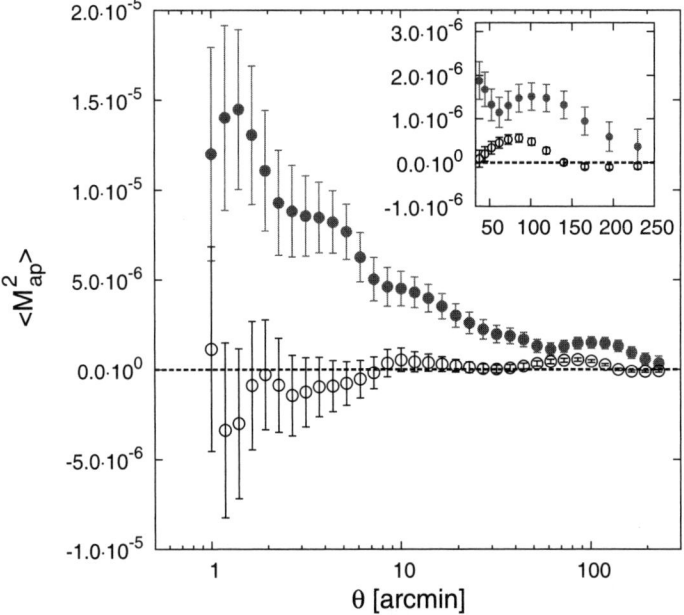

Figure 6.3 The variance of the aperture-mass (in circular apertures) measured from the combined 57 pointings of the CFHTLS (Fu et al., 2008). The error bars of the E-mode include statistical noise added in quadrature to the non-Gaussian cosmic variance. Only statistical uncertainty contributes to the error budget for the B-mode. Filled points show the E-mode, open points the B-mode. The enlargement in the panel shows the signal in the angular range 35′–230′.

et al., 2008). This is probably a consequence of the finite range of angular separations used in the analysis, or the particular estimators used, and not a generic advantage of the aperture mass statistics (Hirata, 2009). Figure 6.4 shows the constraints on σ_8 (the amplitude of matter density fluctuations averaged on the scale of 8 h^{-1} Mpc) and Ω_m, using the aperture-mass statistics $\langle M_{ap}^2 \rangle$ measured from the CFHTLS (Fu et al., 2008).

Figures 6.5 and 6.6 shows the constraints on (σ_8, Ω_m) and (w_0, Ω_m) from the earlier set of the CFHTLS data, using the top-hat variance $\langle \gamma^2 \rangle$ measured from CFHTLS weak lensing survey over 22 (deg)^2 (31 pointings) (Hoekstra et al., 2006). Figure 6.6 assumes a constant dark energy equation of state $w = w_0$.

Comparison of Figures 6.4 and 6.5 shows that the cosmological constraints from the CFHTLS data have not improved notably with the increase of survey area from 22 (deg)^2 to 57 (deg)^2. This is due to the limited knowledge of the source redshifts.

Due to the lack of tomographic measurements, most weak lensing results only constrain a combination of Ω_m and σ_8. The current lack of tomographic measurements from large area surveys also severely limits the constraints on dark energy from weak lensing data (Hoekstra and Jain, 2008).

Figure 6.7 shows how the measurement of σ_8 has evolved from 2001 to 2007 (Hetterscheidt et al., 2007). Figures 6.4–6.7 illustrate both the advantage and the

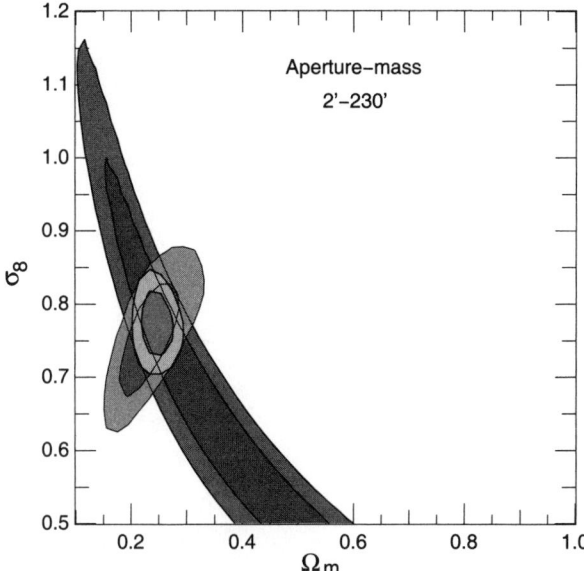

Figure 6.4 Comparison (1σ, 2σ) between WMAP3 (Spergel *et al.*, 2007) (medium contours) and the CFHTLS $\langle M_{ap}^2 \rangle$-results between 2 and 230 arcmin (large contours) by Fu *et al.* (2008). The combined contours of WMAP3 and CFHTLS Wide are shown in small contours.

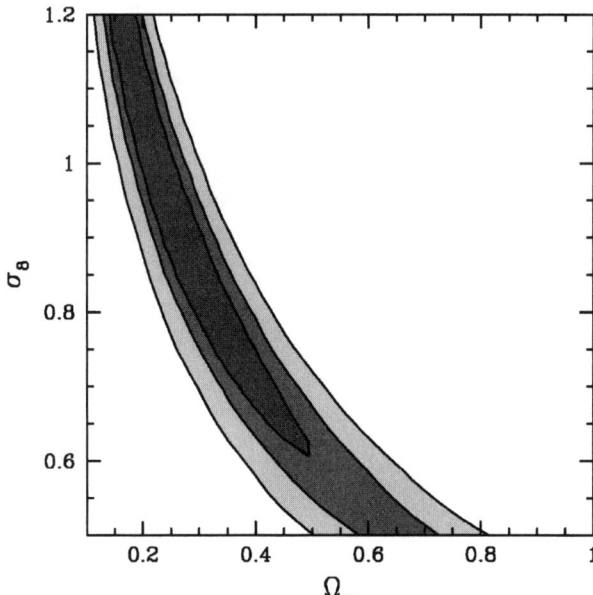

Figure 6.5 Joint constraints on Ω_m and σ_8 constraints from the CFHTLS Wide data over 22 (deg)² (31 pointings) from Hoekstra *et al.* (2006). The contours indicate the 68.3%, 95.4%, and 99.7% confidence limits on two parameters jointly. The Hubble constant and source redshift distribution were marginalized over.

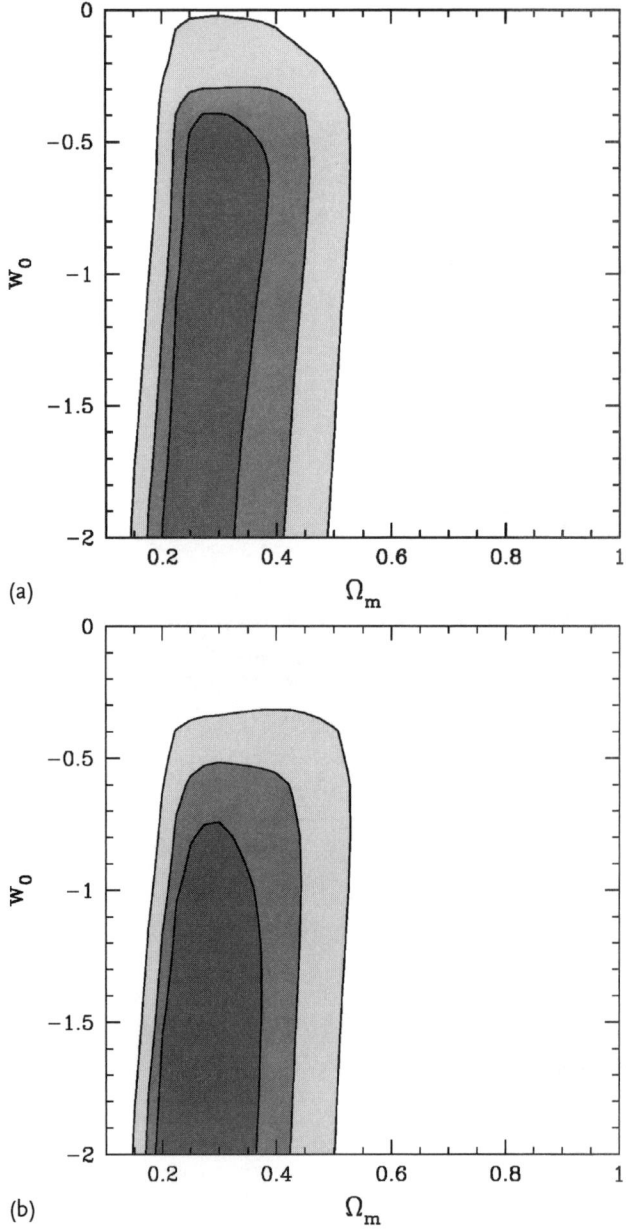

Figure 6.6 Joint constraints on Ω_m and w_0 constraints from the CFHTLS data from Hoekstra et al. (2006). (a) Using the CFHTLS Wide data over 22 (deg)^2 (31 pointings). The contours indicate the 68.3%, 95.4%, and 99.7% confidence limits on two parameters jointly. These are marginalized over $\sigma_8 \in [0.7, 1.0]$, $h \in [0.6, 0.8]$, and the source redshift distribution. (b) Same as (a), but including additional data from the deep component of the CFHTLS (Semboloni et al., 2006).

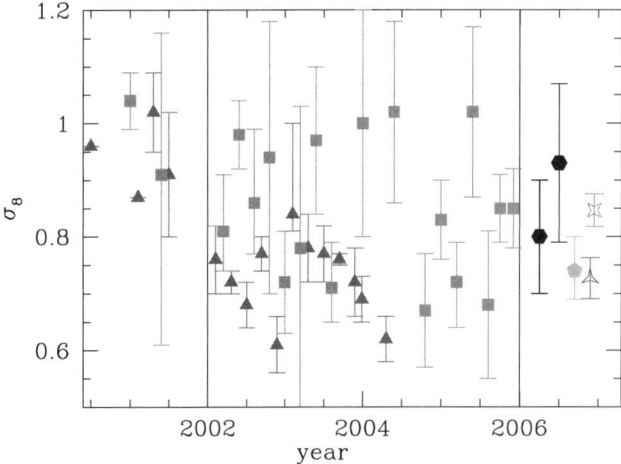

Figure 6.7 The measurements of σ_8 ($\Omega_m = 0.3$) from the analysis of clusters of galaxies (triangles) and cosmological weak lensing (squares) over the last several years (Hetterscheidt et al., 2007). The hexagon with bold and light error bars denote the measurements by Hetterscheidt et al. (2007) from M_{ap} and $\xi_{\pm}$, respectively. The WMAP three-year result is denoted by the pentagon (the σ_8 value would be larger if Ω_m is fixed to 0.3). The open triangle and open star on the right are the average of all 8 determinations between 2002 and 2006 (indicated by the vertical lines) using clusters and cosmic shear, respectively (error bars of single measurements are not taken into account).

disadvantage of weak lensing as a dark energy probe. Weak lensing has the advantage of being sensitive to the clustering of matter, which can potentially allow us to differentiate between dark energy and modified gravity (Knox, Song, and Tyson, 2006). However, if the clustering of matter is not accurately measured, the uncertainty is propagated into the measurement of dark energy parameters (see Figures 6.5 and 6.6).

6.3
Systematics of Weak Lensing

The unbiased measurement of galaxy shapes is perhaps the most difficult task faced by observational cosmology today. The effectiveness of weak lensing as a dark energy probe depends on whether the systematic uncertainties can be tightly controlled, which places the most stringent requirements on instrumentation, and provides severe challenges for theoretical modeling as well. Here we will focus on the current work on modeling and removal of the known systematic uncertainties.

The weak lensing systematic uncertainties can be studied by parametrizing the estimated shear as (Huterer et al., 2006)

$$\hat{\gamma}(z_s, \mathbf{n}) = \gamma_{\text{lens}}(z_s, \mathbf{n}) \left[1 + \phi_{\text{sys}}^{\text{mult}}(z_s, \mathbf{n})\right] + \gamma_{\text{sys}}^{\text{add}}(z_s, \mathbf{n}). \tag{6.39}$$

This includes two types of systematic error contributions, which modify the measured shear via additive and multiplicative terms. In addition, the bin redshift z_s and its width may also be in error, leading to biases in cosmological parameters (Ma, Hu, and Huterer, 2006). Note that the multiplicative shear bias, $\phi_{\text{sys}}^{\text{mult}}$, is not a shear, but a scalar. It can also vary with position, which in turn affect the E and B mode measurements (Guzik and Bernstein, 2005).

Since cosmological information is extracted from shear correlation functions, the true systematic effects modify these correlation functions. The errors that affect individual galaxy shapes but are uncorrelated between galaxy pairs only lead to additional statistical errors. The intrinsic shape noise of galaxies is likely the dominant statistical error. The modeling and reduction of weak lensing systematic effects is an area of active study (see, e.g., Amara and Refregier (2007)).

6.3.1
Point Spread Function Correction

Basic Idea of Point Spread Function Correction
The most difficult and most important step in the analysis of weak lensing data is to correct the observed galaxy shape for the convolution by the point spread function (PSF). In principle, this requires deconvolution in the presence of noise that may not be accurately modeled. In practice, one can convolve model galaxies with the PSF, and find the model galaxy that best matches the observations. Adequate PSF correction requires both knowledge of the PSF, and an optimized method for deconvolution (or its equivalent) that is unbiased and minimizes the loss of information relevant to cosmic shear measurement.

The finite size of the telescope mirror leads to a minimum size for the PSF. The PSF can vary spatially and changes with time, due to pointing jitter and mechanical deformation of the optics. Spatial variation of the PSF is often caused by optical aberrations, with additional contributions from the atmosphere for ground observations. For ground weak lensing observations, atmospheric refraction also leads to time dependence of the PSF. In addition, the PSF depends on detector properties. Pixel size, charge transfer inefficiencies, or other detector nonlinearities will affect the PSF.

Requiring a good knowledge of the PSF drives the complexity and cost of a space weak lensing experiment. For ground-based weak lensing experiment, seeing (caused by turbulence in the atmosphere) plays a dominant role in the PSF, and reduces the number density of galaxies with well-measured shapes. The combination of seeing and the intrinsic size of the PSF leads to a circularization of the observed images; this lowers the amplitude of the observed lensing signal. PSF

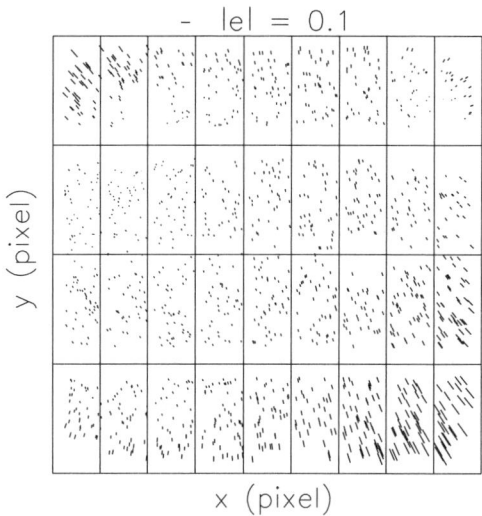

Figure 6.8 The pattern of the PSF anisotropy in an example pointing W3 + 2 + 0 − CFHTLS_W_i_ 143023 + 543031 (Fu et al., 2008). Ticks represent the observed ellipticities at stellar locations. On top of the figure a 10% ellipticity modulus is shown for comparison.

anisotropy leads to coherent alignments in the galaxy shapes. PSF correction has been a major focus of current research in weak lensing.

The measurement of ellipticities at a sufficiently large number of stellar locations is critical for PSF correction. Assuming that the PSF changes smoothly across the field, the PSF anisotropy vector **p** at the galaxy positions can be estimated using a polynomial fit to the measured values of **p** at stellar positions.

Figure 6.8 shows the pattern of the PSF anisotropy in an example pointing from the CFHTLS Wide survey (Fu et al., 2008). This indicates that the PSF anisotropy varies significantly across the whole camera, thus the PSF anisotropy fit would need to be performed in each CCD separately. Since each CCD covers 7×4 arcmin^2 and contains an average of 43 stars, a second order polynomial function provides an accurate mapping of the PSF (Fu et al., 2008).

The KSB+ Method for PSF Correction

The KSB+ method for PSF correction was developed by Kaiser, Squires, and Broadhurst (1995), Luppino and Kaiser (1997), and Hoekstra et al. (1998). It is currently the most widely used method for analyzing weak lensing data. This method requires the definition of a different *complex ellipticity* from Eq. (6.27) to quantify the galaxy image:

$$\chi \equiv \frac{Q_{11} - Q_{22} + 2iQ_{12}}{Q_{11} + Q_{22}}. \tag{6.40}$$

If the image has elliptical isophotes with axis ratio $r \leq 1$, then $\chi = (1 - r^2)(1 + r^2)^{-1} \exp(2i\vartheta)$, where the phase of χ is twice the position angle ϑ of the major axis. This definition assures that the complex ellipticity is unchanged if the galaxy image is rotated by π, since this rotation leaves an ellipse unchanged (Bartelmann and Schneider, 2001).

If the PSF distortion can be described as a small but highly anisotropic distortion convolved with a large circularly symmetric seeing disk (the "smear"), then the ellipticity of a PSF corrected galaxy is given by (Kaiser, Squires, and Broadhurst, 1995)

$$\chi_\alpha^{\text{cor}} = \chi_\alpha^{\text{obs}} - P_{\alpha\beta}^{\text{sm}} p_\beta, \qquad (6.41)$$

where p is a vector that measures the PSF anisotropy, and P^{sm} is the smear polarizability tensor given in Hoekstra et al. (1998). The PSF anisotropy vector $p(\theta)$ can be estimated from the observed stellar ellipticities. Requiring that stars have zero ellipticity after PSF correction ($\chi_\alpha^{*\text{cor}} = 0$), we find

$$p_\mu = \left(P^{\text{sm}*}\right)^{-1}_{\mu\alpha} \chi_\alpha^{*\text{obs}}, \qquad (6.42)$$

where $\chi_\alpha^{*\text{obs}}$ is the observed ellipticity of stars. The isotropic effect of the atmosphere and weight function can be accounted for by applying the pre-seeing shear polarizability tensor correction P^γ, as proposed by Luppino and Kaiser (1997), such that

$$\chi_\alpha^{\text{cor}} = \chi_\alpha^s + P_{\alpha\beta}^\gamma \gamma_\beta, \qquad (6.43)$$

where χ^s is the intrinsic source ellipticity and γ is the pre-seeing gravitational shear. Luppino and Kaiser (1997) showed that

$$P_{\alpha\beta}^\gamma = P_{\alpha\beta}^{\text{sh}} - P_{\alpha\mu}^{\text{sm}} \left(P^{\text{sm}*}\right)^{-1}_{\mu\delta} P_{\delta\beta}^{\text{sh}*}, \qquad (6.44)$$

where P^{sh} is the shear polarizability tensor given in Hoekstra et al. (1998), and $P^{\text{sm}*}$ and $P^{\text{sh}*}$ are the stellar smear (assumed to be isotropic) and shear polarizability tensors, respectively. Combining the PSF correction, Eq. (6.41), and the P^γ seeing correction, the final KSB+ shear estimator $\hat{\gamma}$ is given by

$$\hat{\gamma}_\alpha = (P^\gamma)^{-1}_{\alpha\beta} \left[\chi_\beta^{\text{obs}} - P_{\beta\mu}^{\text{sm}} p_\mu\right]. \qquad (6.45)$$

The KSB method assumes that the PSF can be described as the convolution of a compact anisotropic kernel and a large isotropic kernel. This may be a reasonable assumption for ground based data, but incorrect for space based data (Hoekstra et al., 1998). A suitable set of basis functions can be used to expand the object surface brightness distribution, and relax the assumptions made by the KSB approach about the PSF and galaxy profiles (Bernstein and Jarvis, 2002; Refregier, 2003; Refregier and Bacon, 2003; Nakajima and Bernstein, 2007).

Fu et al. (2008) used the "KSB+" method to analyze the CFHTLS Wide survey weak lensing data over 57 (deg)^2. Figure 6.9 shows the calibration bias m and the residual offset c of the pipeline used by Fu et al. (2008) estimated using the Shear TEsting Programme (STEP) simulations. Note that different PSF models lead to different calibration bias and residual offset.

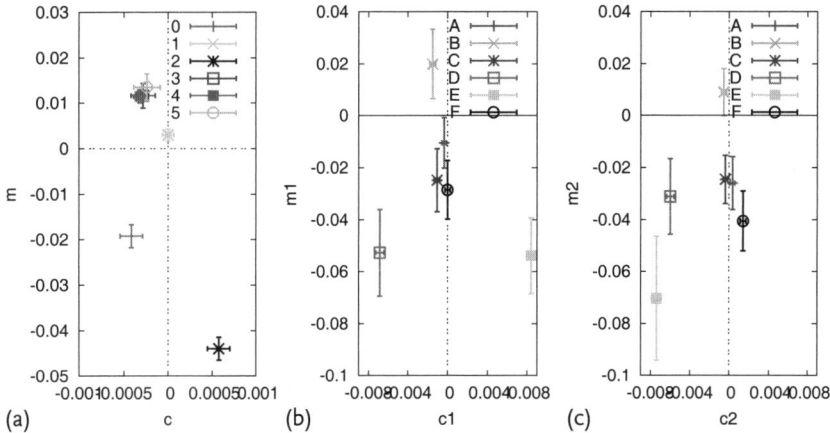

Figure 6.9 The calibration bias m and the residual offset c of the pipeline used by Fu et al. (2008) estimated using the Shear TEsting Programme (STEP) simulations (Fu et al., 2008). (a) The results of STEP1 for the first component of shear. PSF models are labeled from 0 to 5. (b) and (c) The results of STEP2 for the two shear components. PSF models are labeled from A to F.

The Shapelet Method for PSF Correction

The shapelet method aims for a full deconvolution of each galaxy from the PSF. The method has been developed by Bernstein and Jarvis (2002), Refregier (2003), Refregier and Bacon (2003), and Massey and Refregier (2005). It is still under further development to address a number of technical issues in its practical implementation.

An image $f(\mathbf{x})$ is decomposed into shapelets by fitting it to the shapelet expansion

$$f(\mathbf{x}) = \sum_{n=0}^{\infty} \sum_{m=-n}^{n} f_{n,m} \chi_{n,m}(r, \theta; \beta), \qquad (6.46)$$

where $f_{n,m}$ ($m \le n$) are the shapelet coefficients from fitting the image, and the Gauss–Laguerre basis functions are

$$\chi_{n,m}(r, \theta; \beta) = \frac{C_{n,m}}{\beta} \left(\frac{r}{\beta}\right)^{|m|} L_{\frac{n-|m|}{2}}^{|m|}\left(\frac{r^2}{\beta^2}\right) e^{\frac{-r^2}{2\beta^2}} e^{-im\theta}, \qquad (6.47)$$

where $C_{n,m}$ is a normalizing constant, and β is the scale size.

The shapelet model for the PSF contains substructure, skewness and chirality. In general, the ellipticity of the isophotes varies as a function of radius. For computational efficiency, the shapelet series is typically truncated at order $n_{\max} = 12$ (Massey et al., 2007). A high shapelet order is required to accurately model real galaxies, since the profiles of real galaxies are more exponential than Gaussian (Miller et al., 2007).

To simulate the process of real photons incident upon a CCD detector, the shapelet basis functions are first convolved with the PSF in shapelet space, then integrated analytically within the pixels. The convolved basis functions are then fitted to the data in a likelihood analysis, with the shapelet coefficients as free parameters. A deconvolved reconstruction of each galaxy is achieved by resumming *unconvolved* basis functions with the fitted shapelet coefficients.

Some information is lost during convolution, and cannot be recovered through deconvolution. However, the galaxy model in shapelet space can be restricted to the range of physical scales on which information is expected to survive. An iterative algorithm can be designed to optimize the scale size of the shapelets, such that it captures the maximum range of available scales for each individual galaxy (Massey and Refregier, 2005).

In principle, if a sufficiently large range of scales are modeled to lower the residual χ^2_{reduced} to exactly unity, the deconvolved galaxy model can be made to be minimal in noise. In practice, this is made difficult by correlated background noise and the expensive computational cost of incorporating the noise covariance matrix.

Once a deconvolved model of galaxy images, $\tilde{f}(\mathbf{x})$, is obtained, an unbiased estimator of shear is given by (Massey et al., 2007)

$$\tilde{\gamma} = \frac{\iint \tilde{f}(\mathbf{x})r^2 \left(\cos(2\theta), \sin(2\theta)\right) d^2\mathbf{x}}{\iint \tilde{f}(\mathbf{x})r^2 d^2\mathbf{x}}. \tag{6.48}$$

Current Status of PSF Correction

In preparation for the next generation of wide-field weak lensing surveys, a large collaborative project, the Shear TEsting Programme (STEP) has been launched to improve the accuracy and reliability of all weak lensing measurements. STEP is a community effort to test and compare shear measurement algorithms on a common set of simulations (Heymans et al., 2006). The GRavitational lEnsing Accuracy Testing 2008 (GREAT08) Challenge has been issued to continue the progress made through STEP (Bridle et al., 2009).

Weak lensing data can be simulated by applying shear to a galaxy catalog using Eq. (6.30), and convolve the model galaxy and star images with a simulated PSF (Heymans et al., 2006). The performance of a PSF correction method can be evaluated by fitting to (Heymans et al., 2006)

$$\gamma_i - \gamma_i^{\text{true}} = q(\gamma_i^{\text{true}})^2 + m_i \gamma_i^{\text{true}} + c_i, \tag{6.49}$$

where γ_i^{true} ($i = 1, 2$) are the two components of the external shear applied to each image. The absence of calibration bias corresponds to $m_i = 0$. The absence of PSF systematics and shot noise corresponds to $c_i = 0$. For a linear response of the method to shear, $q = 0$.

For clarity, we use Figure 6.10 from Heymans et al. (2006) to illustrate the basic points in PSF correction, although it does not reflect the latest progress that has been made. Figure 6.10 shows the performance of various methods (Heymans et al., 2006). Here $\gamma_2^{\text{true}} = 0$, thus m_2 is absent, and $m = m_1$ (see Eq. (6.49)). The parameter σ_c is the variance of c_1 and c_2 as measured from the six different PSF models used in the STEP project.

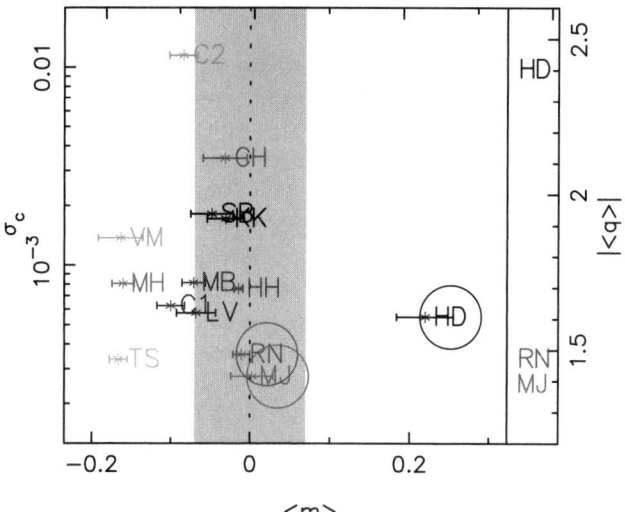

Figure 6.10 Measurements of calibration bias $\langle m \rangle$, PSF residuals σ_c and nonlinearity $\langle q \rangle$ for each author participating in the STEP project (Heymans et al. (2006), MNRAS, Wiley-Blackwell). For the nonlinear cases where $\langle q \rangle \neq 0$ (points enclosed within a large circle), $\langle q \rangle$ is shown with respect to the right hand scale. The lower the value of σ_c, the more successful the PSF correction is at removing all types of PSF distortion. The lower the absolute value of $\langle m \rangle$, the lower the level of calibration bias. The higher the q value the poorer the response of the method to stronger shear. Note that for weak shear $\gamma < 0.01$, the impact of this quadratic term is negligible. Results in the shaded region suffer from less than 7% calibration bias.

The dominant source of error in Figure 6.10 is the correction for the size of the PSF, which leads to an overestimate of the shear by a multiplicative factor $(1 + m)$ (Heymans et al., 2006). Massey et al. (2007) presented a blind analysis of more complicated galaxies, and included an improvement in the statistical accuracy of the test as well. The STEP studies in Heymans et al. (2006) and Massey et al. (2007) showed that current weak lensing analysis pipelines can recover the weak lensing signal with a precision better than 7%, within the statistical errors of current weak lensing analyses. The most successful methods were shown to achieve 1–2% level accuracy, but are not linear in response to shear (see Figure 6.10).

Note that most of the shape measurement techniques can be implemented either in real space or in Fourier space. In particular, if a shear γ is applied to an image, it is equivalent to shearing the Fourier transform $\tilde{I}(k)$ by $-\gamma$. For example, if one stretches the image along the x-axis in real space, it is equivalent to compressing it along the x-axis in Fourier space. The difference is that in Fourier space the PSF convolution is a multiplication instead of a convolution. Fourier-space algorithms are not yet in common use but are very promising (Hirata, 2009).

Impact of PSF Correction on Dark Energy constraints

Figure 6.11 shows the degradation in cosmological parameter accuracy due to shear calibration errors for a survey covering 5000 (deg)^2 with median redshift $z = 0.8$ (Huterer et al., 2006). Note that this figure shows a regime of self-calibration: the increase in the degradation of cosmological parameter accuracy slows with increasing shear calibration errors, because the redshift dependence of the lensing signal allows for joint measurements of parameters describing both shear calibration and cosmology. Note also that for the degradation estimates shown in Figure 6.11, no assumptions are made about the functional forms of the shear calibration errors, and only the shear power spectra are used. When the shear bispectra are combined with the shear power spectra, the self-calibration becomes even more effective (Huterer et al., 2006).

6.3.2
Other Systematic Uncertainties

In addition to PSF correction, there are other known important systematic uncertainties for the weak lensing method. These include uncertainty in the centroids of photometric redshift bins, the intrinsic alignment of galaxies, and nonlinear and baryonic effects (Peacock et al., 2006). Nonlinear and baryonic effects can be minimized by improving theoretical modeling using numerical ray-shooting of realistic cosmological N-body simulations. We discuss photometric redshifts and intrinsic alignments below.

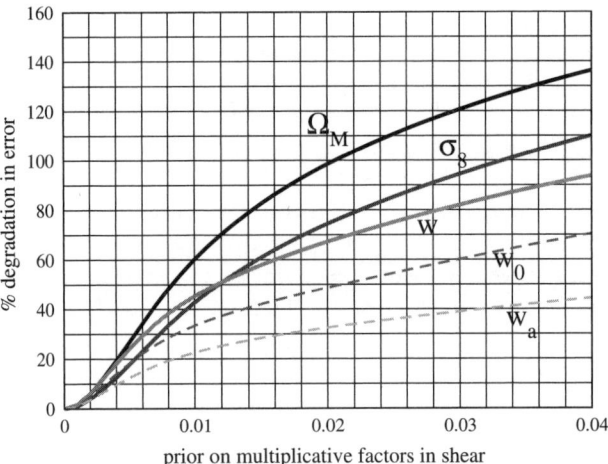

Figure 6.11 Degradation in cosmological parameter accuracy due to shear calibration errors for a survey covering 5000 (deg)^2 with median redshift $z = 0.8$ (Huterer et al. (2006), MNRAS, Wiley-Blackwell). No assumptions are made about the functional forms of the shear calibration errors, and only the shear power spectra are used.

Photometric Redshifts

Photometric redshifts are approximate redshifts estimated from multi-band imaging data, and calibrated using a spectroscopic sample. The dominant features in galaxy spectra affect the broadband colors, and provide the basis for estimating redshifts photometrically. For star-forming galaxies (galaxies with young stellar populations), the dominant features are the Lyman-α and Balmer breaks. For red galaxies (galaxies with old stellar populations), the dominant feature is the 400 nm break. The accuracy and precision of photometric redshifts are critical to the viability of weak lensing as a dark energy probe.

The use of broadband photometry in multiple filters to estimate redshifts of galaxies has been well-established (Weymann *et al.*, 1999). There are two different approaches in estimating photometric redshifts of galaxies. In the empirical fitting method (Connolly *et al.*, 1995; Wang, Bahcall, and Turner, 1998), a training set of galaxies with measured spectroscopic redshifts are used to derive analytical formulae relating the redshift to colors and magnitudes. In the template fitting technique (see for example Puschell, Owen, and Laing (1982); Lanzetta, Yahil, and Fernandez-Soto (1996); Mobasher *et al.* (1996); Sawicki, Lin, and Yee (1997)), the observed colors are compared with the predictions of a set of galaxy spectral energy distribution templates.

In using weak lensing to probe dark energy, photometric redshifts are used to divide galaxies into redshift bins. In the geometric weak lensing method, the centroids of the photometric redshifts must be known to the accuracy of around 0.1% in order to avoid significant degradation of dark energy constraints (see Figure 6.12) (Bernstein and Jain, 2004; Huterer *et al.*, 2006). Note that only the shear

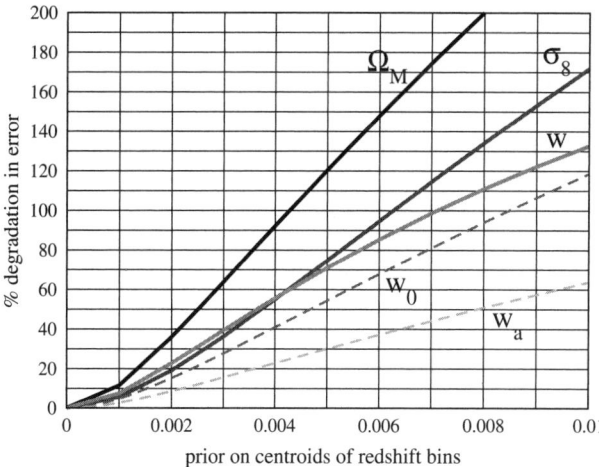

Figure 6.12 Degradation in cosmological parameter accuracy due to the bias in the centroids of the photometric redshifts, for a survey covering 5000 (deg)^2 with median redshift $z = 0.8$ (Huterer *et al.* (2006), MNRAS, Wiley-Blackwell). No assumptions are made about the functional forms of the shear calibration errors, and only the shear power spectra are used.

power spectra are used in Figure 6.12. A self-calibration regime for the centroids of the photometric redshifts arises if shear bispectra are combined with the shear spectra (Huterer et al., 2006).

For fixed observational resources, the bias of the photometric redshifts can be significantly reduced if a full probability distribution in redshift, instead of just the most likely redshift, is used for each galaxy (Wittman, 2009).

The inclusion of imaging in NIR filter bands can dramatically reduce the bias and improve the precision of photometric redshifts. Abdalla et al. (2008) studied the impact of NIR imaging on various proposed weak lensing surveys. They quantified how the photo-z bias $\langle (z_{\text{phot}} - z_{\text{spec}}) \rangle$, and the photo-z rms dispersion, $\langle (z_{\text{phot}} - z_{\text{spec}})^2 \rangle$, decrease in amplitude as functions of redshift with the addition of NIR filter bands.

Intrinsic Alignments

The measurement of cosmic shear is complicated by the correlation in galaxy ellipticities in the absence of lensing. The understanding of these intrinsic alignments in the galaxy shapes is ongoing, and relatively little is known about this effect. Two types of intrinsic alignment effects have been identified so far.

The first type of intrinsic alignment arises from the alignment of galaxy haloes due to tidal gravitational forces (Croft and Metzler, 2000; Crittenden et al., 2001). Our understanding of this effect on cosmic shear measurements is limited by the theoretical uncertainty in modeling the alignment of the haloes themselves, and how well the luminous matter aligns with dark matter. The effect of this type of intrinsic alignment can be minimized by using cross-spectra of galaxies in two different redshift bins (if the photometric redshifts are sufficiently accurate), so that galaxy pairs are separated by large distances (hence the tidal effects are very weak).

The second type of intrinsic alignment is more problematic. The tidal field and the intrinsic shear are both caused by the same matter density field, thus they are correlated. These correlations cause the intrinsic shear of nearby galaxies (due to the tidal field) to be correlated with the gravitational shear acting on more distant galaxies (Hirata and Selja, 2004). This leads to a suppression of the weak lensing signal. Since this type of intrinsic alignment affects pairs of galaxies at different redshifts, it is a systematic effect that is very difficult to remove, and can bias cosmic shear results (Bridle and King, 2007). The correlation of intrinsic and gravitational shear can in principle be projected out using the combined redshift dependence, but unfortunately this amplifies other systematic effects and requires those to be even better controlled (Bridle and King, 2007).

The impact of the second type of intrinsic alignment can be constrained using the density-shear correlations, or the scaling of the cross-correlation tomography signal with redshift (Hirata and Selja, 2004). Thus the degree to which this systematic effect can be controlled is determined by the accuracy of photometric redshifts (Bridle and King, 2007). Better understanding of this systematic effect requires more spectroscopic data (Mandelbaum et al., 2006; Hirata et al., 2007), and further progress in numerical simulations (Heymans et al., 2006).

6.4
Future Prospects for the Weak Lensing Method

The published forecasts of dark energy constraints from future weak lensing surveys are based on the Fisher matrix formalism, which gives the smallest possible statistical errors for parameters measured from a given set of data (see Section 2.4.2).

For parameters **p**, the Fisher matrix element for p_i and p_j is

$$F_{ij} = \sum_{l} \left(\frac{\partial \mathbf{C}}{\partial p_i}\right)^T \mathbf{Cov}^{-1} \frac{\partial \mathbf{C}}{\partial p_j}, \qquad (6.50)$$

where **C** is the column matrix of the observed power spectra, and $\mathbf{Cov}^{-1}$ is the inverse of the covariance matrix between the power spectra.

Figures 6.13 and 6.14 illustrate the dark energy constraints expected from a 5000 (deg)^2 weak lensing survey with median redshift $z = 0.8$ (Takada and Jain, 2004). Figure 6.13 shows the expected sensitivity of the weak lensing shear-shear power spectra to a constant dark energy equation of state. Figure 6.14 shows the expected constraints on a constant dark energy equation of state, and a dark en-

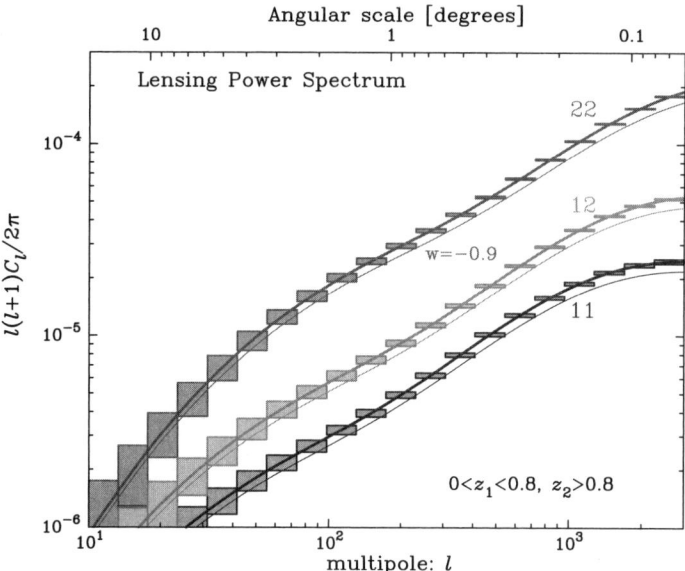

Figure 6.13 The weak lensing shear-shear power spectra constructed from galaxies split into two broad redshift bins, from (Takada and Jain (2004), MNRAS, Wiley-Blackwell). Both of the auto-spectra and one cross-spectrum are shown. The solid curves are predictions for a fiducial ΛCDM model, and include non-linear evolution. The boxes show the expect- ed measurement error due to the sample variance and intrinsic ellipticity errors from a 5000 (deg)^2 survey with median redshift $z = 0.8$. The thin curves are the predictions for a dark energy model with $w = -0.9$. Note that at least four or five redshift bins are expected to be useful from such a survey, leading to many more measured power spectra.

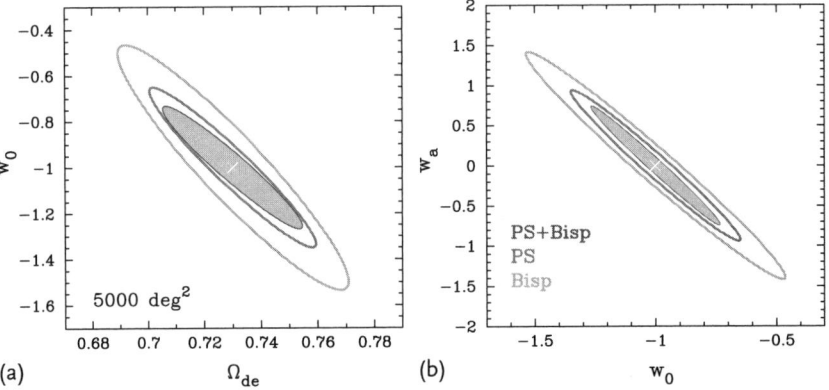

Figure 6.14 Dark energy contours (68.3% confidence level) from (a) lensing power spectra and (b) bispectra from a 5000 (deg)² survey with median redshift $z = 0.8$ (same as in Figure 6.13) from Takada and Jain (2004), MNRAS, Wiley-Blackwell. These forecasts assume Planck priors and do *not* include systematic errors.

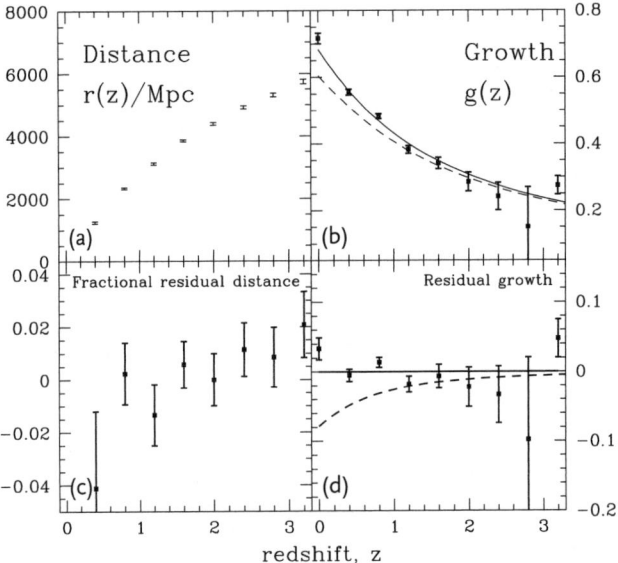

Figure 6.15 Reconstructed distances (a) and (c), and growth factors (b) and (d) from a LSST-like weak lensing survey (figure reprinted with permission from Knox, Song, and Tyson, Phys. Rev. D, 74, 023512 (2006). Copyright (2006) by the American Physical Society). (c) shows the fractional residual distances, $[r(z) - r_{fid}(z)]/r_{fid}(z)$, where $r(z)$ are the reconstructed distances and $r_{fid}(z)$ are the distances in the fiducial DGP model. (d) shows the residual growth factor, $g(z) - g_{fid}(z)$. The curves in (b) and (d) are $g_{fid}(z)$ in the fiducial DGP model (solid) and $g(z)$ for the Einstein gravity model (dashed) with the same $H(z)$ and ρ_m as the DGP model. Although these two models have the same $r(z)$ they are distinguishable by their significantly different growth factors. Note that $g(z) = G(z) \propto D_1(t)$ (see Eqs. (1.20) and (1.27)).

ergy equation of state linear in the cosmic scale factor a, assuming Planck priors. The parameter constraints in Figure 6.14 include the estimated non-Gaussian covariances between the power spectra and bispectra, but do *not* include systematic uncertainties.

If the systematic effects are properly modeled, a weak lensing survey can potentially allow us to differentiate between dark energy and modified gravity (see Figure 6.15 (Knox, Song, and Tyson, 2006)). Figure 6.15 corresponds to a "2π" deep and wide survey of 20 000 (deg)^2 in six wavelength bands from 0.4–1.1 μm to be undertaken by the Large Synoptic Survey Telescope (LSST). Such a survey would yield the shear and photometric redshift of 3 billion source galaxies over a redshift range of 0.2 to 3 (Knox, Song, and Tyson, 2006).

Weak lensing can be used to probe dark energy in two different ways: weak lensing tomography or weak lensing cross-correlation cosmography. Weak lensing tomography requires accurate and precise photometric redshifts to map the clustering of matter, see e.g., Knox, Song, and Tyson (2006). Weak lensing cross-correlation cosmography (Jain and Taylor, 2003) is also known as "the geometric method" (Zhang, Hui, and Stebbins, 2005). The geometric weak lensing method to probe dark energy minimizes the sensitivity to the clustering of matter (which can be a source of systematic uncertainty). We discuss it in detail below.

6.5
The Geometric Weak Lensing Method

The basic idea of the geometric weak lensing method is to construct a map of the foreground galaxies, from which an estimated map of the foreground mass can be made. This foreground mass slice induces shear on all the galaxies in the background. The amplitude of the induced shear as a function of the background redshift is measured, from which a weighted sum of the ratios of angular diameter distances between the source slice and lens slice, and between the lens slice and observer is estimated (Jain and Taylor, 2003; Zhang, Hui, and Stebbins, 2005). Note that this marks an important difference of the weak lensing method from the supernova and baryon acoustic oscillation methods: the weak lensing method gives highly correlated measurements of the cosmic expansion history $H(z)$ in redshift bins through the measurement of distance ratios, while the supernova and baryon acoustic oscillation methods can give uncorrelated measurements of $H(z)$ (Wang and Tegmark, 2005; Seo and Eisenstein, 2003) through uncorrelated distance measurements or the direct measurement of $H(z)$.[11]

---

11) A small correlation in the measured $H(z)$ can arise in the supernova and baryon acoustic oscillation (BAO) methods due to the uncertainties in the supernova peak luminosity calibration and the BAO scale calibration from CMB.

6.5.1
Linear Scaling and Off-Linear Scaling

The geometric weak lensing method can be used with linear scaling (Jain and Taylor, 2003), or off-linear scaling (Zhang, Hui, and Stebbins, 2005).

In the linear scaling approach (Jain and Taylor, 2003), the foreground galaxy distribution is approximated by a delta function at a distance $\hat{\chi}_f$, thus the ratio of $C_{g_i \kappa_j}(l)$ for two different background distributions (labeled b and b') but the same foreground (labeled f) is

$$\frac{C_{g_f \kappa_b}(l)}{C_{g_f \kappa_{b'}}(l)} \approx \frac{\hat{\chi}_f^{-1} - \chi_{eff}(b)^{-1}}{\hat{\chi}_f^{-1} - \chi_{eff}(b')^{-1}}, \qquad (6.51)$$

where

$$\frac{1}{\chi_{eff}(b)} \equiv \frac{\int dz_b\, n(z_b)/\chi(z_b)}{\int\limits_0^\infty dz'\, n(z')}, \qquad (6.52)$$

where $n(z) = dN(z)/dz$ is the galaxy number density distribution. Cosmological parameters can be estimated using Eq. (6.51), by measuring the left hand side and finding the parameter values for the right hand side to match the left hand side.

In the off-linear approach (Zhang, Hui, and Stebbins, 2005), the foreground distribution and the background distribution are assumed to have no overlap. Then the cross-correlation power spectra exhibit an offset-linear scaling:

$$C_{g_f \kappa_b}(l) \approx F_f(l) + G_f(l)/\chi_{eff}(b),$$

$$C_{\gamma_f \gamma_b}(l) \approx A_f(l) + B_f(l)/\chi_{eff}(b). \qquad (6.53)$$

For a fixed foreground distribution of galaxies, as one varies the background redshift distribution, $C_{g_f \kappa_b}(l)$ and $C_{\gamma_f \gamma_b}(l)$ scale linearly through factor $1/\chi_{eff}(b)$, but with an offset given by F_f or A_f.

Note that the Eq. (6.51) and Eq. (6.53) assume a flat universe. For a universe with spatial curvature, the scalings with $1/\chi_{eff}$ can be generalized by replacing χ_{eff} with $\tann(\chi_{eff}/R)$, where R is the curvature radius. For a closed universe, $\tann(x) = \tan(x)$. For an open universe, $\tann(x) = \tanh(x)$. Since this dependence is different from the sinelike function that arises in distance-redshift relations measured from supernovae and transverse-BAO, the combination of geometric weak lensing with SNe or/and transverse-BAO could in principle be used to geometrically determine Ω_k (Hirata, 2009).

Clearly, the offset-linear scaling approach to geometric weak lensing is more robust and conservative than the linear scaling approach, as it involves weaker assumptions about the foreground galaxy distribution. The linear scaling approach has the advantage of being simpler conceptually, and we discuss its implementation below.

6.5.2
Implementation of the Linear Scaling Geometric Method

The observed ellipticity of background galaxy k is (Bernstein and Jain, 2004)

$$e_k^{obs} = \sum_l \mathbf{d}_l(\mathbf{x}_k) g_{lk} + \mathbf{d}_{k,sys} + \mathbf{e}_k, \qquad (6.54)$$

where $\mathbf{d}_l$ is the distortion field imparted on the source by the mass in redshift shell l, $\mathbf{d}_{k,sys}$ is the instrumental systematic distortion, and $\mathbf{e}_k$ is the intrinsic galaxy shape. The dark energy dependence enters primarily through the geometric factor

$$g_{ls} \equiv \frac{r(\chi_s - \chi_l)}{r(\chi_s)}, \qquad (6.55)$$

where $r(\chi) = \chi(z) = \int_0^z \frac{dz'}{H(z')}$ in a flat universe.

The measured data vector in redshift bin s is

$$X_{ls} \equiv \frac{1}{N_s} \sum_{k \in s} e_k^{obs} \cdot \hat{\mathbf{d}}_l(\mathbf{x}_k), \qquad (6.56)$$

where s refers to "source". N_s is the number of source galaxies in bin s. And $\hat{\mathbf{d}}_l$ is a template of the distortion field; it is an estimate of $\mathbf{d}_l$ derived from the galaxies identified near z_l from the photometric/spectroscopic data (without using the background galaxy shape information). The fidelity of the template is parametrized by β_l, defined by

$$\langle \mathbf{d}_l \cdot \hat{\mathbf{d}}_l \rangle = \beta_l \langle d_l^2 \rangle. \qquad (6.57)$$

β_l is related to the bias factor b and the correlation coefficient r between galaxies and mass.

The dark energy and cosmological parameters can be derived through a likelihood analysis comparing Eq. (6.54) with the observed data vector. Typically, the dark energy parameter uncertainties scale roughly as (Bernstein and Jain, 2004)

$$\sigma_p \propto \left(\frac{\sigma_e^2}{n f_{sky} \beta^2} \right)^{0.5} z_{med}^{-0.6} \qquad (6.58)$$

where σ_e is the rms variance of the intrinsic galaxy ellipticity, $Var(e_+) = Var(e_\times) = \sigma_e^2$, n is the source galaxy density, f_{sky} is the fraction of sky covered by the survey, β is the fidelity of the distortion field template, and z_{med} is the median redshift of the survey.

7
Observational Method IV: Clusters as Dark Energy Probe

7.1
Clusters and Cosmology

Clusters of galaxies are the largest structures in the universe that have undergone gravitational collapse. They correspond to the highest density fluctuations in the early universe. It has been argued that because galaxy clusters are so large, their matter content should provide a fair sample of matter content of the universe (White and Frenk, 1991).

There is a long history in the use of galaxy clusters as cosmological probes. Zwicky applied the virial theorem to galaxy motions in the Coma cluster, and estimated a total mass more than 100 times larger than expected by adding up all of the mass in stars. Based on this, he proposed the existence of invisible matter in 1933.

Given the cosmological model, the abundance of galaxy clusters as a function of their mass, $dN/(dM dV)$, can be estimated analytically, or more precisely predicted through gravitational N-body simulations. This prediction can be compared with the observed abundance, $dN/(dM d\Omega dz)$, to constrain the cosmological model. The observed abundance of galaxy clusters first revealed that we live in a low matter density universe (Bahcall, Fan, and Cen, 1997). In combination with the first definitive evidence for a flat universe from the BOOMERANG CMB data (de Bernardis *et al.*, 2000), this constituted the first strong indirect evidence for the existence of dark energy.

Clusters of galaxies can be used to probe dark energy in two different ways: (1) using the cluster number density and its redshift evolution, as well as cluster distribution on large scales (see, e.g., Haiman, Mohr, and Holder (2001); Vikhlinin *et al.* (2003); Schuecker *et al.* (2003)), (2) using clusters as standard candles by assuming a constant cluster baryon fraction (see, e.g., Allen *et al.* (2004)), or using combined X-ray and SZ measurements for absolute distance measurements (see, e.g., Molnar *et al.* (2004)).

7.2
Cluster Abundance as a Dark Energy Probe

The cluster abundance (also known as the *cluster mass function*), $dN/(dMdV)$, can probe dark energy in two ways. First,

$$\frac{dV}{d\Omega dz} \propto \frac{D_A^2(z)}{H(z)}, \tag{7.1}$$

thus the cluster counts depend on the dark energy density $\rho_X(z)$. Second, the cluster abundance function itself, $dN/(dMdV)$, depends on the amplitude of matter density fluctuations; $dN/(dMdV)$ is exponentially sensitive to the growth factor $G(z)$ for fixed mass M.

Mantz *et al.* (2008) gave constraints on dark energy using the X-ray luminosity function of the largest known galaxy clusters at redshift $z < 0.7$. The tightest cosmological parameter constraints at present come from the *Chandra* Cluster Cosmology Project (CCCP) (Vikhlinin *et al.*, 2009a,b). We will discuss the work of Vikhlinin *et al.* (2009) in detail below.

7.2.1
Theoretical Cluster Mass Function

The observed cluster mass function has to be compared with theoretical models in order to extract cosmological constraints. Since the theory of nonlinear gravitational collapse is insensitive to the background cosmology, the mapping of a linear matter power spectrum to the cluster mass function can be described in a model-independent manner (Jenkins *et al.*, 2001).

Jenkins *et al.* (2001) proposed an empirical halo mass function model based on extensive numerical simulations, motivated by the Press–Schechter model (Press and Schechter, 1974). In this model, the quantity $\ln \sigma^{-1}(M, z)$ is used as the mass variable instead of M, where $\sigma^2(M, z)$ is the variance of the linear density field, extrapolated to the redshift z at which haloes are identified, after smoothing with a spherical top-hat filter which encloses mass M in the mean:

$$\sigma^2(M, z) = \frac{G^2(z)}{2\pi^2} \int_0^\infty k^2 P_m(k) W^2(k; M) dk, \tag{7.2}$$

where $G(z) \propto D_1(z)$ is the growth factor of linear perturbations normalized so that $G(z=0) = 1$, $P_m(k)$ is the linear matter power spectrum at $z = 0$, and $W(k; M)$ is the Fourier-space representation of a real-space top-hat filter enclosing mass M at the mean density of the universe. The use of $\ln \sigma^{-1}(M)$ as the "mass" variable factors out most of the difference in the mass functions between different epochs, cosmologies and power spectra.

The halo mass function $f(\sigma, z; s)$ is defined as (Jenkins *et al.*, 2001)

$$f(\sigma, z; s) \equiv \frac{M}{\rho_0} \frac{dn_s(M, z)}{d \ln \sigma^{-1}}, \tag{7.3}$$

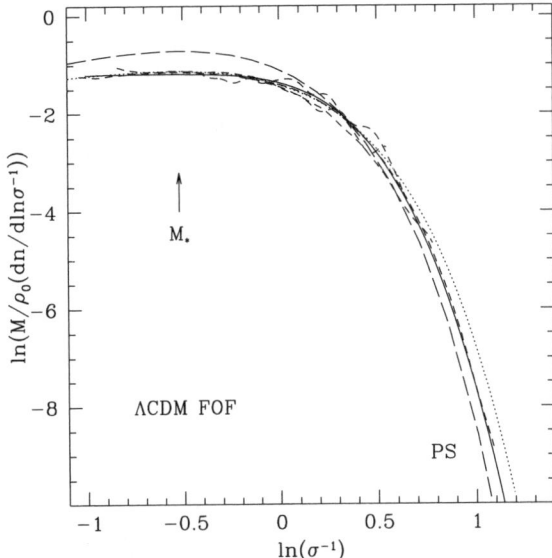

Figure 7.1 The halo mass function at $z = 0$ in a ΛCDM model with $\Omega_m = 0.3$, $\Omega_\Lambda = 0.7$, $\Gamma = \Omega_m h = 0.21$, and $\sigma_8 = 0.9$ (Jenkins et al. (2001), MNRAS, Wiley-Blackwell). Halos were found using the friends-of-friends (FOF) algorithm of Davis et al. (1985) with $b = 0.164$. The short dashed lines represent results from the individual simulations, the solid curve is the fit of Eq. (7.4) to the combined results of the simulations, while the long dashed line shows the Press–Schechter prediction for $\delta_c = 1.675$. The arrow marks the characteristic mass scale, M_*, where $\sigma(M_*) = \delta_c$, and corresponds to the position of the peak in the Press–Schechter mass function.

where s is a label identifying the cosmological model and halo finder under consideration, $n(M, z)$ is the abundance of haloes with mass less than M at redshift z, and $\rho_0(z)$ is the mean density of the universe at that time.

Figure 7.1 shows the halo mass functions from the N-body simulations for a ΛCDM model with $\Omega_m = 0.3$, $\Omega_\Lambda = 0.7$, $\Gamma = \Omega_m h = 0.21$, and $\sigma_8 = 0.9$ (Jenkins et al., 2001). Halos were found using the friends-of-friends (FOF) algorithm of Davis et al. (1985) with $b = 0.164$ (b is the linking parameter in the FOF algorithm). The short dashed lines show results from the individual simulations, the solid curve is the fit to the combined results of the simulations given by

$$f(M) = 0.301 \exp\left[-|\ln \sigma^{-1} + 0.64|^{3.88}\right], \tag{7.4}$$

in the range $-0.96 \leq \ln \sigma^{-1} \leq 1.0$. The difference between these fitting functions and the actual mass functions is typically less than 10%. The long dashed line shows the Press–Schechter prediction using $\delta_c = 1.675$ (δ_c is a threshold parameter). The arrow marks the characteristic mass scale, M_*, where $\sigma(M_*) = \delta_c$, and corresponds to the position of the peak in the Press–Schechter mass function.

Jenkins et al. (2001) computed the halo mass functions from the N-body simulations of various cosmological models, including ΛCDM, OCDM, SCDM, and τCDM. They found that all the mass functions are fitted with an accuracy of better than about 20% over the entire range by the following formula:

$$f(M) = 0.315 \exp\left[-|\ln \sigma^{-1} + 0.61|^{3.8}\right], \tag{7.5}$$

valid over the range $-1.2 \leq \ln \sigma^{-1} \leq 1.05$.

7.2.2
Cluster Mass Estimates

Since cluster mass is not an observable, it must be inferred using observables such as the average X-ray temperature, T, and the X-ray gas mass fraction, f_g. This requires the establishment of cluster mass and observable relations.

Cluster mass is not a well-defined quantity. It can be defined as the mass within the radius corresponding to a fixed mean overdensity, Δ:

$$M_\Delta \equiv M(< r_\Delta) \equiv \Delta \cdot \rho_c \cdot \frac{4}{3}\pi r_\Delta^3, \tag{7.6}$$

where the critical density $\rho_c(z) \equiv 3H^2(z)/8\pi G$. Vikhlinin et al. (2009b) chose $\Delta = 500$ – the radius within which the clusters are relatively relaxed, and the largest radius at which the intra-cluster medium (ICM) temperature can be reliably measured with *Chandra* and *XMM-Newton* (Vikhlinin et al., 2009a). Using significantly lower Δ dramatically increases observational uncertainties; at significantly higher values of Δ, the theoretical uncertainties start to increase while there is no crucial gain on the observational side (Vikhlinin et al., 2009a).

To calibrate the cluster mass and observable relations, we need to know the masses of a sample of clusters. The hydrostatic mass measurements of clusters are usually used for calibration. The hydrostatic estimate of total cluster mass within a radius r is given by

$$M(r) = -\frac{kT(r)r}{G\mu m_H}\left(\frac{d\ln \rho_g}{d\ln r} + \frac{d\ln T}{d\ln r}\right), \tag{7.7}$$

where m_H is the Hydrogen mass, and $\mu = 0.592$ is the mean molecular weight of the fully ionized H-He plasma. Given $M(r)$, r_{500} and hence M_{500} can be found by solving Eq. (7.6).

The Cluster Mass and Gas Mass Relation

The cluster total mass is often estimated using the X-ray derived hot gas mass M_{gas}:

$$M_{tot} = f_g^{-1} M_{gas}, \tag{7.8}$$

where f_g is the *cluster gas-to-mass ratio*, or the *cluster gas fraction*. The radius at which the cluster mass is defined, r_{500}, can be found by solving (see Eq. (7.6))

$$\frac{M_{gas}(r) f_g^{-1}}{(4/3)\pi r^3 \rho_c(z)} = \Delta = 500. \tag{7.9}$$

This gives $M_{gas}(r_{500})$, and Eq. (7.8) gives the cluster total mass $M_{tot} \equiv M_{500}$.

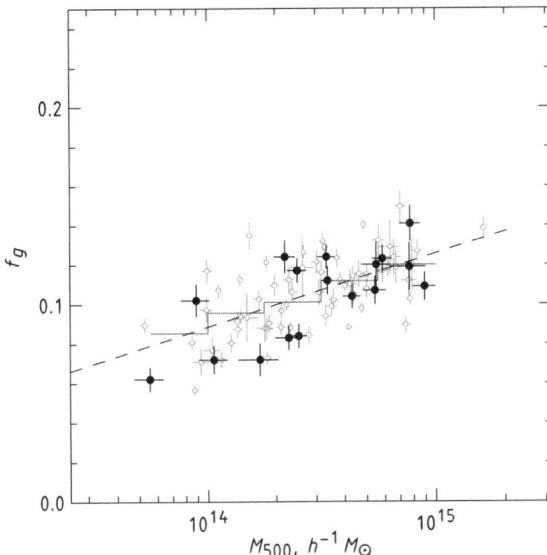

Figure 7.2 Trend of f_g within $r = r_{500}$ with cluster mass derived from X-ray observations (Vikhlinin et al., 2009a). The solid black circles show the results from direct hydrostatic mass measurements of 17 clusters (Vikhlinin et al., 2006, 2009a). Gray circles show approximate estimates using the $M_{tot} - T_X$ correlation. The scatter is consistent with mass measurement uncertainties, either from hydrostatic estimates (Nagai, Vikhlinin, and Kravtsov, 2007) or from $M_{tot} - T_X$ correlation (Kravtsov, Vikhlinin, and Nagai, 2006). The error bars indicate only the formal measurement uncertainties.

If all cluster baryons are in the ICM, the ICM follows the distribution of dark matter, and clusters contain the same baryon-to-matter ratio as the global value, then $f_g = \Omega_b/\Omega_m$, a ratio accurately determined by CMB measurements. However, f_g depends on both cluster mass and redshift.

The observed gas fraction in clusters is significantly lower than the cosmic average, with most of the "missing" component in stars. The observed f_g also depends on cluster mass (Mohr, Mathiesen, and Evrard, 1999; Vikhlinin et al., 2006; Zhang et al., 2006), likely due to baryon cooling and galaxy formation (Kravtsov, Nagai, and Vikhlinin, 2005), energy feedback from the central AGNs (Bode et al., 2007), evaporation of supra-thermal protons (Loeb, 2007), and so on.

Figure 7.2 shows the trend of f_g within $r = r_{500}$ with cluster mass derived from X-ray observations (Vikhlinin et al., 2009a). The solid black circles show the results from direct hydrostatic mass measurements (Vikhlinin et al. (2006) sample with 7 additional clusters), which are more accurate than the approximate estimates using the $M_{tot} - T_X$ correlation (gray circles). The results from direct hydrostatic mass measurements give a linear fit of

$$f_g(h/0.72)^{1.5} = 0.125 + 0.037 \log M_{15} , \qquad (7.10)$$

where M_{15} is the cluster total mass, M_{500}, in units of $10^{15} \, h^{-1} \, M_\odot$. Extrapolation of this trend to lower masses describes the observed f_g for galaxy groups (Sun et al.,

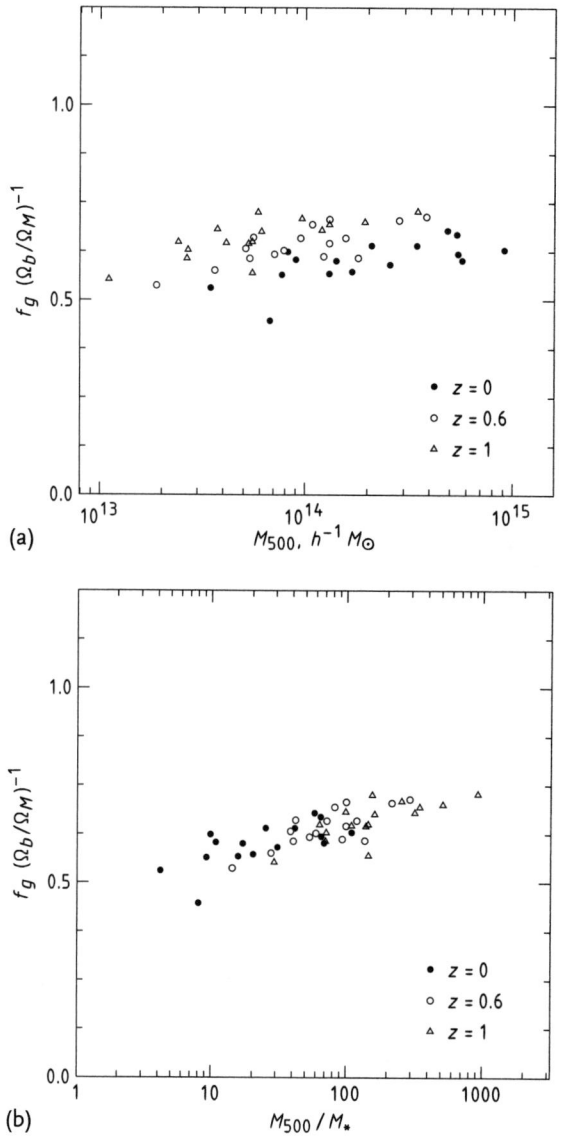

Figure 7.3 (a) The dependence of f_g within r_{500} on the cluster mass observed in high-resolution cosmological simulations with cooling, star formation, and feedback (Kravtsov, Nagai, and Vikhlinin, 2005; Nagai, Vikhlinin, and Kravtsov, 2007; Vikhlinin et al., 2009a). The simulated clusters show qualitatively the same trend as that observed at $z = 0$ (Figure 7.2), but there is a clear evolution of f_g for the given M. (b) The z-dependence is almost completely removed if we scale the cluster masses by the characteristic nonlinear mass scale, M_*.

2009). Across the useful mass range, 10^{14}–$10^{15}\,h^{-1}\,M_\odot$, f_g is determined to 4–5%. The systematic uncertainties are dominated by those of the hydrostatic mass estimates.

The $f_g(M)$ trend for high-z clusters cannot be established observationally independent of the underlying cosmology. Therefore, hydrodynamic simulations must be relied upon to study the redshift dependence of $f_g(M)$. Figure 7.3 shows the dependence of f_g within r_{500} on the cluster mass observed in high-resolution cosmological simulations with cooling, star formation, and feedback (Kravtsov, Nagai, and Vikhlinin, 2005; Nagai, Vikhlinin, and Kravtsov, 2007). The simulated clusters show qualitatively the same trend as that observed at $z = 0$ (see Figure 7.2), but there is a clear evolution of f_g for the given M. The redshift dependence is almost completely removed (see Figure 7.3b), if we scale the cluster masses by M_*, the mass scale corresponding to a linear matter density fluctuation amplitude of 1.686 (Vikhlinin et al., 2009a). Thus M_* is determined by $\sigma(M_*) = 1.686$, where $\sigma(M)$ is the rms fluctuation of the matter density field smoothed with a top hat filter containing mass M (see Eq. (7.2)). Therefore, $f_g(M, z)$ can be estimated by fitting it at $z = 0$ to the observed f_g, and evolved to $z > 0$ assuming that $f_g(M/M_*)$ is independent of redshift (Vikhlinin et al., 2009a).

The Cluster Mass and Temperature Relation

The average X-ray temperature is one of the most widely used cluster mass indicators. The M–T relation expected in self-similar theory is given by

$$M_{500} \propto T^{3/2} E(z)^{-1}, \quad E(z) \equiv H(z)/H_0. \tag{7.11}$$

This relation can be understood easily since the ICM temperature is expected to scale with the depth of gravitational potential, $T \propto M/R$, and the cluster mass definition of Eq. (7.6) yields $R \propto M^{1/3}$. In addition, Eq. (7.11) generally describes the ICM temperatures found in the cosmological numerical simulations (Evrard, Metzler, and Navarro, 1996; Mathiesen and Evrard, 2001; Borgani et al., 2004; Kravtsov, Vikhlinin, and Nagai, 2006). It is most practical to define the average cluster temperature as the average X-ray spectral temperature, T_X – the value derived from a single temperature fit to the total cluster spectrum integrated within a given radial range.

Vikhlinin et al. (2009a) found that for the 17 low-redshift relaxed clusters with very high quality *Chandra* observations, the temperature profiles extend sufficiently far to permit hydrostatic mass estimates at $r = r_{500}$. The mass and temperature measurements for these 17 clusters (see Figure 7.4; note that the error bars are symmetrized for simplicity) are fit to the power law,

$$M = M_0 E(z)^{-1} (T/5\,\text{keV})^\alpha, \tag{7.12}$$

normalized at $T = 5$ keV because this is approximately the median temperature for this sample, and therefore the estimates for M and α should be uncorrelat-

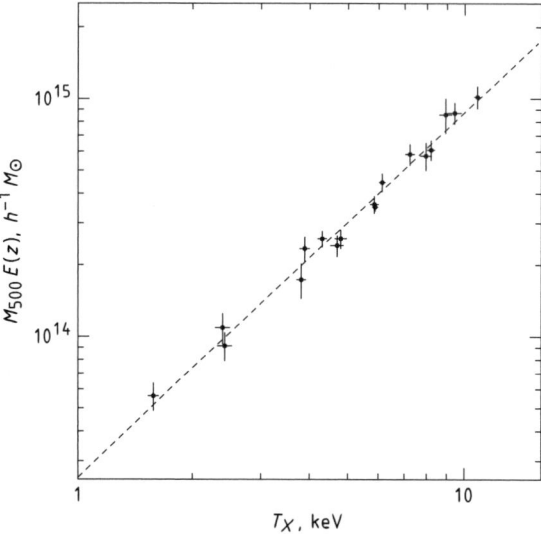

Figure 7.4 Calibration of the M–T relation using X-ray hydrostatic mass measurements for a sample of 17 relaxed *Chandra* clusters with the temperature profile measurements extending to $r = r_{500}$ (Vikhlinin *et al.*, 2006, 2009a). The dashed line shows the best-fit power law relation.

ed. The dashed line in Figure 7.4 shows the best-fit power law relation with a slope of 1.53 ± 0.08 (consistent with the expectation of the self-similar theory, see Eq. (7.11)).

Note that there is a systematic offset in the normalization of the $M_{\text{tot}} - T_X$ relation for relaxed and unrelaxed clusters; merging clusters tend to have lower temperatures for the same mass (Mathiesen and Evrard, 2001; Kravtsov, Vikhlinin, and Nagai, 2006; Ventimiglia *et al.*, 2008). This systematic offset cannot be measured directly using the X-ray data, and will ultimately be measured with a weak lensing analysis of a large sample (not yet available at present). A correction must be made to transfer the $M_{\text{tot}} - T_X$ calibration to the entire data set that contains both relaxed and merging clusters. Numerical simulations indicate that the offset is $(17 \pm 5)\%$ in mass for a fixed T_X; the T_X-based mass estimates for clusters identified as mergers should be corrected upwards by a factor of 1.17, with statistical uncertainties of ± 0.05 (Kravtsov, Vikhlinin, and Nagai, 2006). There is no obvious trend of this offset with redshift, or the difference in the slope of the relations for relaxed and merging clusters.

Observed clusters can be classified as relaxed or unrelaxed using the morphology of their X-ray images, equivalent to the classification scheme used in simulations. "Unrelaxed" clusters are those with secondary maxima, filamentary X-ray structures, or significant isophotal centroid shifts (Vikhlinin *et al.*, 2009a). Ventimiglia *et al.* (2008) showed that the deviations from the mean $M_{\text{tot}} - T_X$ relation in the simulated clusters are correlated with the quantitative substructure measures.

The Cluster Mass and Y_X Relation

The tightest cluster mass and observable relation is the $M_{tot} - Y_X$ relation, proposed by Kravtsov, Vikhlinin, and Nagai (2006), with Y_X defined as

$$Y_X \equiv T_X M_{gas,X} , \qquad (7.13)$$

where T_X is the temperature derived from fitting the cluster X-ray spectrum integrated within the projected radii $0.15 r_{500} - 1 r_{500}$, and $M_{gas,X}$ is the hot gas mass within the sphere r_{500}, derived from the X-ray image.

Y_X approximates the total thermal energy of the ICM within r_{500}, and also the integrated low-frequency Sunyaev–Zeldovich flux (Sunyaev and Zeldovich, 1972). Hydrodynamic simulations have found that the total thermal energy, Y, is a very good indicator of the total cluster mass (daSilva et al., 2004; Motl et al., 2005; Hallman et al., 2006; Nagai, 2006). In the simplest self-similar model (Kaiser, 1986), Y scales with the cluster mass as

$$M_{tot} \propto Y^{3/5} E(z)^{-2/5} , \qquad (7.14)$$

which follows from Eqs. (7.8), (7.11), and (7.13). This scaling is a consequence of the expected evolution in the $M_{tot} - T$ relation (Eq. (7.11)) and the assumption of the self-similar model that f_g is independent of cluster mass. This scaling has been verified by hydrodynamic simulations (Kravtsov, Vikhlinin, and Nagai, 2006).

The main reason for the validity of Eq. (7.14) is that the total thermal energy of the ICM is not strongly disturbed by cluster mergers (Poole et al., 2007), unlike T_X or X-ray luminosity (Ricker and Sarazin, 2001). In addition, the $M_{tot} - Y$ scaling also appears to be not very sensitive to the effects of gas cooling, star formation, and energy feedback (Nagai, 2006).

The X-ray proxy of Y, Y_X, is potentially even more stable with respect to cluster mergers than the "true" Y (Kravtsov, Vikhlinin, and Nagai, 2006). After a merger, the cluster temperature (thus the total thermal energy Y) is biased somewhat low because of incomplete dissipation of bulk ICM motions. Meanwhile, the same bulk motions cause gas density fluctuations, which leads to an overestimation of M_{gas} from the X-ray analysis (Mathiesen, Evrard, and Mohr, 1999). Thus the merger-induced deviations of the average temperature and derived M_{gas} are anti-correlated and hence partially canceled out in Y_X. Hydrodynamic simulations found that there is no detectable systematic offset in the normalization of the $M_{tot} - Y_X$ relations for relaxed and unrelaxed clusters, with an upper limit of 4% in the difference in M_{tot} for fixed Y_X within a typical simulated sample (Kravtsov, Vikhlinin, and Nagai, 2006).

Note that in using Y_X as a mass indicator, Y_X should be determined within r_{500}, which is itself unknown. Therefore r_{500} and hence M_{500} can be found by solving (see Eq. (7.6))

$$C_Y (T_X M_{gas}(r))^{\alpha_Y} E(z)^{-2/5} = 500 \times (4/3)\pi r^3 \rho_c(z) , \qquad (7.15)$$

where C_Y and α_Y are the parameters of the power law approximation to the $M_{tot} - Y_X$ relation:

$$M_{tot} = C_Y Y_X^{\alpha_Y} E(z)^{-2/5} . \qquad (7.16)$$

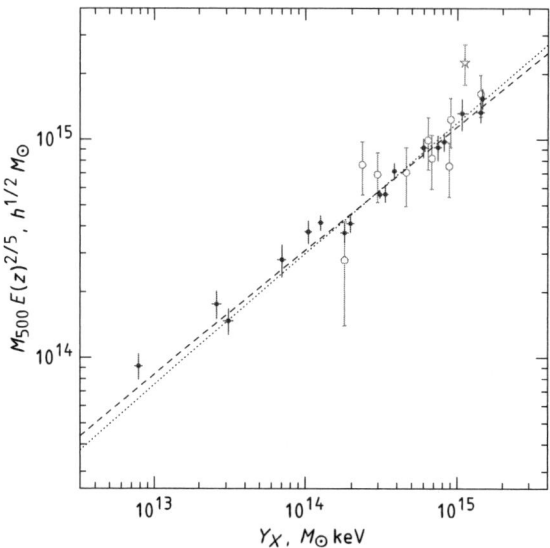

Figure 7.5 Calibration of the $M_{tot} - Y_X$ relation. Solid points with error bars represent *Chandra* results for 17 clusters (Vikhlinin et al., 2006, 2009a). Dashed line shows a power law fit (excluding the lowest-mass cluster) with the free slope. Dotted line shows the fit with the slope fixed at the self-similar value, 3/5. Open points show weak lensing measurements from Hoekstra (2007) (these data are not used in the fit; the strongest outlier is A1689 (open star), a known case of large scale structures superposed along the line of sight.

Figure 7.5 shows the calibration of the $M_{tot} - Y_X$ relation. Solid points with error bars show *Chandra* results for 17 clusters (Vikhlinin et al., 2006, 2009a). The total mass and Y_X measurements are most uncertain for the lowest-temperature cluster. Dashed line shows a power-law fit (excluding the lowest-temperature cluster)

$$M E(z)^{2/5} \propto Y_X^{0.57 \pm 0.05} . \tag{7.17}$$

consistent with the self-similar relation (shown by a dotted line). Open points show weak lensing measurements from Hoekstra (2007); these data are not used in the fit but are shown for verification of mass estimates.

7.2.3
Cluster Abundance Estimation

The cumulative cluster abundance can be computed as

$$N(> M) = \sum_{M_i > M} V(M_i)^{-1} , \tag{7.18}$$

where $V(M_i)$ is the survey volume that corresponds to the cluster with mass M_i. The CCCP samples span similar mass at low and high redshifts, which is very important for controlling systematic uncertainties (Vikhlinin et al., 2009a).

The sample volumes can be computed as a function of X-ray luminosity:

$$V(L_X) = \int_{z_1}^{z_2} A(f_x, z) \frac{dV}{dz} dz , \tag{7.19}$$

where f_x is the X-ray flux corresponding to the object with luminosity L_X at redshift z, dV/dz is the cosmological volume-redshift relation, and $A(f_x, z)$ is the effective survey area for such objects. The cluster luminosity and flux are related by

$$f_x = \frac{L_X}{4\pi d_L(z)^2} K(z) , \tag{7.20}$$

where $d_L(z)$ is the bolometric distance, and $K(z)$ is the K-correction factor (Jones et al., 1998).

To fit cluster abundance models to data, we need to know the survey volume as a function of mass, not luminosity. If there is a well-defined relation between the cluster mass and its X-ray luminosity, then

$$\frac{dV(M)}{dz} = \int_{L_X} \frac{dV(L)}{dz} P(L|M, z) dL , \tag{7.21}$$

where $P(L_X|M, z)$ is the probability for a cluster with mass M to have a luminosity L_X at redshift z. Thus the volume in the given redshift interval is given by

$$V(M) = \int_{z_1}^{z_2} dz \int_{L_X} A(f_x, z) \frac{dV}{dz} P(L|M, z) dL . \tag{7.22}$$

The simplest model that seems to adequately describe the observed $M_{\text{tot}} - L_X$ relations is given by

$$P(\ln L|M) \propto \exp\left[-\frac{(\ln L - \ln L_0)^2}{2\sigma^2}\right], \tag{7.23}$$

where

$$L_0 = C(z) M^\alpha . \tag{7.24}$$

The evolution factor $C(z)$ can be determined empirically for each assumed cosmological model (Vikhlinin et al., 2009a). It is often assumed to be of the following forms:

$$C(z) = C_0(1+z)^\gamma \quad \text{or} \quad C(z) = C_0 E(z)^\gamma . \tag{7.25}$$

The $L_X - M$ relation is characterized by four parameters, C_0, α, γ, and σ, which can be determined using mass estimates for clusters in the sample (Vikhlinin et al.,

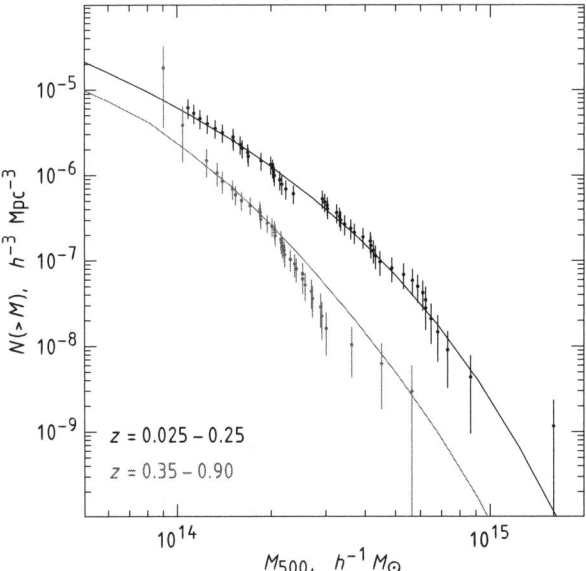

Figure 7.6 Cluster mass functions for the CCCP low-z and high-z samples (upper and lower set of data points) from Vikhlinin et al. (2009a). The masses were estimated using the Y_X method. The error bars show the Poisson uncertainties. Solid lines show the model predictions for the adapted cosmological model $\Omega_m = 0.3$, $\Omega_\Lambda = 0.7$, $h = 0.72$, with only σ_8 fitted to the cluster data. The evolution of the mass function is nonnegligible within either redshift range. To take this into account, the model number densities for each mass were weighted with $dV(M)/dz$ (Eq. (7.21)) within the redshift bin.

2009a). The best fit to Eqs. (7.24) and (7.25) for the CCCP sample is (Vikhlinin et al., 2009a)

$$\ln L_X = (47.392 \pm 0.085) + (1.61 \pm 0.14) \ln M_{500} + (1.850 \pm 0.42) \ln E(z)$$
$$- 0.39 \ln(h/0.72) \pm (0.396 \pm 0.039) ,$$

(7.26)

where the last term on the right hand side indicates the observed scatter in L_X for fixed M.

Figure 7.6 shows the cluster abundances for the CCCP low-z and high-z samples, with cluster masses estimated using the Y_X method (Vikhlinin et al., 2009a). Solid lines show the model predictions for the fiducial cosmological model $\Omega_m = 0.3$, $\Omega_\Lambda = 0.7$, $h = 0.72$, with only σ_8 fitted to the cluster data. The strongest observed deviation of the data from the model is a marginal deficit of clusters in the distant sample near $M_{500} = 3 \times 10^{14} \, h^{-1} \, M_\odot$ – 4 clusters were observed where 9.5 are expected, a 2σ deviation. The differential cluster abundances from the CCCP low-z and high-z samples indicate that this deficit is consistent with the Poisson noise expected in the data (Vikhlinin et al., 2009a).

7.2.4
Cosmological Parameters Constraints

Figure 7.7 illustrates the sensitivity of the cluster abundance measured by the CC-CP to the cosmological model, with the two panels corresponding to two different cosmological models. Figure 7.7a corresponds to a flat model with $\Omega_\Lambda = 0.75$, while Figure 7.7b corresponds to an open model with the same Ω_m, but with $\Omega_\Lambda = 0$. In each panel, the upper and lower curves and data points correspond to the low-z and high-z samples, respectively. The overall model normalization is adjusted to the low-z mass function in both panels. Both the model and the data at high redshifts differ for the two cases. The measured mass function depends on the distance-redshift relation (which differs for the two cosmological models). The model mass function depends on the growth of structure and overdensity thresholds corresponding to $\Delta_{\text{crit}} = 500$ (which differ for the two cosmological models). While the flat model with $\Omega_\Lambda = 0.75$ appears in excellent agreement with data, the open model with the same Ω_m but $\Omega_\Lambda = 0$ fails to reproduce the number density of $z > 0.55$ clusters (Vikhlinin et al., 2009b).

Quantitative constraints on cosmological parameters require a likelihood analysis. Since the number of clusters in a current sample is limited, Poisson statistics applies. The mass intervals can be divided into narrow bins, ΔM, so that the probability to observe a cluster with an estimated mass in this bin is small, $p(M^{\text{est}}, z)\Delta M \ll 1$, and there is at most one cluster per bin. The likelihood function in this case can be written as (Cash, 1979)

$$\ln L = \sum_i \ln\left(p(M_i^{\text{est}}, z_i)\Delta M_i\right) - \iint_{M,z} p(M^{\text{est}}, z) \mathrm{d}M^{\text{est}} \mathrm{d}z \,, \tag{7.27}$$

where the summation is over the clusters in the sample and the integration is over pre-selected $z_{\min} - z_{\max}$ and $M_{\min} - M_{\max}$ intervals. Note that the estimated masses depend on the background cosmology. When M_i^{est} is changed because of the variation in the cosmological parameters, we should correspondingly stretch the mass interval, $\Delta M = \Delta M^{(0)} M/M^{(0)}$, where $M^{(0)}$ and $\Delta M^{(0)}$ are the estimated mass and width of the interval for some fixed reference cosmological model. Hence

$$\ln L = \sum_i \ln p(M_i^{\text{est}}, z_i) + \sum_i \ln M_i^{\text{est}} - \iint_{M,z} p(M^{\text{est}}, z) \mathrm{d}M^{\text{est}} \mathrm{d}z \,, \tag{7.28}$$

where the constant term containing $M^{(0)}$ and $\Delta M^{(0)}$ have been dropped.

The probability density distribution of the observed masses is given by convolution of the model distribution of the true masses and the scatter between M^{est} and M^{true} (Vikhlinin et al., 2009a):

$$p(M^{\text{est}}, z) = \left(\frac{\mathrm{d}n}{\mathrm{d}M^{\text{true}}} \frac{\mathrm{d}V(M^{\text{true}}, z)}{\mathrm{d}z}\right) \otimes \text{scatter}(M^{\text{est}}, M^{\text{true}}) \,, \tag{7.29}$$

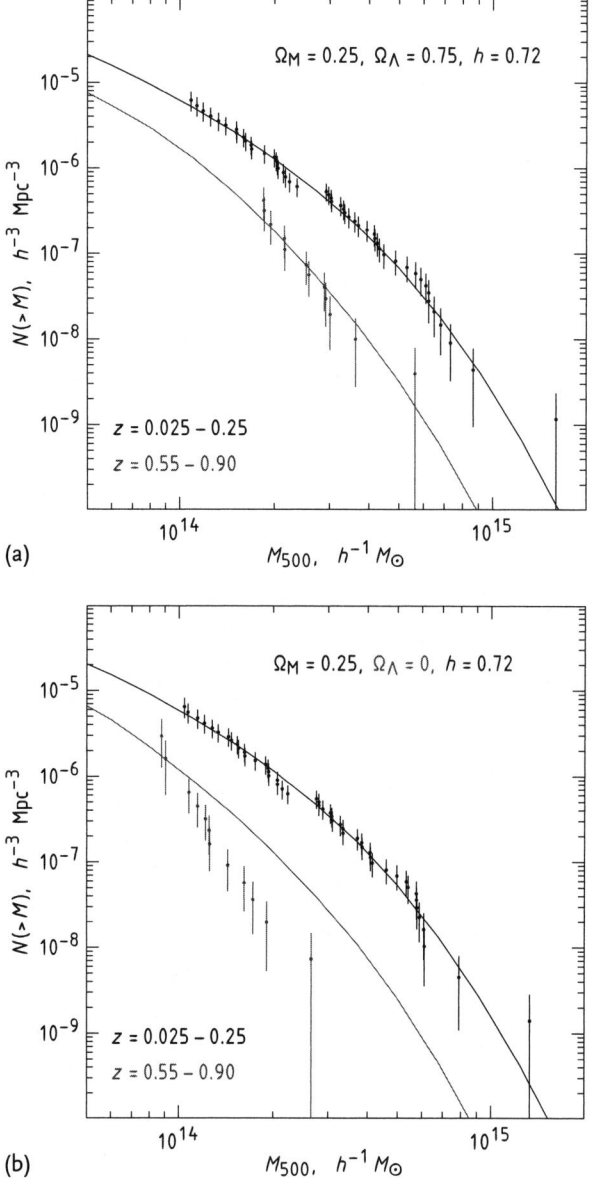

Figure 7.7 Illustration of the sensitivity of the cluster abundance measured by the CCCP to the cosmological model (Vikhlinin et al., 2009b). (a) shows the measured cluster abundance and model predictions (with only the overall normalization at $z = 0$ adjusted) computed for a cosmology which is close to the best-fit model. The low-z mass function is the same as in Figure 7.6, while for the high-z clusters only the most distant subsample ($z > 0.55$) is shown. (b) shows the measured cluster mass function and model predictions for a cosmology with $\Omega_\Lambda = 0$.

where the model distribution of the true masses is given by the product of the theoretical mass function, dn/dM^{true}, and the survey volume at this redshift, $dV(M^{\text{true}}, z)/dz$ (see Eq. (7.21)). A log-normal distribution is a good approximation for the scatter in the mass estimates, and so the convolution in Eq. (7.29) can be written as (Vikhlinin et al., 2009a)

$$p(M^{\text{est}}, z) = \frac{1}{M^{\text{est}}} \frac{1}{\sqrt{2\pi} \, \sigma^{\text{est}}} \int_{-\infty}^{\infty} \frac{dn}{d\ln M^{\text{true}}} \frac{dV(M^{\text{true}}, z)}{dz}$$
$$\cdot \exp\left(-\frac{(\ln M^{\text{est}} - \ln M^{\text{true}})^2}{2(\sigma^{\text{est}})^2}\right) d\ln M^{\text{true}}. \tag{7.30}$$

The function $p(M^{\text{est}}, z)$ enters the expression for the likelihood function in the summation over observed clusters (first term in Eq. (7.28)) and in the integral over the observed range (second term in the same equation). The term $dn/d\ln M^{\text{true}}$ is the differential cluster mass function at the given redshift. Cosmological parameters enter the calculation of $dV(M^{\text{true}}, z)$ through the volume-redshift relation and the evolving cluster $L_X - M$ relation which is derived using L_X and M_{tot} estimated in the assumed cosmological model.

The likelihood function implicitly depends on the cosmological parameters through the model of cluster mass function (reflecting the growth, normalization, and shape of the matter density perturbation power spectrum $P_m(k)$), through the cosmological volume-redshift relation which determines the survey volume, and through the distance-redshift and $E(z)$ relations which affect the cluster mass estimates. The best-fit parameters are obtained by maximizing the likelihood function to estimate uncertainties of the model parameters. The advantage of this approach is that no binning in either mass or redshift is used.

The local sample of CCCP gives strong constraints on $\Omega_m h$ and σ_8, the amplitude of linear perturbations at the scale of $8\,h^{-1}$ Mpc. Note that σ_8 measures the variance of linear matter density fluctuations in spheres of radius $8\,h^{-1}$ Mpc. Since

$$\left(\frac{\delta\rho}{\rho}\right)^2 = \int_0^\infty \frac{dk}{k} \Delta^2(k), \quad \Delta^2(k) \equiv \frac{k^3 |\delta_k|^2}{2\pi^2} \equiv \frac{k^3 P(k)}{2\pi^2}, \tag{7.31}$$

σ_8 is given by

$$\sigma_8^2 = \int_0^\infty \frac{dk}{k} \Delta^2(k) \left[\frac{3 j_1(kr)}{kr}\right]^2, \tag{7.32}$$

with $r = 8\,h^{-1}$ Mpc, and $j_1(x)$ is spherical Bessel function.

The fit to the shape of the CCCP local cluster mass function gives $\Omega_m h = 0.184 \pm 0.035$ (including uncertainties in the L–M power law slope α, see Eq. (7.24)), with the primordial matter power spectrum power law index n_S fixed at $n_S = 0.95$. The normalization of the cluster mass function is exponentially sensitive to σ_8, but

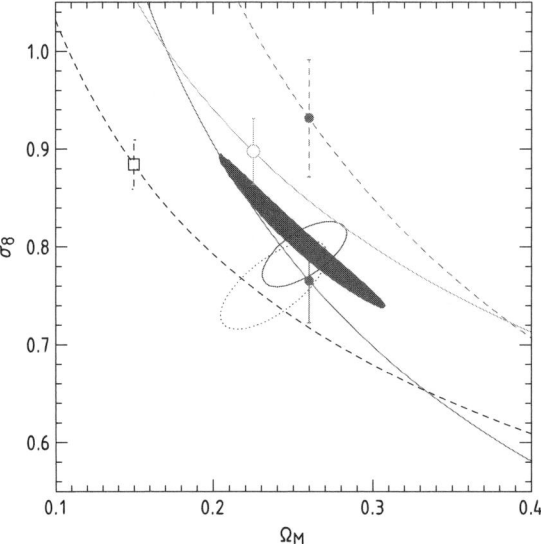

Figure 7.8 Comparison of the σ_8 measurement from clusters with other measurements (Vikhlinin et al., 2009b). Solid region is the 68% confidence level region (corresponding to $\Delta\chi^2 = 1$) from the CCCP cluster data. Hollow oval contours show the WMAP three and five-year results (dotted and solid contours, respectively) (Spergel et al., 2007; Dunkley et al., 2008). For other measurements, the general direction of degeneracy is shown as a solid line and a 68% uncertainty in σ_8 at a representative value of Ω_m is indicated. Filled circles show the weak lensing shear results from Hoekstra et al. (2006) and Fu et al. (2008) (dashed and solid lines, respectively). Open circle shows results from a cluster sample with galaxy dynamics mass measurements (Rines, Diaferio, and Natarajan, 2007). Finally, open squares represents the results from Reiprich and Böhringer (2002), approximately the lower bound of recently published X-ray cluster measurements.

this dependence is degenerate with that of Ω_m, since the mass to length scale correspondence is mainly determined by Ω_m. Figure 7.8 shows the joint constraints on σ_8 and Ω_m from a likelihood analysis of the CCCP local cluster mass function, assuming a flat universe, with the HST prior on h ($h = 0.72 \pm 0.08$ (Freedman et al., 2001)) to help break the degeneracy between Ω_m and h, and then marginalized over h (Vikhlinin et al., 2009b). Figure 7.8 also shows other σ_8 measurements for comparison.

Figure 7.9 shows constraints on Ω_X and a constant dark energy equation of state parameter w_0 derived from cluster mass function evolution, assuming a flat universe, and with the HST prior on h (Vikhlinin et al., 2009b). The marginalized constraints are $\Omega_X = 0.75 \pm 0.04$ and $w_0 = -1.14 \pm 0.21$. The cluster results compare favorably with those from other individual methods (SNe Ia, BAO, and CMB). Combining cluster data with SNe Ia, BAO, and CMB data gives $w_0 = -0.991 \pm 0.045$ (± 0.04 systematic) and $\Omega_X = 0.740 \pm 0.012$ (Vikhlinin et al., 2009b). Figure 7.10 shows constraints on $w_X(z) = w_0 + w_a z/(1+z)$ in a flat universe (Vikhlinin et al., 2009b).

7.3 X-Ray Gas Mass Fraction as a Dark Energy Probe

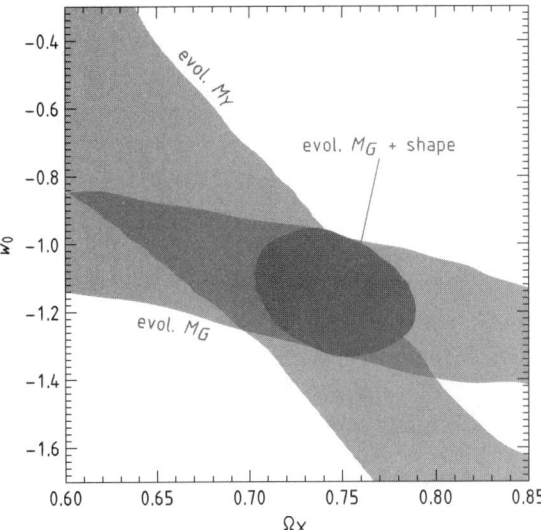

Figure 7.9 Constraints on Ω_X and a constant dark energy equation of state parameter w_0 derived from cluster mass function evolution in a spatially flat universe (Vikhlinin et al., 2009b). The results for M_{gas} and Y_X-based total mass estimates are shown separately. The inner dark region shows the effect of adding the mass function shape information to the evolution of the M_{gas}-based mass function.

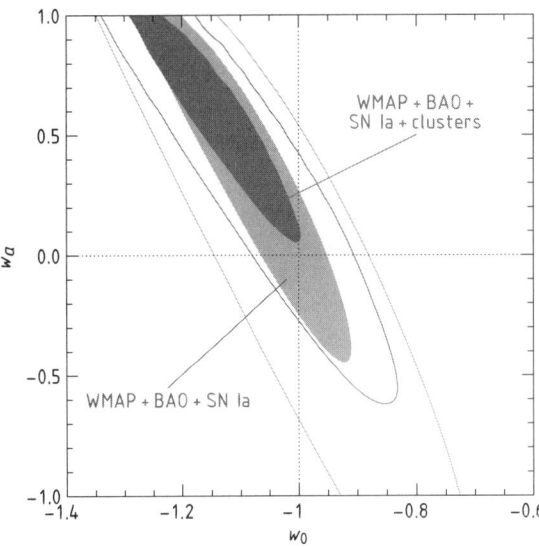

Figure 7.10 Constraints on dark energy equation of state, $w_X(z) = w_0 + w_a z/(1+z)$, including cluster data (Vikhlinin et al., 2009b). A flat universe is assumed.

7.3
X-Ray Gas Mass Fraction as a Dark Energy Probe

The measured X-ray gas mass fraction, f_{gas}, can be modeled and used to derive constraints on dark energy (Allen et al., 2008; Samushia and Ratra, 2008). Figure 7.11 shows the X-ray gas mass fraction profiles for six low redshift clusters ($z \lesssim 0.15$), assuming a ΛCDM reference cosmology with $\Omega_m = 0.3$, $\Omega_\Lambda = 0.7$, and $h = 0.7$ (Allen et al., 2008). The radial axes are scaled in units of r_{2500} in the reference ΛCDM cosmology, corresponding to an angle $\theta_{2500}^{\Lambda CDM}$, for which the mean enclosed mass density is 2500 times the critical density of the universe at the redshift of the cluster. Note that the profiles tend to a common value at r_{2500}, which corresponds to about one quarter of the virial radius. Thus the r_{2500} scale is sufficiently large to provide small systematic scatter, but not too large such that systematic uncertainties in the background modeling become important.

Figure 7.12 shows the apparent variation of the X-ray gas mass fraction measured within r_{2500} as a function of redshift for the reference ΛCDM (Figure 7.12a) and reference SCDM ($\Omega_m = 1.0$, $\Omega_\Lambda = 0.0$, $h = 0.5$) (Figure 7.12b) cosmologies (Allen et al., 2008). The cosmological constraints are derived assuming that f_{gas} is approximately constant with redshift.

The cosmological constraints are obtained in a likelihood analysis by fitting the reference ΛCDM measurements of f_{gas} to a model given by

$$f_{gas}^{\Lambda CDM}(z) = \frac{KA\gamma b_d(z)}{1 + s_*(z)}\left(\frac{\Omega_b}{\Omega_m}\right)\left[\frac{d_A^{\Lambda CDM}(z)}{d_A(z)}\right]^{1.5}, \qquad (7.33)$$

where $d_A(z)$ and $d_A^{\Lambda CDM}(z)$ are the angular diameter distances to the clusters in the current test model and the reference cosmology. The factor K is a "calibration" constant that parameterizes the residual uncertainty in the accuracy of the instrument calibration and X-ray modeling, assumed to be $K = 1.0 \pm 0.1$. The factor A accounts for the change in angle subtended by r_{2500} as the underlying cosmology

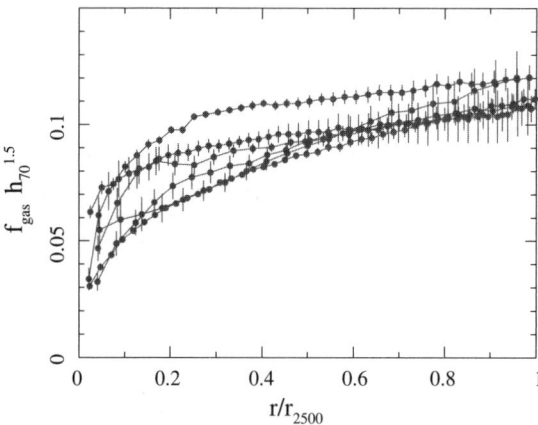

Figure 7.11 The X-ray gas mass fraction profiles for six low-redshift clusters ($z \lesssim 0.15$), assuming a ΛCDM reference cosmology with $\Omega_m = 0.3$, $\Omega_\Lambda = 0.7$, $h = 0.7$ (Allen et al. (2008), MNRAS, Wiley-Blackwell). The radial axes are scaled in units of r_{2500}.

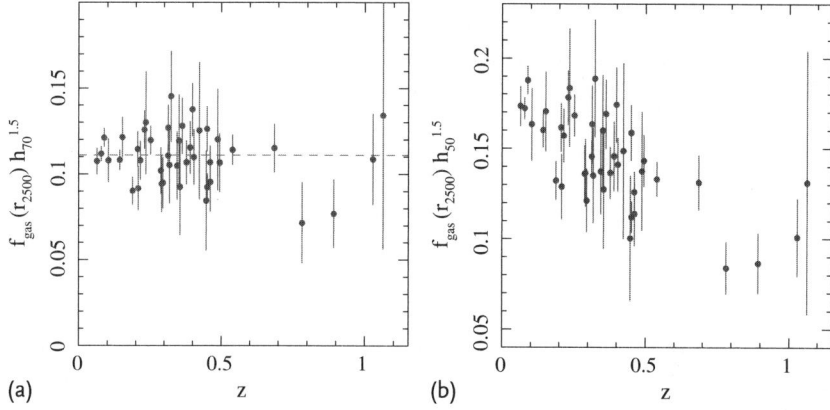

Figure 7.12 The apparent variation of the X-ray gas mass fraction measured within r_{2500} as a function of redshift for the reference ΛCDM (a) and the reference SCDM ($\Omega_m = 1.0$, $\Omega_\Lambda = 0.0$, $h = 0.5$) (b) cosmologies (Allen et al. (2008), MNRAS, Wiley-Blackwell). The plotted error bars are statistical root-mean-square 1σ uncertainties. The global, absolute normalization of the f_{gas} values should be regarded as uncertain at the ~ 10–15 per cent level due to systematic uncertainties in instrument calibration, modeling and the level of nonthermal pressure support.

is varied,

$$A = \left(\frac{\theta_{2500}^{\Lambda\text{CDM}}}{\theta_{2500}}\right)^\eta \approx \left(\frac{H(z)d_A(z)}{[H(z)d_A(z)]^{\Lambda\text{CDM}}}\right)^\eta, \tag{7.34}$$

where $\eta = 0.214 \pm 0.022$. The parameter γ models nonthermal pressure support in the clusters, assumed to satisfy $1.0 < \gamma < 1.1$.

The factor $b_d(z) = b_0(1 + \alpha_b z)$ in Eq. (7.33) is the "depletion" factor; the ratio by which the baryon fraction measured at r_{2500} is depleted with respect to the universal mean. Such depletion is a natural consequence of the thermodynamic history of the gas. The nonradiative simulations of hot, massive clusters give $b_0 = 0.83 \pm 0.04$ at r_{2500}, and are consistent with no redshift evolution in b_d for $z < 1$ (Eke, Navarro, and Frenk, 1998; Crain et al., 2007). A range of $-0.1 < \alpha_b < 0.1$ encompasses a range of evolution allowed by recent simulations including various approximations to the detailed baryonic physics (Kay et al., 2004; Ettori et al., 2006; Crain et al., 2007; Nagai, Vikhlinin, and Kravtsov, 2007).

The parameter $s_*(z) = s_0(1 + \alpha_s z)$ in Eq. (7.33) models the average baryonic mass fraction in stars (in galaxies and intra-cluster light combined), with $s_0 = (0.16 \pm 0.05)(h/0.7)^{0.5}$ (Lin and Mohr, 2004), and $-0.2 < \alpha_s < 0.2$ (to allow for evolution in the stellar baryonic mass fraction of $\pm 20\%$ per unit redshift interval).

Using the method outlined here, Allen et al. (2008) analyzed the Chandra f_{gas} data, and found that the cluster f_{gas} data give competitive constraints on dark energy compared to other methods. Future prospects for using f_{gas} to probe dark energy can be found in Rapetti, Allen, and Mantz (2008).

7.4
Systematic Uncertainties and Their Mitigation

The dominant systematic uncertainties of clusters as dark energy probe are the uncertainties in the cluster mass estimates that are derived from observed properties, such as X-ray or optical luminosities and temperature (Majumdar and Mohr, 2003, 2004; Lima and Hu, 2004). Section 7.2.2 describes the three methods that have been used to estimate cluster masses, and illustrates the systematic uncertainties involved in each method.

One unique aspect in the cluster method for probing dark energy is the dependence on hydrodynamic simulations to assess calibration relations that cannot be established by observational data without assuming a cosmological model. For example, the redshift dependence of the cluster gas fraction $f_g(M, z)$ can only be estimated using hydrodynamic simulations (Kravtsov, Nagai, and Vikhlinin, 2005; Vikhlinin *et al.*, 2009a). On the other hand, the best proxy for cluster mass, $Y_X \equiv T_X M_{gas,X}$ (both the cluster X-ray temperature T_X and the cluster gas mass $M_{gas,X}$ can be inferred from X-ray observations), was found through hydrodynamic simulations (Kravtsov, Vikhlinin, and Nagai, 2006). Figure 7.13 shows the total cluster mass versus the proxy for cluster mass, Y_X, based on simulated X-ray data (Kravtsov, Vikhlinin, and Nagai, 2006). This mass proxy has subsequently been validated by observational data, see Figure 7.5 (Vikhlinin *et al.*, 2009a). Thus the dependence on hydrodynamic simulations is both a source of systematic errors, and a tool with which cluster mass estimates can be improved.

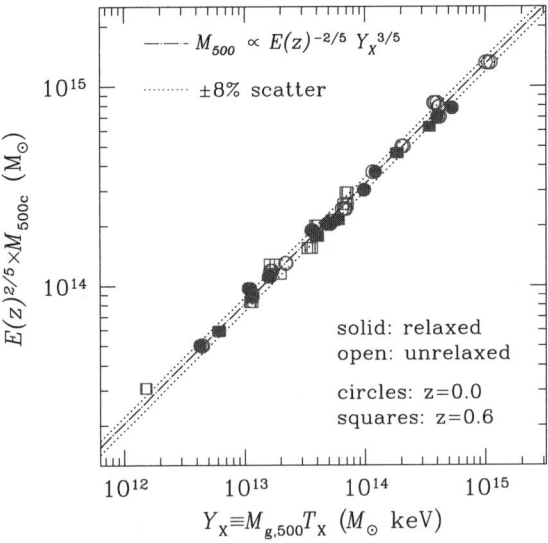

Figure 7.13 Total cluster mass versus $Y_X \equiv T_X M_{gas,X}$, a proxy based on the cluster gas mass and temperature (both of which can be inferred from X-ray observations), based on simulated X-ray data (Kravtsov, Vikhlinin, and Nagai, 2006).

Large, well-defined and statistically complete samples of galaxy clusters are required to derive robust dark energy constraints from cluster data. Future surveys aim to select clusters using data from X-ray satellites with high resolution and wide sky coverage, and multi-band optical and NIR surveys to obtain photometric redshifts for clusters (Haiman, Mohr, and Holder, 2001). In addition, weak lensing observations of a large number of clusters will help calibrate the cluster mass and observable relations.

8
Other Observational Methods for Probing Dark Energy

8.1
Gamma Ray Bursts as Cosmological Probe

Gamma-ray bursts (GRBs) are the most luminous astrophysical events observable today, because they are at cosmological distances (Paczynski, 1995). The duration of a gamma-ray burst is typically a few seconds, but can range from a few milliseconds to several minutes. The initial burst at gamma-ray wavelengths is usually followed by a longer-lived afterglow at longer wavelengths (X-ray, ultraviolet, optical, infrared, and radio). Gamma-ray bursts have been detected by orbiting satellites about two to three times per week. Most observed GRBs appear to be collimated emissions caused by the collapse of the core of a rapidly rotating, high-mass star into a black hole. A subclass of GRBs (the "short" bursts) appear to originate from a different process, the leading candidate being the collision of neutron stars orbiting in a binary system. See Meszaros (2006) for a recent review on GRBs.

GRBs can be used as distance indicators. The main advantage of GRBs over SNe Ia is that they span a much greater redshift range, from low z to $z > 8$; Tanvir et al. (2009) reported the discovery of a GRB at $z \simeq 8.3$. The main disadvantage is that GRBs have to be calibrated for each cosmological model tested. This is in contrast to SNe Ia, where the calibration relations can be established using nearby SNe Ia, and applied to high z SNe Ia to extract cosmological constraints. There are no nearby GRBs that can be used for calibration.[12] Thus, the GRB data must be fitted simultaneously for calibration and cosmological parameters. To avoid circular logic, a likelihood analysis must be performed fitting the observables to a set of calibration parameters *and* cosmological parameters that enter through $d_L(z)$ (Schaefer, 2004).

8.1.1
Calibration of GRBs

Long-duration gamma-ray bursts (GRBs) have eight luminosity relations relating the observable burst properties to luminosity (see, e.g., Schaefer and Collazzi (2007)). Some of these relations are problematic when GRB redshifts are un-

---

12) A nearby GRB may cause extinction of life on Earth.

known (Nakar and Piran, 2005; Li, 2007). Following Schaefer (2007), we only consider five calibration relations for GRBs here. These relate GRB luminosity, L, or the total burst energy in the gamma rays, E_γ, to observables of the light curves and/or spectra: τ_{lag} (time lag), V (variability), E_{peak} (peak of the νF_ν spectrum), and τ_{RT} (minimum rise time) (Schaefer, 2007):

$$\log\left(\frac{L}{1\,\text{erg s}^{-1}}\right) = a_1 + b_1 \log\left[\frac{\tau_{\text{lag}}(1+z)^{-1}}{0.1\,\text{s}}\right], \tag{8.1}$$

$$\log\left(\frac{L}{1\,\text{erg s}^{-1}}\right) = a_2 + b_2 \log\left[\frac{V(1+z)}{0.02}\right], \tag{8.2}$$

$$\log\left(\frac{L}{1\,\text{erg s}^{-1}}\right) = a_3 + b_3 \log\left[\frac{E_{\text{peak}}(1+z)}{300\,\text{keV}}\right], \tag{8.3}$$

$$\log\left(\frac{E_\gamma}{1\,\text{erg}}\right) = a_4 + b_4 \log\left[\frac{E_{\text{peak}}(1+z)}{300\,\text{keV}}\right], \tag{8.4}$$

$$\log\left(\frac{L}{1\,\text{erg s}^{-1}}\right) = a_5 + b_5 \log\left[\frac{\tau_{\text{RT}}(1+z)^{-1}}{0.1\,\text{s}}\right]. \tag{8.5}$$

The lag of a GRB, τ_{lag}, is the time shift between the hard and soft light curves, with the soft photons coming somewhat later than the hard photons. The lag-luminosity relation (Band, 1997; Norris, Marani, and Bonnell, 2000) is thought to be caused by the speed of the radiative cooling of the shocked material (and hence the lag) being determined by the luminosity (Schaefer, 2004).

The variability of a GRB, V, is a quantitative measure of whether its light curve is spiky or smooth. It is a measure of the sharpness of the pulse structure, which is determined by the size of the visible region in the jet, which in turn depends on the bulk Lorentz factor of the jet Γ_{jet}. The variability-luminosity relation (Fenimore and Ramirez-Ruiz, 2000) has the largest scatter among the five GRB calibration relations discussed here, and is thought to be caused by the fact that both the variability and luminosity are functions of Γ_{jet}.

E_{peak} is the peak energy of the νF_ν spectrum of the GRB. Not surprisingly, E_{peak} carries the most distance information (Amati et al., 2002; Schaefer, 2003). The E_{peak}–E_γ relation (Ghirlanda, Ghisellini, and Lazzati, 2004; Bloom, Frail, and Kulkarni, 2003; Amati et al., 2002) is the tightest of the GRB calibration relations. To be included in this relation, the GRB afterglow must have an observed jet break in its light curve, and this means that only a fraction of GRBs with redshifts can contribute to establishing this relation. This relation has been used by various groups (Ghirlanda, Ghisellini, and Lazzati, 2004; GRB and cosmology references). Figure 8.1 shows the E_{peak}–E_γ calibration relation for GRBs from Schaefer (2007); it includes new bursts found by the Swift satellite.

The minimum rise time – luminosity relation (τ_{RT}–L) of GRBs was originally predicted because the luminosity depends on a power law of the bulk Lorentz factor of the jet Γ_{jet}, while the minimum rise time in the GRB light curve also depends on Γ_{jet} through the size of the visible region in the shocked jet (Panaitescu and Kumar, 2002; Schaefer, 2002). This observed calibration relation provides a test and confirmation of the prediction that τ_{RT} is correlated with luminosity as a power law.

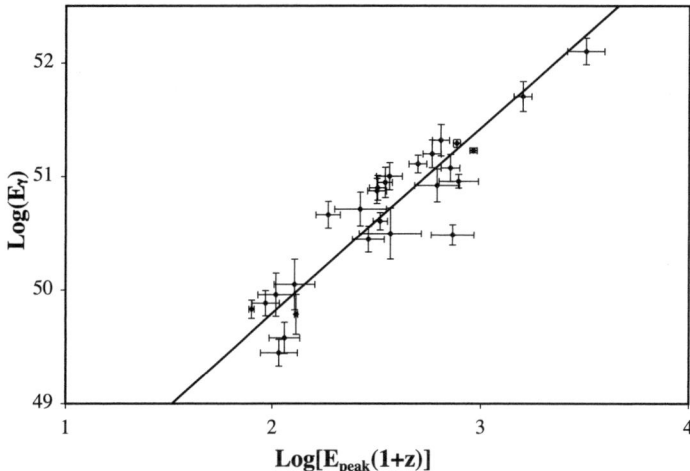

Figure 8.1 $E_{peak} - E_\gamma$ relation of GRBs (Schaefer, 2007). The E_{peak} values for 27 GRBs have been corrected to the restframe of the GRB and plotted versus E_γ (the total burst energy in the γ-rays). The best-fit power law is also shown.

In order to calibrate GRBs, L and E_γ must be related to the observed bolometric peak flux, P_{bolo}, or the bolometric fluence, S_{bolo}:

$$L = 4\pi d_L^2 P_{bolo} \tag{8.6}$$

$$E_\gamma = E_{\gamma,iso} F_{beam} = 4\pi d_L^2 S_{bolo}(1+z)^{-1} F_{beam}, \tag{8.7}$$

where $E_{\gamma,iso}$ is the isotropic energy. Clearly, the calibration of GRBs depends on the cosmological model through the luminosity distance $d_L(z)$.

The cosmological constraints from GRBs are sensitive to how the GRBs are calibrated. Calibrating GRBs using Type Ia supernovae (SNe Ia) gives tighter constraints than calibrating GRBs internally (Liang et al., 2008). However, in order to combine GRB data with any other cosmological data sets, GRBs should be calibrated internally, *without* using any external data sets.

In fitting the five calibration relations, one needs to fit a data array $\{x_i, y_i\}$ with uncertainties $\{\sigma_{x,i}, \sigma_{y,i}\}$, to a straight line

$$y = a + bx \tag{8.8}$$

through the minimization of χ^2 given by (Press et al., 2007)

$$\chi^2 = \sum_{i=1}^{N} \frac{(y_i - a - bx_i)^2}{\sigma_{y,i}^2 + b^2 \sigma_{x,i}^2}. \tag{8.9}$$

It is convenient to define

$$x_i^{(\alpha)} \equiv \log\left(x_{0,i}^{(\alpha)}\right), \tag{8.10}$$

thus

$$x_{0,i}^{(1)} = \frac{\tau_{\text{lag},i}(1+z)^{-1}}{0.1\,\text{s}} \tag{8.11}$$

$$x_{0,i}^{(2)} = \frac{V(1+z)}{0.02} \tag{8.12}$$

$$x_{0,i}^{(3)} = x_{0,i}^{(4)} = \frac{E_{\text{peak},i}(1+z)}{300\,\text{keV}} \tag{8.13}$$

$$x_{0,i}^{(5)} = \frac{\tau_{\text{RT},i}(1+z)^{-1}}{0.1\,\text{s}} \tag{8.14}$$

and

$$y_i^{(1)} = y_i^{(2)} = y_i^{(3)} = y_i^{(5)} = \log\left(\frac{L}{1\,\text{erg s}^{-1}}\right)$$
$$= \log(4\pi P_{\text{bolo},i}) + 2\log\overline{d_L}\,,$$

$$y_i^{(4)} = \log\left(\frac{E_\gamma}{1\,\text{erg}}\right)$$
$$= \log\left[\frac{4\pi S_{\text{bolo},i} F_{\text{beam},i}}{1+z}\right] + 2\log\overline{d_L}\,, \tag{8.15}$$

where we have defined

$$\overline{d_L} \equiv (1+z)H_0 r(z)/c\,. \tag{8.16}$$

Since the absolute calibration of GRBs is unknown, the Hubble constant cannot be derived from GRB data. Thus we have defined the data arrays $\{y_i\}$ such that

$$c/H_0 = 9.2503 \times 10^{27}\,h^{-1}\,\text{cm} \tag{8.17}$$

is absorbed into the overall calibration.

Furthermore, for the E_{peak}–L and E_{peak}–E_γ relations, the measurement error of E_{peak} is asymmetric, thus we need to modify the χ^2 such that

$$\sigma_{x,i} = \sigma_{x,i}^+\,, \quad \text{if } (y_i - a)/b \geq x_i\,;$$

$$\sigma_{x,i} = \sigma_{x,i}^-\,, \quad \text{if } (y_i - a)/b < x_i\,, \tag{8.18}$$

where $\sigma_{x,i}^+$ and $\sigma_{x,i}^-$ are the $\pm$ measurement errors.

As noted by Schaefer (2007), the statistical errors on $\{a_i, b_i\}$ are quite small, but the χ^2's are very large due to the domination of systematic errors. Following Schaefer (2007), we derive the systematic errors by requiring that $\chi^2 = \nu$ (the degrees of freedom), and that $\sigma_{\text{tot}}^2 = \sigma_{\text{stat}}^2 + \sigma_{\text{sys}}^2$.

For illustration on the cosmological parameter dependence of the calibration of GRBs, Table 8.1 shows the systematic errors, as well as the constants $\{a_i, b_i\}$ for

Table 8.1 Systematic errors for the five GRB calibration relations from Eqs. (8.1)–(8.5)

	$\Omega_m = 0.27$	$\Omega_m = 0.2$	$\Omega_m = 0.4$
a_1	-3.901 ± 0.027	-3.848 ± 0.027	-3.979 ± 0.026
b_1	-1.154 ± 0.033	-1.167 ± 0.033	-1.138 ± 0.032
$\sigma_{sys,1}$	0.417	0.427	0.405
a_2	-3.822 ± 0.011	-3.781 ± 0.011	-3.883 ± 0.011
b_2	3.983 ± 0.050	4.017 ± 0.05	3.934 ± 0.049
$\sigma_{sys,2}$	0.927	0.931	0.922
a_3	-3.821 ± 0.010	-3.782 ± 0.010	-3.881 ± 0.010
b_3	1.830 ± 0.027	1.863 ± 0.027	1.787 ± 0.026
$\sigma_{sys,3}$	0.466	0.466	0.467
a_4	-5.612 ± 0.024	-5.574 ± 0.024	-5.672 ± 0.024
b_4	1.452 ± 0.059	1.468 ± 0.060	1.430 ± 0.058
$\sigma_{sys,4}$	0.204	0.200	0.211
a_5	-3.486 ± 0.023	-3.431 ± 0.024	-3.567 ± 0.023
b_5	-1.590 ± 0.044	-1.617 ± 0.044	-1.557 ± 0.042
$\sigma_{sys,5}$	0.591	0.598	0.582

the five calibration relations for $\Omega_m = 0.2, 0.27, 0.4$ for a flat universe with a cosmological constant (Wang, 2008c). Note that the a_i in this table are smaller than the definition of Schaefer (2007) by $2\log(9.2503 \times 10^{27}\, h^{-1})$. Note also that the derived systematic errors change by less than 3% for the different models. Since systematic errors should be independent of the cosmological model, we can take the systematic errors for the $\Omega_m = 0.27$ flat ΛCDM model to be the standard values (Wang, 2008c).

For most of the calibration relations, fitting straight lines using errors in both coordinates does not give significantly different results from fitting straight lines using errors in the y coordinates only. For the V–L relation, assuming the $\Omega_m = 0.27$ flat ΛCDM model, fitting straight lines using errors in the y coordinates only, we find $a_2 = -3.540 \pm 0.003$, $b_2 = 1.649 \pm 0.012$, and $\sigma_{sys,2} = 0.518$. The slope and the systematic error are both significantly smaller than the results shown in Table 8.1. Since the x coordinates have significant measurement errors in all five calibration relations, the latter should be fitted to straight lines using errors in both x and y coordinates (Wang, 2008c).

Note that the calibration relations from Table 8.1 (the parameters a_i and b_i) should *not* be used when we derive model-independent distances from GRBs. Table 8.1 is only used to show that the calibration of GRBs depends on the assumptions about cosmological parameters, but the systematic uncertainties of the calibration parameters a_i and b_i are *not* sensitive to the assumptions about cosmological parameters. Thus, we will derive calibration relations of GRBs for each set of assumed distances, but we will assume that the systematic errors on a_i and b_i are given by the $\Omega_m = 0.27$ flat ΛCDM model.

8.1.2
Model-Independent Distance Measurements from GRBs

Following Schaefer (2007), we weigh the five estimators of distance of GRBs (from the five calibration relations) as follows:

$$\left(\log \overline{d_L}^2\right)_i^{\text{data}} = \frac{\sum_{a=1}^{5} \left(\log \overline{d_L}^2\right)_i^{(a)} / \sigma_{i,a}^2}{\sum_{a=1}^{5} 1/\sigma_{i,a}^2} \tag{8.19}$$

$$\sigma \left(\log \overline{d_L}^2\right)_i^{\text{data}} = \left(\sum_{a=1}^{5} 1/\sigma_{i,a}^2\right)^{-1/2} \tag{8.20}$$

where

$$\left(\log \overline{d_L}^2\right)_i^{(a)} = a_a + b_a x_i^{(a)} - \log(4\pi P_{\text{bolo},i}), \quad a = 1, 2, 3, 5 \tag{8.21}$$

$$\left(\log \overline{d_L}^2\right)_i^{(4)} = a_4 + b_4 x_i^{(4)} - \log \left[\frac{4\pi S_{\text{bolo},i} F_{\text{beam},i}}{1+z}\right], \tag{8.22}$$

and

$$\sigma_{i,a}^2 = \sigma_{a_a}^2 + \left(\sigma_{b_a} x_i^{(a)}\right)^2 + \left(\frac{b_a \sigma\left(x_{0,i}^{(a)}\right)}{x_{0,i}^{(a)} \ln 10}\right)^2 + \left(\frac{\sigma(P_{\text{bolo},i})}{P_{\text{bolo},i} \ln 10}\right)^2 + \left(\sigma_{\text{sys}}^{(a)}\right)^2,$$

$$a = 1, 2, 3, 5 \tag{8.23}$$

$$\sigma_{i,4}^2 = \sigma_{a_4}^2 + \left(\sigma_{b_4} x_i^{(4)}\right)^2 + \left(\frac{b_4 \sigma\left(x_{0,i}^{(4)}\right)}{x_{0,i}^{(4)} \ln 10}\right)^2 + \left(\frac{\sigma(S_{\text{bolo},i})}{S_{\text{bolo},i} \ln 10}\right)^2$$

$$+ \left(\frac{\sigma(F_{\text{beam},i})}{F_{\text{beam},i} \ln 10}\right)^2 + \left(\sigma_{\text{sys}}^{(4)}\right)^2. \tag{8.24}$$

For each cosmological model given by $\overline{d_L}^2$, we can calibrate the GRBs as described in Section 8.1.1, and then derive the distance estimate $\left(\log \overline{d_L}^2\right)_i$ from each GRB. The χ^2 of a model is given by

$$\chi_{\text{GRB}}^2 = \sum_{i=1}^{N_{\text{GRB}}} \frac{\left[\left(\log \overline{d_L}^2\right)_i^{\text{data}} - \log \overline{d_L}^2(z_i)\right]^2}{\left[\sigma \left(\log \overline{d_L}^2\right)_i^{\text{data}}\right]^2}. \tag{8.25}$$

The treatment of the asymmetric errors in E_{peak} is given by Eq. (8.18).

We can summarize the cosmological constraints from GRB data, $\overline{d_L}^2(z_{\text{GRB}})$, in terms of a set of model-independent distance measurements $\{\overline{r_p}(z_i)\}$:

$$\overline{r_p}(z_i) \equiv \frac{r_p(z_i)}{r_p(0.17)}, \quad r_p(z) \equiv \frac{(1+z)^{1/2}}{z} \frac{H_0}{ch} r(z), \tag{8.26}$$

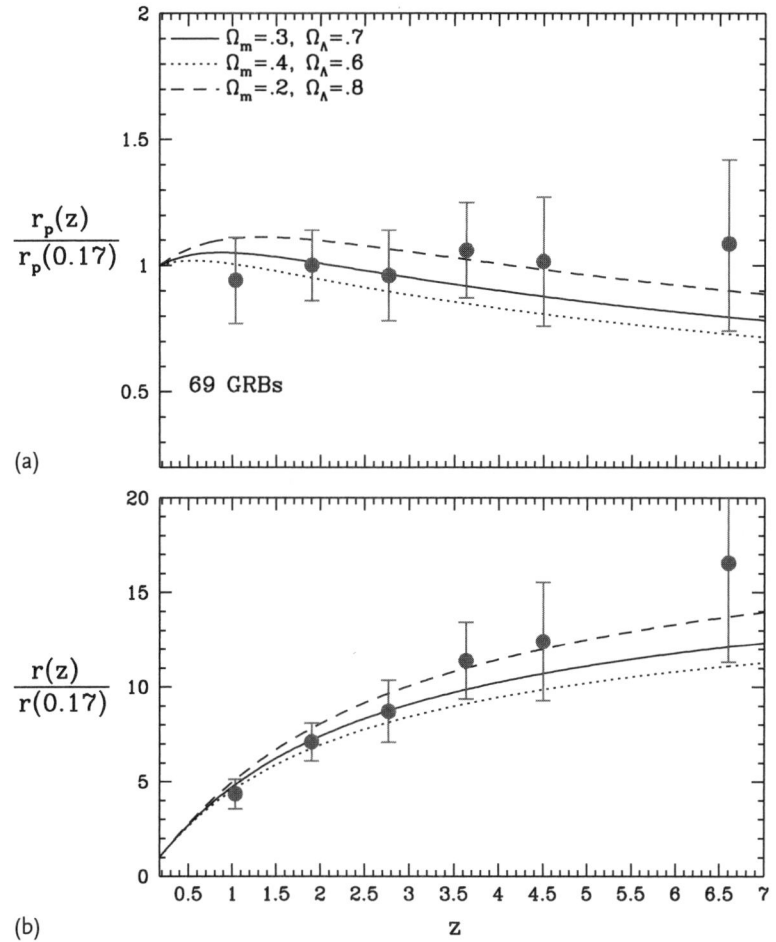

Figure 8.2 The distances measured from 69 GRBs using the five calibration relations in Eqs. (8.1)–(8.5) (Wang, 2008c). The error bars indicate the 68% C.L. uncertainties. (a) shows the scaled distances $r_p(z_i)$ (see Eq. (8.26)); (b) shows the corresponding comoving distances $r(z_i)$.

where $r(z) = d_L(z)/(1+z)$ is the comoving distance at z given by Eq. (2.24). For the 69 GRBs from Schaefer (2007), the lowest redshift GRB has $z = 0.17$, while the highest redshift GRB has $z = 6.6$. There are only four GRBs at $4.5 \leq z \leq 6.6$. We find that the optimal binning is to divide the redshift range between 0.17 and 4.5 into 5 evenly spaced bins, and choose the last bin to span from 4.5 and 6.6 (see Figure 8.2).

Note that the ratio $r_p(z)/r_p(0.17)$ is the most convenient distance parameter choice for the currently available GRB data, since $z = 0.17$ is the lowest redshift GRB in the data set, and the absolute calibration of GRBs is unknown. Using the distance ratio $r_p(z)/r_p(0.17)$ removes the dependence on Hubble constant (which is unknown due to the unknown absolute calibration of GRBs).

Because $\{\overline{r_p}(z_i)\}$ varies very slowly for all cosmological models allowed by current data, the scaled distance $\overline{r_p}(z)$ at an arbitrary redshift z can be found using cubic spline interpolation from $\{\overline{r_p}(z_i)\}$ to $\sim$1–3% percent accuracy for $N_{\text{bin}} = 6$ with our choice of binning.

Note that for a given set of possible values of $\{\overline{r_p}(z_i)\}$ ($i = 1, 2, \ldots, 6$), the luminosity distance at an arbitrary redshift, $d_L(z)$, is given by the accurate interpolation described above. Thus *no* assumptions about cosmological parameters are made. We calibrate the GRBs for each set of possible values of $\{\overline{r_p}(z_i)\}$ ($i = 1, 2, \ldots, 6$), and compute the likelihood of this set of $\{\overline{r_p}(z_i)\}$ in a Markov Chain Monte Carlo analysis. Hence the distances $\{\overline{r_p}(z_i)\}$ are *independent* of assumptions about cosmological parameters.

Figure 8.2 shows the distances $\{\overline{r_p}(z_i)\}$ measured from 69 GRBs using the five calibration relations in Eqs. (8.1)–(8.5). Table 8.2 gives the mean and 68% confidence level errors of $\{\overline{r_p}(z_i)\}$. The normalized covariance matrix of $\{\overline{r_p}(z_i)\}$ is given in Table 8.3. To use these GRB distance measurements to constrain cosmological models, use

$$\chi^2_{\text{GRB}} = \left[\Delta\overline{r_p}(z_i)\right] \cdot \left(\text{Cov}^{-1}_{\text{GRB}}\right)_{ij} \cdot \left[\Delta\overline{r_p}(z_j)\right]$$
$$\Delta\overline{r_p}(z_i) = \overline{r_p}^{\text{data}}(z_i) - \overline{r_p}(z_i) \,, \tag{8.27}$$

Table 8.2 Distances measured from 69 GRBs with 68% C.L. upper and lower uncertainties.

	z	$\overline{r_p}^{\text{data}}(z)$	$\sigma\left(\overline{r_p}(z)\right)^+$	$\sigma\left(\overline{r_p}(z)\right)^-$
0	0.17	1.0000	–	–
1	1.036	0.9416	0.1688	0.1710
2	1.902	1.0011	0.1395	0.1409
3	2.768	0.9604	0.1801	0.1785
4	3.634	1.0598	0.1907	0.1882
5	4.500	1.0163	0.2555	0.2559
6	6.600	1.0862	0.3339	0.3434

Table 8.3 Normalized covariance matrix of distances measured from 69 GRBs.

	1	2	3	4	5	6
1	1.0000	0.7056	0.7965	0.6928	0.5941	0.5169
2	0.7056	1.0000	0.5653	0.6449	0.4601	0.4376
3	0.7965	0.5653	1.0000	0.5521	0.5526	0.4153
4	0.6928	0.6449	0.5521	1.0000	0.4271	0.4242
5	0.5941	0.4601	0.5526	0.4271	1.0000	0.2999
6	0.5169	0.4376	0.4153	0.4242	0.2999	1.0000

where $\bar{r}_p(z)$ is defined by Eq. (8.26). The covariance matrix is given by

$$(\text{Cov}_{\text{GRB}})_{ij} = \sigma(\bar{r}_p(z_i))\sigma(\bar{r}_p(z_j))\left(\overline{\text{Cov}}_{\text{GRB}}\right)_{ij}, \qquad (8.28)$$

where $\overline{\text{Cov}}_{\text{GRB}}$ is the normalized covariance matrix from Table 8.3, and

$$\begin{aligned}\sigma(\bar{r}_p(z_i)) &= \sigma\left(\bar{r}_p(z_i)\right)^+, \quad \text{if } \bar{r}_p(z) \geq \bar{r}_p(z)^{\text{data}}; \\ \sigma(\bar{r}_p(z_i)) &= \sigma\left(\bar{r}_p(z_i)\right)^-, \quad \text{if } \bar{r}_p(z) < \bar{r}_p(z)^{\text{data}},\end{aligned} \qquad (8.29)$$

where $\sigma\left(\bar{r}_p(z_i)\right)^+$ and $\sigma\left(\bar{r}_p(z_i)\right)^-$ are the 68% C.L. errors given in Table 8.2.

Using the distance measurements from GRBs (see Tables 8.2 and 8.3), we find $\Omega_m = 0.247$ (0.122, 0.372) (mean and 68% C.L. range). Using the 69 GRBs directly, we find $\Omega_m = 0.251$ (0.135, 0.365). This demonstrates that our model-independent distance measurements from GRBs can be used as a useful summary of the current GRB data.

8.1.3
Impact of GRBs on Dark Energy Constraints

Figure 8.3 shows the joint confidence contours for $(\Omega_m, \Omega_\Lambda)$, from an analysis of 307 SNe Ia (compiled by Kowalski et al. (2008)) with and without 69 GRBs, as-

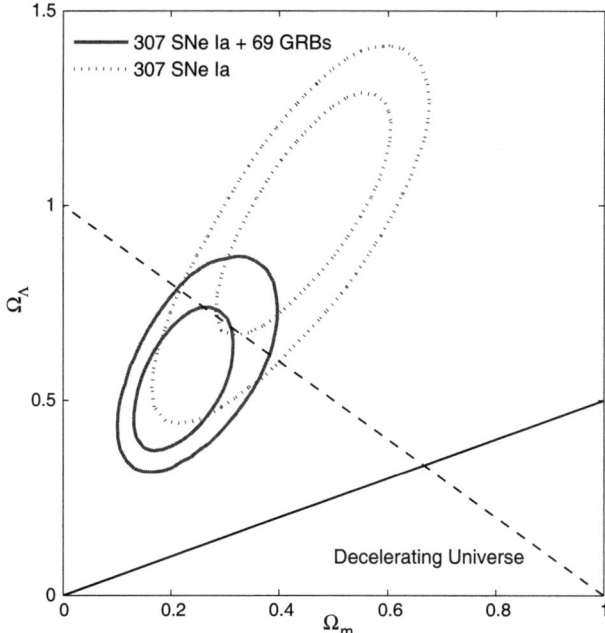

Figure 8.3 The joint confidence contours for $(\Omega_m, \Omega_\Lambda)$, from an analysis of 307 SNe Ia with and without 69 GRBs (Wang, 2008c). A cosmological constant is assumed.

suming a cosmological constant. This shows that the addition of GRB data significantly reduces the uncertainties in (Ω_m, Ω_Λ), and shifts the best-fit parameter values towards a lower matter density universe. Figure 8.4 shows the joint confidence contours for (w_0, Ω_m), from an analysis of 307 SNe Ia with and without 69 GRBs, assuming a flat universe. This shows that SNe Ia alone rules out a cosmological constant at greater than 68% C.L., but the addition of GRB data significantly shifts the best-fit parameter values, and a cosmological constant is consistent with combined SN Ia and GRB data at 68% C.L.

We now consider a dark energy equation of state linear in the cosmic scale factor a. Figure 8.5 shows the joint confidence contours for (w_0, $w_{0.5}$) and (w_0, w_a), from a joint analysis of 307 SNe Ia (Kowalski et al., 2008) with CMB data from WMAP5 (Komatsu et al., 2009), and SDSS BAO scale measurement (Eisenstein et al., 2005), with and without 69 GRBs. HST prior on H_0 has been imposed ($h = 0.72 \pm 0.08$ (Freedman et al., 2001)), and a flat universe is assumed. Note that $w_{0.5} = w_X(z = 0.5)$ in the linear parametrization (Wang, 2008a)

$$w_X(a) = \left(\frac{a_c - a}{a_c - 1}\right) w_0 + \left(\frac{a - 1}{a_c - 1}\right) w_{0.5}$$

$$= \frac{a_c w_0 - w_{0.5} + a(w_{0.5} - w_0)}{a_c - 1} \tag{8.30}$$

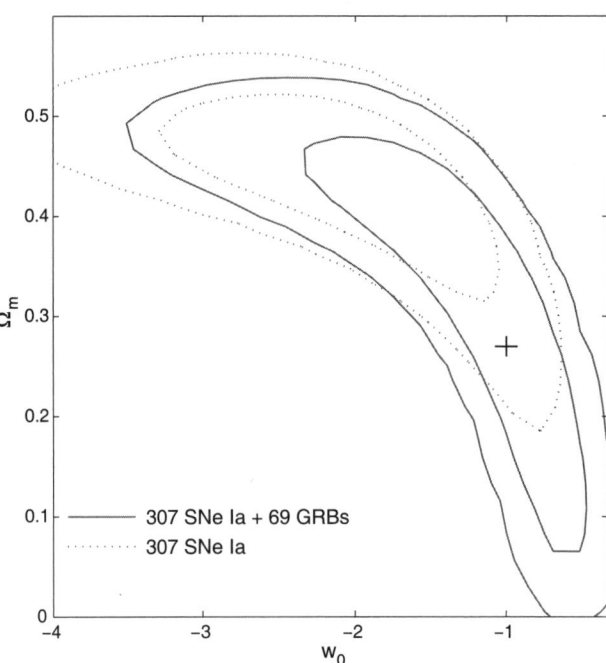

Figure 8.4 The joint confidence contours for (w_0, Ω_m), from an analysis of 307 SNe Ia with and without 69 GRBs (Wang, 2008c). A flat universe and dark energy with constant equation of state are assumed.

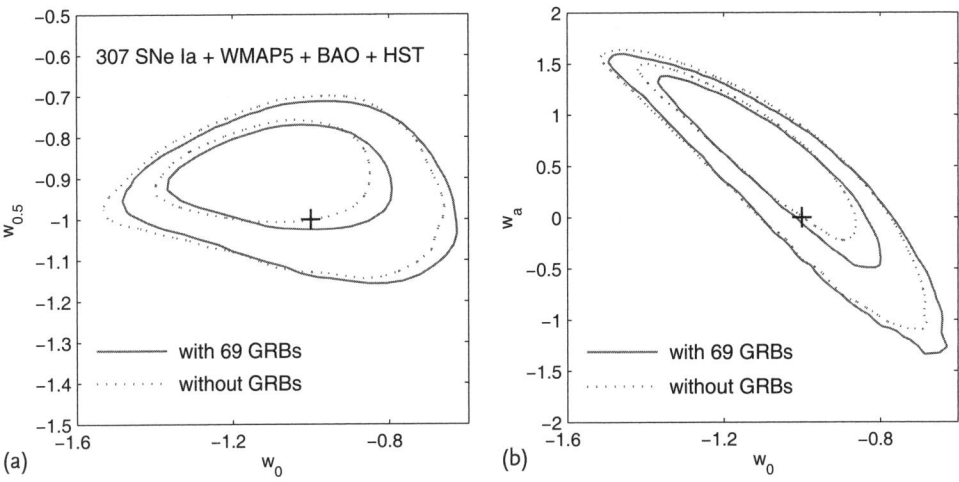

Figure 8.5 The joint confidence contours for $(w_0, w_{0.5})$ (a) and (w_0, w_a) (b), from a joint analysis of 307 SNe Ia with CMB data from WMAP5, and SDSS BAO scale measurements, with and without 69 GRBs (Wang, 2008c). HST prior on H_0 has been imposed and a flat universe is assumed.

with $a_c = 2/3$ (i.e., $z_c = 0.5$). Equation (8.30) corresponds to a dark energy density function

$$X(z) = \frac{\rho_X(z)}{\rho_X(0)} = \exp\left\{3\left[1 + \left(\frac{a_c w_0 - w_{0.5}}{a_c - 1}\right)\right] \ln(1+z) \right.$$
$$\left. + 3\left(\frac{w_{0.5} - w_0}{a_c - 1}\right) \frac{z}{1+z}\right\} \tag{8.31}$$

Equation (8.30) is related to $w_X(z) = w_0 + (1-a)w_a$ (Chevallier and Polarski, 2001) by setting (Wang, 2008a)

$$w_a = \frac{w_{0.5} - w_0}{1 - a_c}, \quad \text{or} \quad w_{0.5} = w_0 + (1 - a_c)w_a. \tag{8.32}$$

Wang (2008a) showed that $(w_0, w_{0.5})$ are much less correlated than (w_0, w_a), thus are a better set of parameters to use. Figure 8.5 shows that the addition of GRB data notably shifts the 68% C.L. contours of $(w_0, w_{0.5})$ and (w_0, w_a) to enclose the cosmological constant model ($w_X(a) = -1$).

8.1.4
Systematic Uncertainties

The most obvious systematic uncertainties of GRBs as cosmological probe are gravitational lensing and Malmquist bias. The biases in estimated cosmological parameters due to these appear small compared to the expected parameter errors for SWIFT GRBs (Oguri and Takahashi, 2006; Schaefer, 2007).

Since the majority of currently observed GRBs are located at $z > 1$, weak gravitational lensing is a main source of systematic uncertainty (Oguri and Takahashi, 2006). The only observable modified by weak lensing is E_{peak}, due to the magnification from cosmic large scale structure. Oguri and Takahashi (2006) found that the bias in the estimates of cosmological parameters due to gravitational lensing is strongly dependent on the shape of the GRB luminosity function.

Malmquist bias leads to two sampling effects. The first Malmquist bias sampling effect occurs for the distance limit. Due to intrinsic scatters and observational errors, GRBs just inside the distance limit will be excluded if the random fluctuations push the apparent brightness below the detection threshold, while the opposite occurs for the GRBs just outside the distance limit. Since there is more volume just outside the distance limit than there is just inside, more overbright GRBs will be included compared to the underbright GRBs that are excluded. The second Malmquist bias sampling effect occurs for the luminosity. There are always more lower luminosity events that are excluded due to random fluctuations compared to the higher luminosity events that are included in the sample.

Schaefer (2007) analyzed the effect of gravitational lensing and Malmquist biases on cosmological constraints from GRBs, and found that the resultant biases in cosmological constraints are surprisingly small, with the average bias of 0.03 mag and the rms scatter of this bias being 0.14 mag. This may have been the consequence of the small range of variation (a factor of two for $0.17 \leq z \leq 6.6$) in the average burst brightness (compared to the experimental threshold), and the various competing effects (from increasing volume, increasing number density, decreasing luminosity function, and decreasing detection probability with increasing distances) all come close to canceling each other out for typical burst properties (Schaefer, 2007).

8.2
Cosmic Expansion History Derived from Old Passive Galaxies

Basic Idea

Jimenez and Loeb (2002) first proposed using relative galaxy ages to constrain cosmological parameters. Simon, Verde, and Jimenez (2005) presented the cosmic expansion history derived from old passive galaxies.

In old passive galaxies (or red galaxies), star formation rate is very small, thus the brightness of these galaxies is dominated by an old stellar population. The aging of stellar populations can be measured by fitting stellar population models to spectroscopic data.

By measuring the age difference between two passively evolving galaxies at different redshifts, one could determine the derivative of redshift with respect to cosmic time, dz/dt. Since

$$H(z) = -\frac{1}{1+z}\frac{dz}{dt}, \tag{8.33}$$

this leads to a measurement of the cosmic expansion rate $H(z)$ at the mean redshift of the two galaxies. Since relative ages, and *not* absolute ages, of galaxies are measured, systematic effects on the absolute scale cancel out (Jimenez and Loeb, 2002).

This method requires first selecting samples of passively evolving galaxies with high-quality spectroscopy. Then synthetic stellar population models are used to constrain the age of the *oldest* stars in each galaxy, after marginalizing over the metallicity and star formation history. Then the *differential* ages of galaxies are computed and used as an estimator for dz/dt, which in turn gives $H(z)$.

Sample Selection

The first sample from Simon, Verde, and Jimenez (2005) consists of 10 early-type field galaxies, chosen from Treu et al. (1999, 2001, 2002). They discarded galaxies for which the spectral fit indicates an extended star formation, defined as the best-fit declining exponential for star formation rate having a decay time in excess of 0.1 Gyr. Their second sample consists of 20 old passive galaxies, obtained from the publicly released Gemini Deep Survey (GDDS) data (Abraham et al., 2004). This is the GDDS subsample of 20 red galaxies for which the most likely star formation history is that of a single burst of star formation of duration less than 0.1 Gyr (in most cases the duration of the burst is consistent with 0 Gyr, that is, the galaxies have been evolving passively since their initial burst of star formation) (McCarthy et al., 2004). Finally, two radio galaxies, 53W091 and 53W069 (Dunlop et al., 1996; Evrard, Metzler, and Navarro, 1996; Spinrad et al., 1997; Nolan et al., 2003), were added to arrive at a total of 32 galaxies (Simon, Verde, and Jimenez, 2005).

Cosmic Expansion History Measurement

Figure 8.6a shows the absolute age for the 32 passively evolving galaxies determined from fitting stellar population models, plotted as a function of redshift. It is apparent that there is a clear age-redshift relation: the lower the redshift the older the galaxies. Figure 8.6b shows the $H(z)$ measurements from the differential ages of galaxies in Figure 8.6a. These $H(z)$ measurements are consistent and complementary to other current observational constraints (Figueroa, Verde, and Jimenez, 2008).

Simon, Verde, and Jimenez (2005) obtained 8 measurements of $H(z)$ (Figure 8.6b) from the 32 galaxy ages (Figure 8.6a) as follows. They first grouped together all galaxies that are within $\Delta z = 0.03$ of each other, and discarded those galaxies that are more than 2σ away from the oldest galaxy in that bin. By averaging over the galaxies in a redshift bin, they obtained an estimate for the age of the universe at the mean redshift of that bin. They then computed age differences only for those redshift bins that are separated by more than $\Delta z = 0.1$ (to ensure that age evolution between the two bins is larger than the error in age determination) but no more than $\Delta z = 0.15$. They found that the resultant dz/dt depends only weakly on the choice of binning.

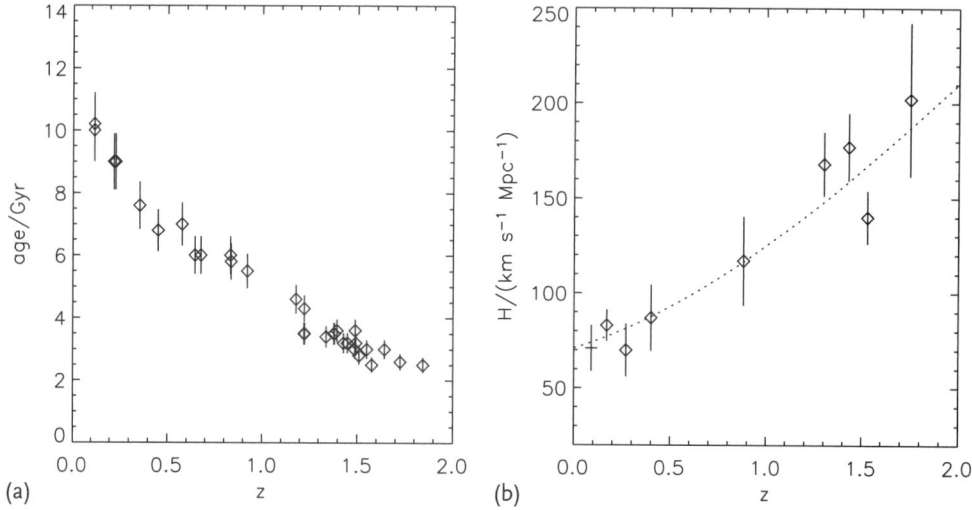

Figure 8.6 The absolute age for the 32 passively evolving galaxies determined from fitting stellar population models (a), and the $H(z)$ derived from the differential ages of these galaxies (b) (reprinted with permission from Simon, Verde, and Jimenez, Phys. Rev. D, 71, 123001 (2005). Copyright (2006) by the American Physical Society). In (b), the lowest redshift data point is the Hubble constant determination from Jimenez et al. (2003). The dotted line is the value of $H(z)$ for the ΛCDM model.

Systematic Effects

In the use of old passive galaxies as cosmological probes, galaxy ages are obtained assuming a single-burst stellar population model, and a single metallicity model (Simon, Verde, and Jimenez, 2005). The single metallicity approximation affects the estimated ages at a level below the statistical errors. The single-burst stellar population approximation may not be a good approximation in general, although it appears to work well for the subset of 20 old galaxies from the GDDS (McCarthy et al., 2004). Massive elliptical galaxies are known to experience some level of star formation activity at low redshift. Thus the assumption of a single-burst in star formation modeling can lead to a bias in the galaxy age estimate towards younger ages. Thus this method relies on obtaining a fair sample of the old passive galaxy population, so that there are enough single-burst galaxies in each redshift bin to define the "red envelope".

8.3
Radio Galaxies as Cosmological Probe

Basic Idea

The use of extended radio galaxies for cosmology was introduce by Daly (1994), and developed by Guerra and Daly (1998), Guerra, Daly, and Wan (2000), Daly and Guerra (2002), Podariu et al. (2003), and Daly et al. (2009). It is interesting to note that Daly, Guerra, and Wan (1998) used a sample of radio galaxies with $0 < z < 2$ to show that the universe has a low mean mass density. This was among the first observational evidence for dark energy.

In the standard unified model for radio galaxies and radio loud quasars, the sources are intrinsically the same, but radio galaxies lie close to the plane of the sky, and radio loud quasars are oriented along the line of sight to the observer. Thus radio galaxies rather than radio loud quasars are used to minimize projection effects (Daly et al., 2009).

The properties and structure of the subset of classical double radio galaxies are well-described by the standard "twin jet" model (e.g., Blandford and Rees (1974); Scheuer (1974); Begelman, Blandford, and Rees (1984); Begelman and Cioffi (1989); Daly (1990)). Particles are accelerated and material is channeled away from the vicinity of a massive black hole along relatively narrow, oppositely directed jets, and is deposited in the radio hot spot. Here particles are re-accelerated to relativistic energies and produce synchrotron radiation in the presence of a local magnetic field. Relativistic plasma flows from the radio hot spots, and, as time goes on, the location of the radio hot spot moves further from the central black hole, leaving behind a "radio bridge" or "radio lobe" of relativistic material. These galaxies form a very homogeneous population (Daly et al., 2009), and are referred to as the extended radio galaxies (ERG).

In the ERG method for distance estimation, the ratio

$$R_*(\beta, y(z)) \equiv \langle D \rangle / D_* \equiv \exp(\kappa_{RG}) \tag{8.34}$$

is assumed to be a constant. The parameter $\langle D \rangle$ denotes the average size of the full population of ERGs at a given redshift, while the parameter D_* denotes the average size of a given radio galaxy at that redshift. The ratio R_* is given by (Guerra and Daly, 1998; Daly and Djorgovski, 2003)

$$R_* = k_0 y^{(6\beta-1)/7} (k_1 y^{-4/7} + k_2)^{(\beta/3-1)}, \tag{8.35}$$

where k_0, k_1, and k_2 are observed quantities (Guerra, Daly, and Wan, 2000), β is the ERG model parameter, and $y = r/[c H_0^{-1}]$ is the dimensionless comoving distance.

Distance Measurement

In fitting Eq. (8.34) to data, the difference between $\ln(\langle D \rangle / D_*)$ and κ_{RG} is minimized (Guerra and Daly, 1998), to obtain the best-fit values of β, κ_{RG}, and cosmological parameters (including dark energy parameters) through $r(z)$. Figure 8.7 shows the size D_* and the ratio $\langle D \rangle / D_*$ obtained for the best-fit values of cosmological parameters and β, and normalized to unity using the best-fit value of κ_{RG}, assuming a quintessence model (Daly et al., 2009).

The ERG method is complimentary to the SN Ia method, and gives cosmological constraints that are consistent to that from SNe Ia. Figure 8.8 shows the dimensionless comoving distances to 30 radio galaxies and 192 supernovae published by Davis et al. (2007). These SNe Ia include the ESSENCE SNe Ia from Wood-Vasey et al. (2008), the SNLS SNe from Astier et al. (2006), and the high redshift HST SNe Ia from Riess et al. (2007).

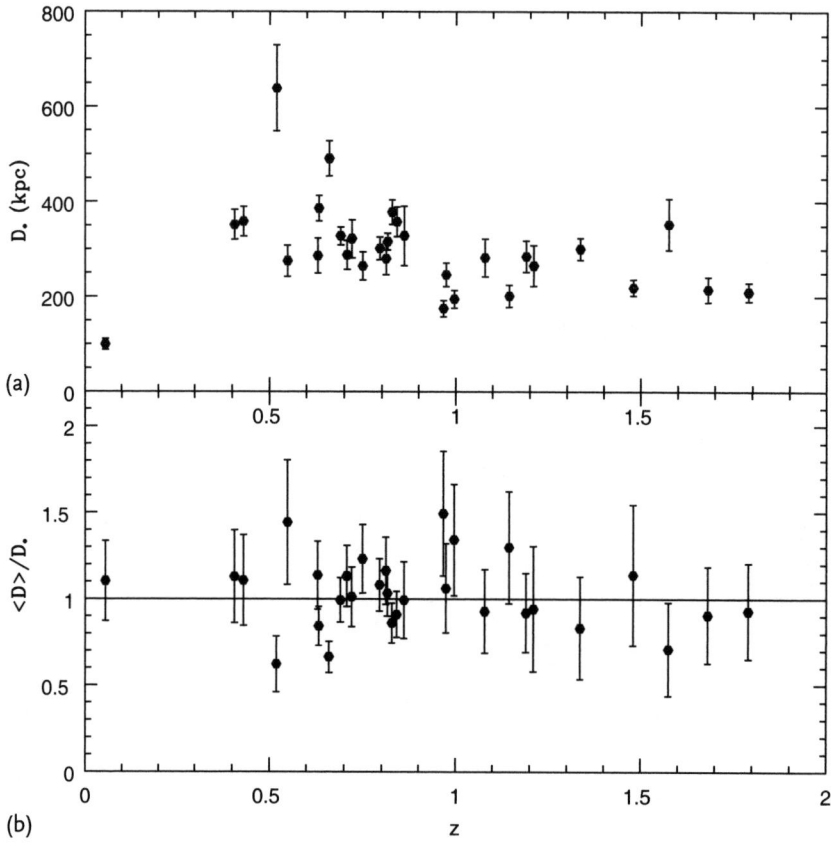

Figure 8.7 The size D_* (a) and the ratio $\langle D \rangle / D_*$ (b) obtained for the best-fit values of cosmological parameters and β, and normalized to unity using the best-fit value of κ_{RG}, assuming a quintessence model (Daly et al., 2009).

Systematic Effects

With powerful radio sources, the potential systematic effects are radio power selection effects and projection effects (Daly, 2009). These effects, including their impact on D_*, were studied in detail by Wan and Daly (1998a,b). The sources used to date for cosmological studies using the ERG method have been drawn from the parent population of 3CR radio galaxies, known for radio power selection effects, that is, an increase in radio power of the sample with redshift.

The ERG method does not require any assumptions about the source environments or the relation between the beam power and radio power of the source. The only assumptions required are that strong shock physics applies at the extremities of the source, and that the total source lifetime is given by a power law of the beam power (see, e.g., Daly et al. (2009)). Both of these assumptions have now been carefully studied and appear to be well-justified. The key observables that go into the determination of D_* are the source width, pressure, and lobe prop-

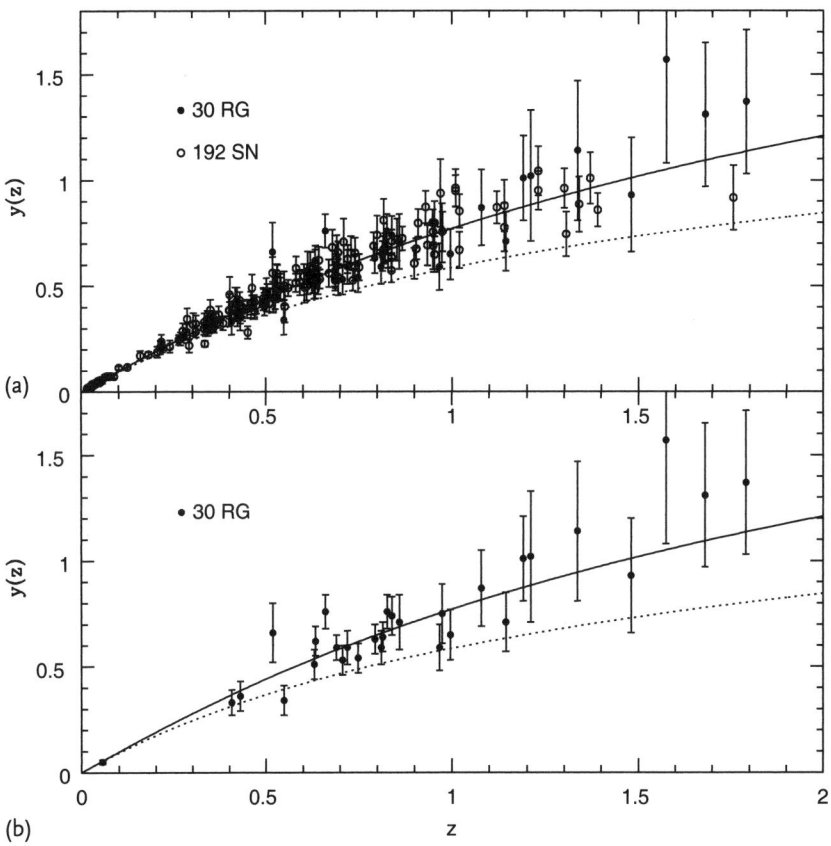

Figure 8.8 Dimensionless comoving distances to 30 radio galaxies (b) compared with those to the 192 supernovae (a) (Daly et al., 2009). The SNe Ia are from Davis et al. (2007). The dotted curves indicate y(z) expected in a flat matter-dominated universe with $\Omega_m = 1$, and the solid curves indicate that expected in a flat ΛCDM model, with $\Omega_m = 0.3$ and $\Omega_\Lambda = 0.7$.

agation velocity; the lobe propagation velocity enters quite weakly, and there is a growing consensus that determinations of this quantity have been demonstrated to be reasonably accurate (O'Dea et al., 2009). The source width and pressure are straightforward to determine observationally, and in determining the pressure, the method accounts for offsets of the radio emitting region from minimum energy conditions. One of the primary determinants of D_*, the beam power, which is a combination of the source width, pressure, and velocity, has been shown to be independent of offsets from minimum energy conditions for these sources (O'Dea et al., 2009).

Since the ERG method relies on the ratio of $\langle D \rangle$ to D_*, both of which are subject to precisely the same selection effects (see Daly and Guerra (2002)), a prediction of the evolution of D_* is not required. Thus this method is relatively unencumbered by selection effects.

8.4
Solar System Tests of General Relativity

Parametrized Post-Newtonian Approach

The most common approach to testing general relativity in the solar system is the "parametrized post-Newtonian" (PPN) approach first introduced by Eddington (1957). For recent reviews on this subject, see Reynaud and Jaekel (2008), and Will (2006).

In the simplest model of the solar system, we reduce the gravity sources to the Sun as a point-like mass M at rest. Using Eddington's gauge convention, we can write the general form of an isotropic and stationary metric:

$$ds^2 = g_{00} c^2 dt^2 + g_{rr} \left[dr^2 + r^2 \left(d\theta^2 + \sin^2\theta \, d\phi^2 \right) \right]. \tag{8.36}$$

In this simple model, the exact solution to Einstein's equations can be written in terms of the reduced Newtonian potential ϕ:

$$g_{00} = \left(\frac{1 + \phi/2}{1 - \phi/2} \right)^2, \quad g_{rr} = -(1 - \phi/2)^4, \quad \phi \equiv -\frac{GM}{rc^2}. \tag{8.37}$$

Since $GM/c^2 \sim 1.5\,\text{km}$, $\phi \sim 10^{-8}$ on Earth due to the Sun. Therefore, we can expand Eq. (8.37) into a power series:

$$g_{00} = 1 + 2\phi + 2\phi^2 + \cdots$$

$$g_{rr} = -1 + 2\phi + \cdots \tag{8.38}$$

The family of PPN metrics results from inserting constants in the coefficients of the terms in the expansion:

$$g_{00} = 1 + 2\alpha\phi + 2\beta\phi^2 + \cdots$$

$$g_{rr} = -1 + 2\gamma\phi + \cdots. \tag{8.39}$$

The parameter α can be set to unity by fixing Newton's gravitational constant G to its effective value in the solar system. The parameters β and γ then characterize families of PPN models, with general relativity corresponding to $\beta = \gamma = 1$.

Since the predicted geodesics associated with the metric in Eq. (8.39) depend on the PPN parameters β and γ, experiments and observations can be used to constrain these parameters. The current bound on γ is given by the experiment performed through radar ranging of the Cassini probe during its 2002 solar occultation (Bertotti, Iess, and Tortora, 2003):

$$\gamma - 1 = (2.2 \pm 2.3) \times 10^{-5}. \tag{8.40}$$

Lunar laser ranging gives the current bound on a linear combination of β and γ (Williams, Turyshev, and Boggs, 2004)

$$\eta \equiv 4(\beta - 1) - (\gamma - 1) = (4.4 \pm 4.5) \times 10^{-4}. \tag{8.41}$$

Assuming that the measurements of γ and η are uncorrelated, we obtain

$$\beta - 1 = (1.2 \pm 1.1) \times 10^{-4} \, . \tag{8.42}$$

Modified Gravity Models

Modified gravity models generically predict that the locally measured Newtonian gravitational constant varies with cosmic time. For scalar-tensor theories of gravity, the predictions for $\dot{G}/G$ can be written in terms of the derivatives of the asymptotic scalar field. If G does vary with cosmic time, its rate of variation should be around the same order of magnitude as the expansion rate of the universe, that is, $\dot{G}/G \sim H_0 = 1.02 \times 10^{-10} \, h \, \text{yr}^{-1}$. The HST key project gives $h = 0.72 \pm 0.08$ (Freedman et al., 2001).

In solar system measurements, bounds on $\dot{G}/G$ can be obtained phenomenologically by replacing G with $G_0 + \dot{G}_0(t - t_0)$ in Newton's equations of motion. Lunar laser ranging gives the tightest current bound of (Williams, Turyshev, and Boggs, 2004)

$$\dot{G}/G = (4 \pm 9) \times 10^{-13} \, \text{yr}^{-1} \, . \tag{8.43}$$

The above uncertainty is $\sim 1\%$ of H_0, the expected variation rate of G if it does vary with cosmic time.

Future Prospects for Solar System Tests of General Relativity

The planned ESA mission BepiColombo (http://www.rssd.esa.int/BepiColombo/), with a current launch date in 2013, will explore the planet Mercury with instrumentation that enables extremely accurate tracking. Milani et al. (2002) estimated that BepiColombo can realistically reach an accuracy of $\sim 2 \times 10^{-6}$ for γ, and an accuracy of $\lesssim 10^{-5}$ for η (implying an accuracy of $\sim 2 \times 10^{-6}$ for β). The improvement of the bound on the time variation of G from BepiColombo is expected to be modest, because of the problem of absolute calibration of the ranging transponder (Milani et al., 2002).

9
Basic Instrumentation for Dark Energy Experiments

Based on the previous chapters, it is clear that optical to NIR wide-field imaging and massive multi-object spectroscopy are the most important capabilities for future dark energy projects. These imply specific preferences in telescope and optical design, and choices in detectors and spectroscopic masks.

9.1
Telescope

Wide-field imaging requires a telescope that accesses a wide angle with an extremely low-distortion focal plane. The main aberrations resulting from the shape of reflecting surfaces are: spherical aberration, coma, astigmatism, and field curvature.

It is easier, and thus cheaper, to grind lenses and mirrors into spheroids. But spherical surfaces do not focus a parallel beam of light to a single point. This results in *spherical aberration*, which can be removed by modifying the surface shapes from spherical to parabolic.

Parabolic mirrors work perfectly only when the beam is parallel to the optical axis. Off-axis beams suffer from coma and astigmatism. *Coma* produces elongated images of point sources that lie off the optical axis, because the focal lengths of paraboloids vary with the angle between the direction of an incoming light ray and the optical axis. *Astigmatism* derives from having different parts of a lens or mirror converge an image at slightly different locations on the focal plane. When a system of lenses or mirrors is designed to correct for astigmatism, field curvature can be a problem. *Field curvature* is due to the focusing of images on a curve rather than on a plane. In addition, if the plate scale depends on the distance from the optical axis, *field distortion* results.

The Ritchey–Chrétien telescope is designed to eliminate spherical aberration and coma through the use of hyperbolic primary and secondary mirrors. The well known Schmidt telescope is designed to provide a wide-angle and low distortion field of view. To minimize coma, a spheroidal primary mirror is used in combination with a correcting lens to help remove spherical aberration. The SDSS 2.5 m telescope uses a modified Ritchey–Chrétien design with two mirrors and two correctors (see Figure 9.1).

Dark Energy. Yun Wang
Copyright © 2010 WILEY-VCH Verlag GmbH & Co. KGaA, Weinheim
ISBN: 978-3-527-40941-9

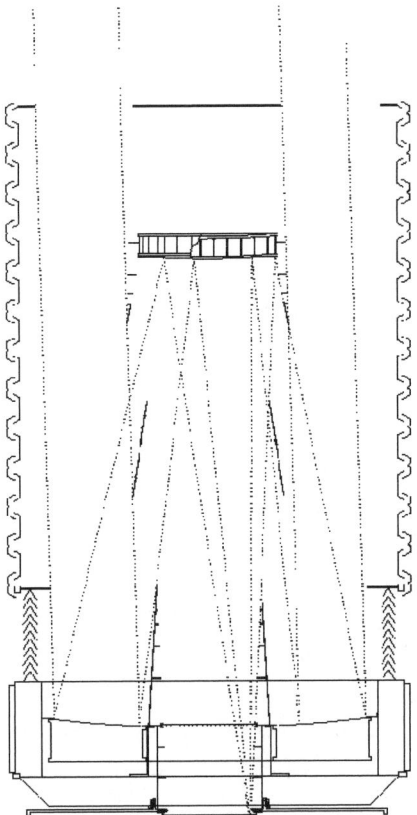

Figure 9.1 SDSS 2.5 m telescope optical configuration, showing the light baffles and a set of rays from the extreme field edge (Gunn et al. (2006), SDSS Collaboration, http://www.sdss.org/). The first transmissive corrector (the Gascoigne or common corrector) is nearly coincident with the vertex of the primary mirror. The second transmissive corrector, which forms the main structural element of the SDSS mosaic camera (Gunn et al., 1998), is just before the focal service. In spectroscopic mode, the imaging camera, including this corrector, is dismounted and replaced with the spectroscopic corrector and the spectroscopic plug plate and fiber harnesses. The baffle system consists of the wind baffle (outside the beam to the primary mirror) and the secondary, conical, and primary baffles (inside). The conical baffle is suspended about halfway between the primary and secondary. Light enters the wind baffle through the annular opening at the top. The outer interlocking C-shaped baffles that form the upper tube are carried by an independently mounted and driven wind baffle mechanism and take the major wind loads on the telescope.

There are a number of three-mirror telescopes that have been proposed for wide-field imaging, including the Baker–Paul three-mirror design (Paul, 1935; Baker, 1969), and the Korsch three-mirror anastigmatic telescope (Korsch, 1977).

The use of Korsch's three-mirror design, usually referred to as "the three-mirror anastigmat (TMA)", has become widespread in space instrumentation, because of its excellent performance, and the flexibility it affords. Figure 9.2 shows a popular configuration of the Korsch three-mirror anastigmatic telescope. The TMA is cor-

9.1 Telescope | 207

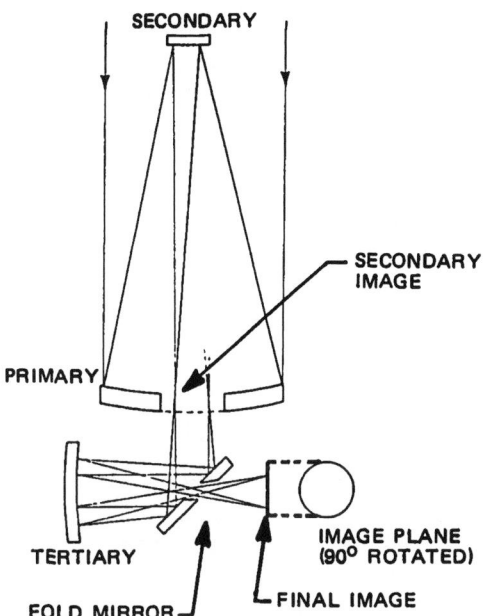

Figure 9.2 A configuration of the Korsch anastigmatic three-mirror telescope (Korsch, 1977).

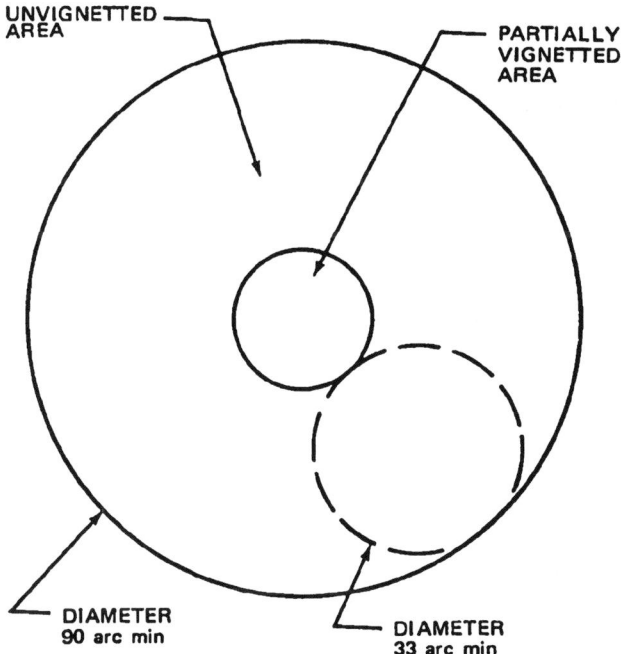

Figure 9.3 Image area in the telescope focal plane for the Korsch anastigmatic three-mirror telescope (Korsch, 1977).

rected for four aberrations: spherical aberration, coma, astigmatism, and field curvature (Korsch, 1977). Korsch's original TMA (a 1.5 m telescope) had a very small central obscuration (see Figure 9.3), an easily accessible flat image field of 1.5° in diameter with a geometric rms spot size smaller than 0.07″ throughout the field (far superior compared to a Ritchey–Chrétien telescope (Korsch, 1977)), and excellent stray light suppression (see Figure 9.4). Figure 9.5 shows an isometric view of the configuration of the Korsch TMA telescope from Figure 9.2.

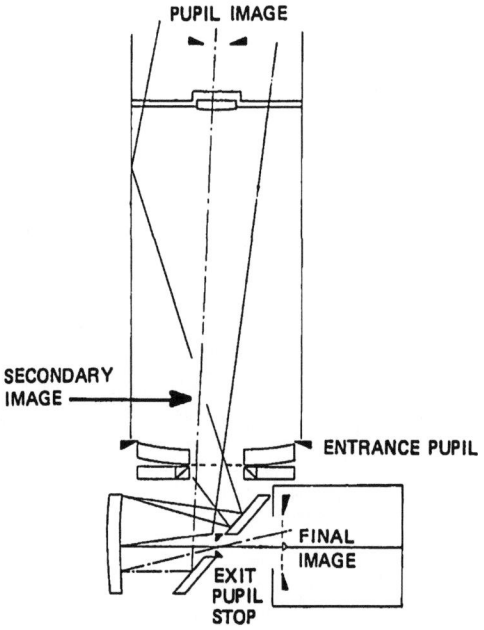

Figure 9.4 Stray light path for the configuration of the Korsch anastigmatic three-mirror telescope shown in Figure 9.2 (Korsch, 1977).

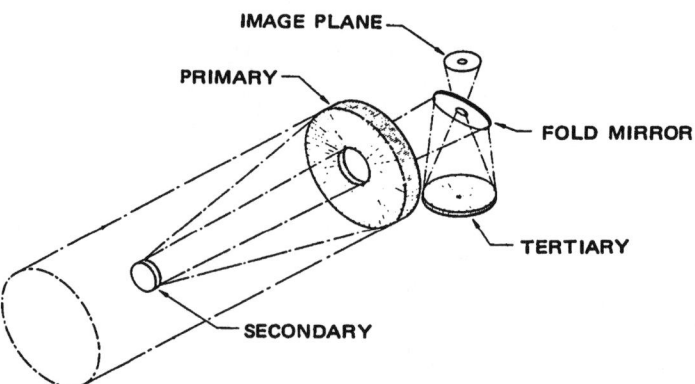

Figure 9.5 Isometric view of the configuration of the Korsch anastigmatic three-mirror telescope shown in Figure 9.2 (Korsch, 1977).

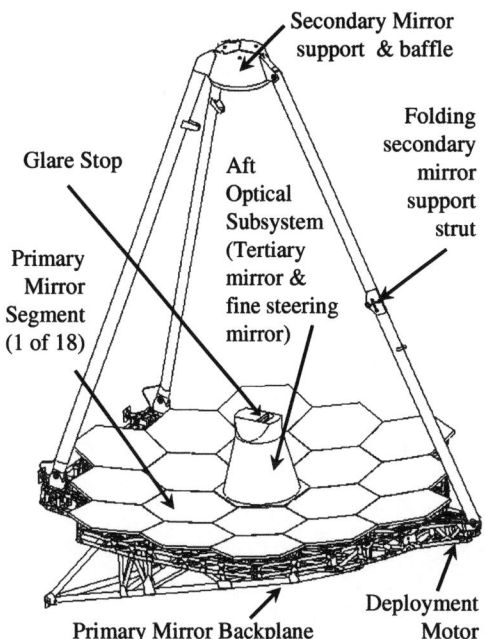

Figure 9.6 Isometric view of the JWST telescope (Gardner et al., 2006). The JWST telescope is a three-mirror anastigmat with a primary mirror made up of 18 hexagonal segments.

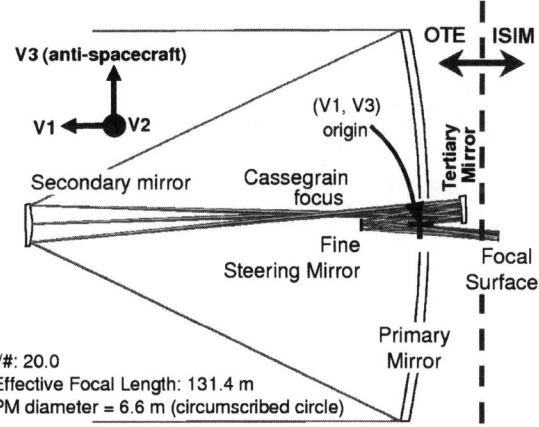

Figure 9.7 The JWST telescope optical layout (Gardner et al., 2006). The JWST optical telescope has an effective f/number of 20 and an effective focal length of 131.4 m.

The James Webb Space Telescope (JWST) uses a TMA with a primary mirror made up of 18 hexagonal segments (Gardner et al., 2006). Figure 9.6 shows an isometric view of the JWST telescope (Gardner et al., 2006), and Figure 9.7 shows the JWST telescope optical layout (Gardner et al., 2006). The TMA has also been used in proposed mission concepts for the Joint Dark Energy Mission (JDEM), see for example, Aldering *et al.* (2004), Wang *et al.* (2004), and Crotts *et al.* (2005).

9.2
NIR Detectors

Panoramic detectors sensitive to IR radiation and with the low noise required by astronomical applications have been developed using both Indium Antimonide (InSb) and Mercury Cadmium Telluride (HgCdTe) materials. HgCdTe detectors are now used at the major telescopes in the world as well as on the Hubble Space Telescope (NIRCAM and WFC3). Figure 9.8 shows an example of the $2K \times 2K$ HgCdTe focal planes from Teledyne Imaging Sensors that will be used by the JWST NIRCam. HgCdTe references contain recent reviews on HgCdTe infrared detectors.

HgCdTe is an alloy of CdTe and HgTe. Both CdTe and HgTe are mixtures of trivalent and pentavalent atoms, yielding an effectively quadrivalent lattice, as in Si or Ge. CdTe is a semiconductor with a bandgap of ~ 1.5 eV at room temperature, while HgTe is a semimetal with zero bandgap energy. A HgCdTe alloy with any bandgap between zero and 1.5 eV can be obtained by mixing CdTe and HgTe in the appropriate proportions. Since HgCdTe is transparent at wavelengths corresponding to photon energies below the energy gap, the exact alloy composition of the HgCdTe determines the longest wavelength at which the detector can absorb light.

Figure 9.8 An example of the $2K \times 2K$ HgCdTe focal planes from Teledyne Imaging Sensors that will be used by the JWST NIRCam (Chuh *et al.*, 2006).

Thus a HgCdTe detector can be "tuned" in wavelength band to minimize thermal noise.

In a HgCdTe detector, a photon is detected if it has enough energy to kick a charge carrier (electron or hole) from the valence band to the conduction band. This charge carrier is then collected by an external readout integrated circuit and transformed into an electric signal.

Thermal agitation may also push a charge carrier in the conduction band, creating a spurious "dark current" signal with its associated noise. This intrinsic noise is temperature-dependent, and can be greatly reduced by cooling the detector at very low temperatures. Detectors with narrower bandgaps, that is, sensitive to longer wavelengths, usually require lower temperatures to control dark current. Typical HgCdTe detectors with 2.5 µm cutoff can be operated at liquid N_2 temperature (77 K). Figure 9.9 shows the dark current versus temperature for MBE HgCdTe from Teledyne Imaging Sensors (Chuh *et al.*, 2006).

A key enabler for focal plane arrays (FPAs) is the SIDECAR ASIC (Chuh *et al.*, 2006). Figure 9.10 shows the SIDECAR ASIC micrograph (Figure 9.10a), and the SIDECAR functional block diagram with ROIC and external interfaces (Figure 9.10b). SIDECAR (System for Image Digitization, Enhancement, Control and Retrieval) is an FPA controller-on-a-chip (Loose *et al.*, 2002). It replaces multiple conventional PCB boards, thereby dramatically reducing power, size and weight of

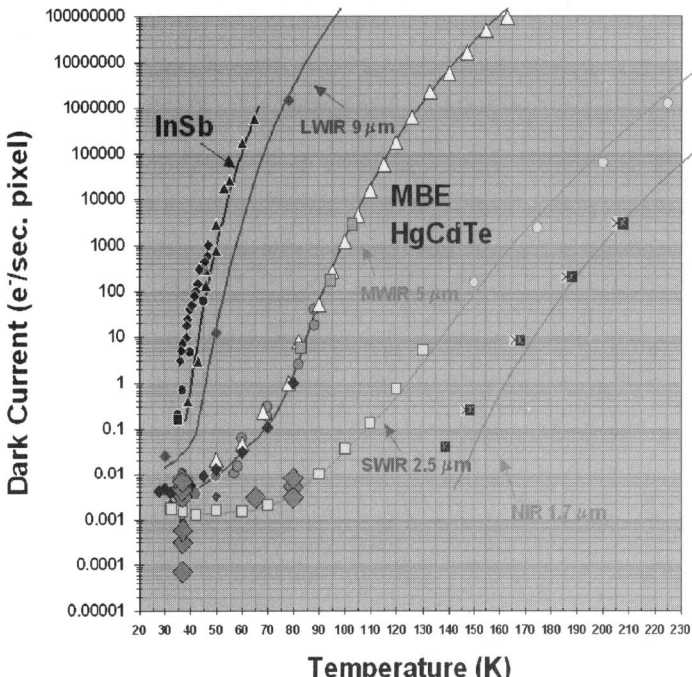

Figure 9.9 Dark current versus temperature for MBE HgCdTe from Teledyne Imaging Sensors (Chuh *et al.*, 2006).

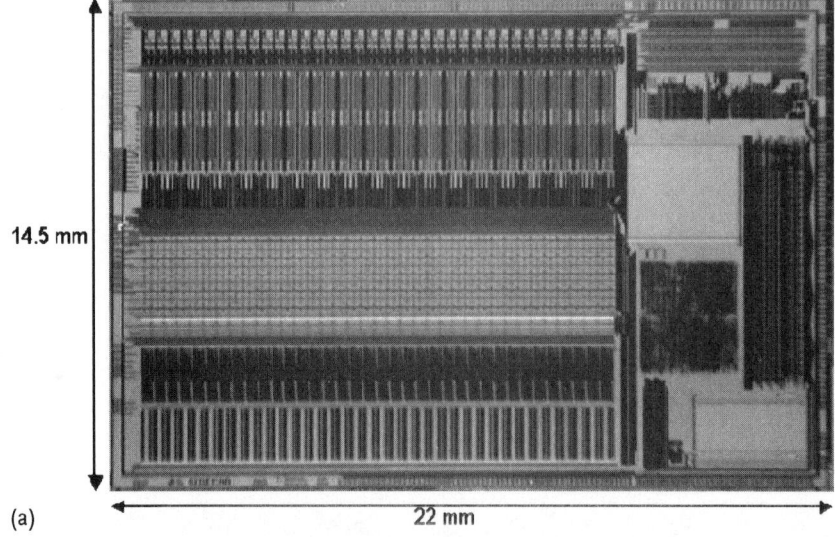

SIDECAR: System for Image Digitization, Enhancement, Control and Retrieval

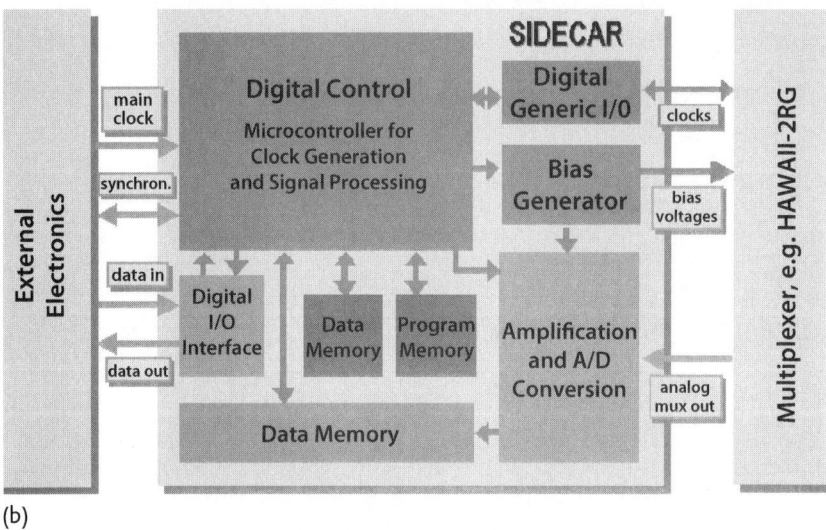

Figure 9.10 SIDECAR (Chuh et al. (2006), Teledyne Imaging Sensors). (a) The SIDECAR ASIC micrograph. (b) The SIDECAR functional block diagram with ROIC and external interfaces.

the support electronics, while increasing the ease to design and integrate an imaging subsystem. Multiple FPAs can be operated simultaneously by a single ASIC. For mosaic FPAs, SIDECAR offers significant improvements because it eliminates the complexity and performance risk associated with acquiring tens or hundreds of millions of pixels of data per frame (Chuh et al., 2006). ASICs have been developed for JWST, and have been used already in space instrumentation: the ACS (the Advanced Camera for Surveys) repair on the HST has been done using ASICs.

9.3
Multiple-Object Spectroscopic Masks

Traditional multiple-object spectroscopic masks use fibers (see, e.g. Limmongkol et al. (1993)). This can enable simultaneous spectroscopy of hundreds of faint galaxies. For example, the SDSS spectrograph uses custom drilled aluminum plug-plates to hold the fibers in the telescope focal plane. These plug-plates are installed in a fixture called a fiber cartridge, and manually stuffed with fibers before nightfall. The fibers are brought to the slithead, which is incorporated in the cartridge, and connected to the spectrograph. During the night, a number of different cartridges can be swapped out to observe different fields. Each cartridge contains single strand object fibers, and also a number of coherent fiber bundles capable of imaging a few arcseconds of the sky. These are usually placed on preselected guide stars and feed a CCD camera mounted on the spectrograph. The guide stars are used to center the telescope on the plug-plate field, adjust the plate scale of the telescope, control focus, and guide the exposure. One larger (30 arcsec) bundle can be used to measure the sky brightness. Image quality and photometric data from these guide and sky bundles can be used to estimate the exposure time required to complete the

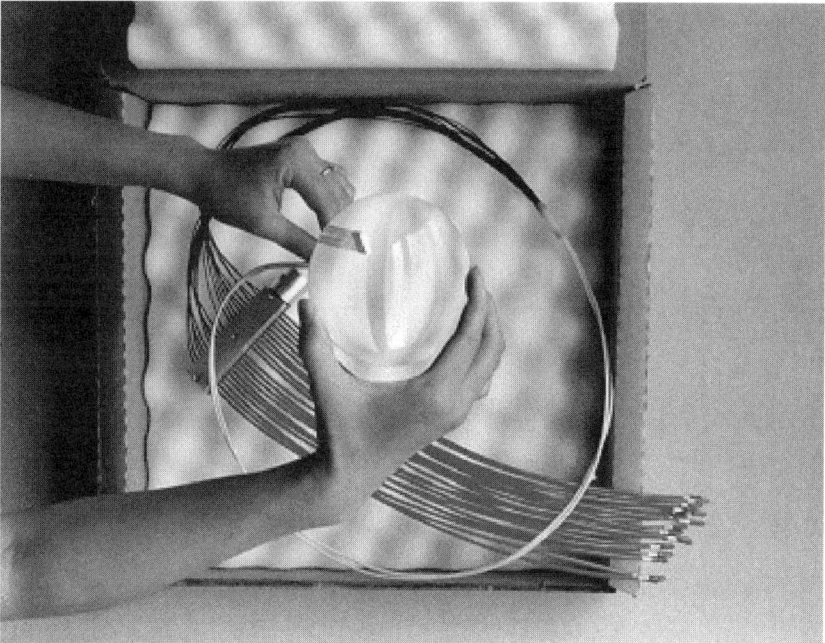

Figure 9.11 The SDSS prototype fiber optic cartridge with 20 fibers (J. Gunn and SDSS Collaboration, http://www.sdss.org/). The ends, which are plugged and unplugged during operations, are protected by tough nylon tubing. The lens enlarges the v-groove block termination at the slit end.

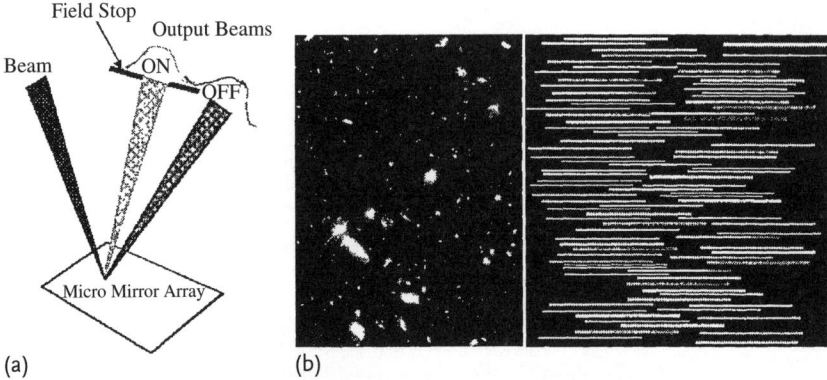

Figure 9.12 The concept of a programmable spectroscopic mask using micromirror arrays (MacKenty and Stiavelli, 2000). (a) Illustration of how a micromirror array can be used to define slits. (b) As an example, a simulated $R = 300$ observation of the Hubble Deep Field with a multi-object spectrograph using micromirrors on the HST.

observation. Figure 9.11 shows the SDSS prototype fiber optic cartridge with 20 fibers.

Next generation multiple-object spectroscopic masks use micro-electro-mechanical-systems (MEMS), following the approach pioneered for NIRSpec on JWST (MacKenty and Stiavelli, 2000; Moseley et al., 2000). MEMS technology enables a rapidly reconfigurable slit device. Instead of punched aperture plates or movable fibers, digital electronics are used to control microscopic mechanical elements to form slits. Figure 9.12 illustrates how a micromirror array can be used to define slits (MacKenty and Stiavelli, 2000).

MEMS technology allows the integration of microscopic mechanical elements together with their control electronics on a common silicon substrate. The control electronics is fabricated using standard integrated circuit processes. The mechanical components are fabricated using "micromachining" processes that selectively etch away parts of the silicon wafer or add new structural layers.

Two concepts were studied for the JWST programmable spectroscopic masks, microshutter arrays (MSAs), and micromirror devices (DMDs). MSAs were chosen for the JWST final design. Figure 9.13 shows the schematic layout of the JWST NIRSpec slit mask (using MSAs) overlaid on the detector array and projected to the same angular scale (Gardner et al., 2006).

The JWST MSAs consist of microshutters that are tiny cells measuring 100×200 μm. The microshutters are arranged in a waffle-like grid that contains 171×365 shutters. The microshutter cells have lids that open and close when a magnetic field is applied. Each cell can be controlled individually, allowing it to be opened or closed to view or block a portion of the sky. Figure 9.14 shows two closeup views of the microshutters. MSA have been developed and optimally matched to the JWST NIRSpec, and can operate at low temperatures (Moseley et al., 2000).

DMDs consist of an array of up to 2.21 million aluminum micromirrors fabricated on top of a complementary metal oxide semiconductor (MOS) static ran-

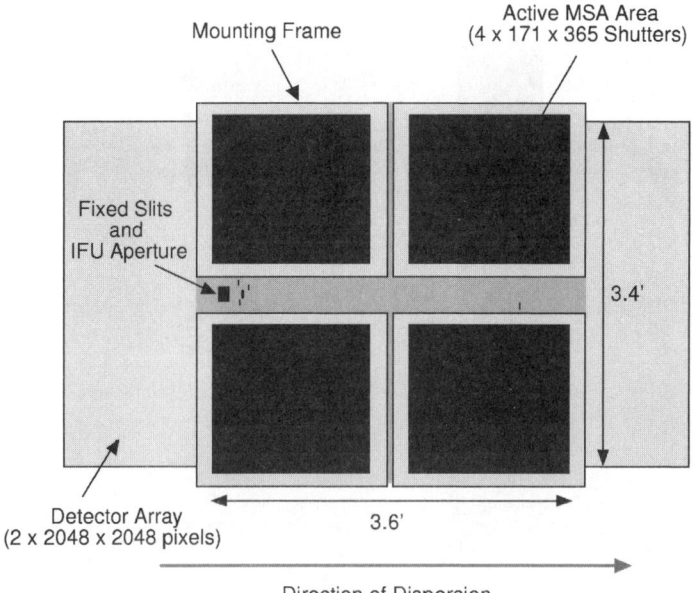

Figure 9.13 Schematic layout of the JWST NIRSpec slit mask overlaid on the detector array and projected to the same angular scale (Gardner *et al.*, 2006).

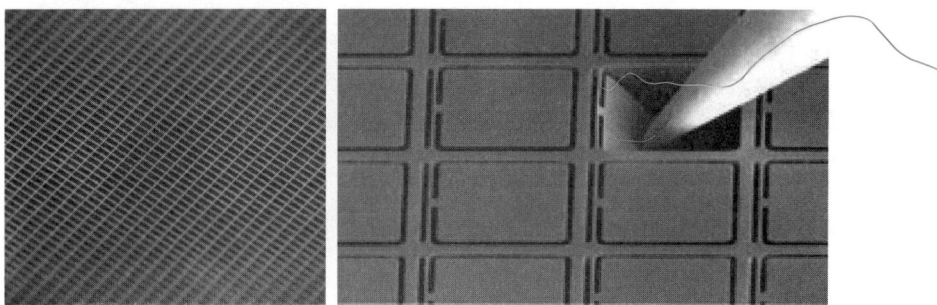

Figure 9.14 Two closeup views of the microshutters (MSA reference, 2009).

dom access memory (SRAM) array. Each micromirror, independently controlled, can switch along its diagonal thousands of times per second as a result of electrostatic attraction between the mirror structure and the underlying electrodes. One advantage of DMDs is that they are commercially available. Since their invention by L. Hornbeck at Texas Instruments in 1988, DMDs have been massively produced and their qualities continuously improved. At present, DMDs represent one of the leading technologies in digital imaging, widely used in video projectors and home television systems. Commercial DMDs do not operate at low temperatures and this limits their use to wavelengths shorter than approximately 2 μm.

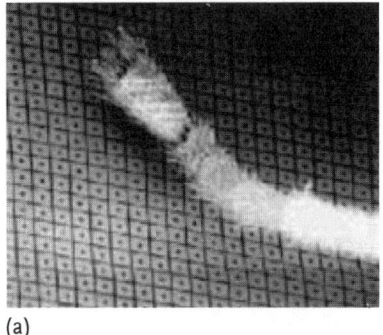 

Figure 9.15 DMD array (Cimatti et al., 2009). (a) DMD array with an ant leg for comparison. (b) Packaged DMD CINEMA (2048 × 1080) device.

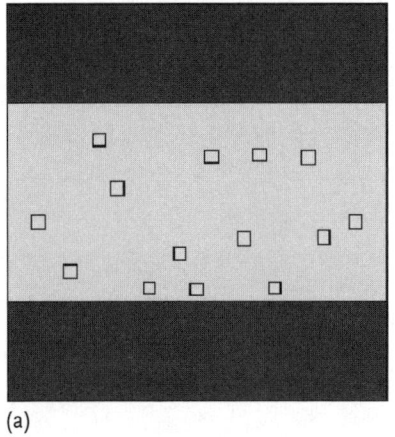 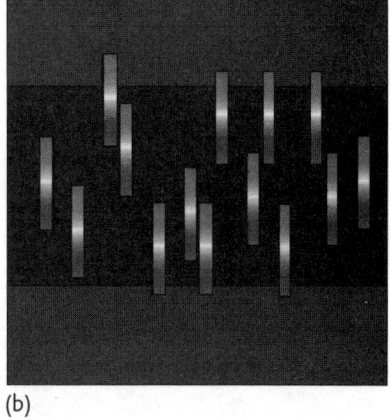

Figure 9.16 Sketch of the data acquisition and observing procedure (Cimatti et al., 2009). (a) the DMD field is projected onto the detector (dark background) in broadband imaging mode. The light color indicates the high background, targets are represented by small squares. (b) All DMDs are turned off except those of the targets, and the prism is inserted. The background is low (dark color) and spectra are produced.

Texas Instruments produces several types of DMDs, including the CINEMA device with more than 2 million pixels (2048 × 1080 pixels) and 13.68 μm pixel pitch (see Figure 9.15b), and the 1080p High Definition TV device with 1920×1080 pixels and 10.6 μm pixel pitch.

DMDs are already in use in astronomy. MacKenty et al. developed an Infrared Multi-Object Spectrometer (IRMOS) using DMDs (MacKenty et al., 2006). IRMOS contains ∼500 000 micromirrors. IRMOS was delivered to Kitt Peak National Observatory (KPNO) in September 2004, and is now available for proposals at KPNO (MacKenty et al., 2006).

The current generation of commercial DMDs appear to be comparable or superior to MSAs in the figures of merit such as filling factor, contrast, reliability, and cosmetic defects. Thus DMDs are an excellent choice as slit mask for NIR spectrographs that aim to obtain of order one hundred millions galaxy redshifts to probe dark energy. Figure 9.16 illustrates the data acquisition and observing procedure using DMDs on a proposed dark energy space mission (Cimatti *et al.*, 2009).

10
Future Prospects for Probing Dark Energy

Ongoing dark energy projects use pre-existing instrumentations, not all of which are optimal for probing dark energy. For the next-generation dark energy projects, the instrumentation should be dictated by the science goals by design. This gives us a precious opportunity to optimize the instrumentation for efficiency and control of systematic effects, and maximize the scientific return of planned dark energy projects.

We will first discuss some general considerations in designing the optimal dark energy experiment, and give two examples of proposed dark energy space mission concepts to illustrate the critical roles of optical to NIR wide-field imaging and massive multi-object spectroscopy. We will then discuss the big picture of proposed future dark energy projects in the context of recommendations made to the national and international funding agencies.

10.1
Designing the Optimal Dark Energy Experiment

The challenge to solving the mystery of cosmic acceleration will not be the statistics of the data obtained, but the tight control of systematic effects inherent in the data. Thus the optimal experiment to probe the nature of cosmic acceleration should employ multiple techniques. The feasibility of this was first demonstrated by JEDI (Joint Efficient Dark-energy Investigation), a mission concept proposed for JDEM (Wang *et al.*, 2004; Crotts *et al.*, 2005; Cheng *et al.*, 2006).

JEDI can probe the nature of cosmic acceleration in three independent ways: (1) using Type Ia supernovae (SNe Ia) as cosmological standard candles over a range of distances, with $\sim$14 000 SNe Ia with redshifts ranging from 0 to 2; (2) using baryon acoustic oscillations (BAO) as a cosmological standard ruler over a range of cosmic epochs, with a spectroscopic redshift survey of 100 million galaxies over the redshift range of 0.5 to 2 over $\sim$10 000 (deg)^2, and (3) mapping the weak gravitational lensing distortion (WL) by foreground galaxies of the images of background galaxies at different distances, with a weak lensing survey over $\sim$1000–10 000 (deg)^2 with a median redshift of 1 to 1.5.

JEDI uses three powerful independent probes of cosmic acceleration to provide the redundancy critical for detecting and controlling systematic errors, and placing accurate and precise constraints on the nature of cosmic acceleration. These include the measurement of the cosmic expansion history $H(z)$ as a function of cosmic time in three different ways (SNe Ia, BAO, and WL), to determine whether dark energy density varies with cosmic time; and the measurement of the growth history of cosmic large scale structure $f_g(z)$ in two different ways (galaxy clustering from the galaxy redshift survey and WL), to differentiate between dark energy and modified gravity as causes for the observed cosmic acceleration. In addition, JEDI can use galaxy clusters selected by their weak lensing shear as an additional cosmic acceleration probe to provide cross-checks and further tighten constraints.

The science goals of JEDI are enabled by its instrumental capabilities. JEDI is a 2 m aperture space telescope capable of simultaneous wide-field imaging (0.8–4.2 microns) and multiple object spectroscopy (1–2 or 0.8–3.2 microns) with a field of view of $\sim 1\,(\text{deg})^2$. The JEDI spectrograph fields are adjacent to the imaging fields, allowing simultaneously imaging and spectroscopy in adjacent fields. JEDI uses the microshutter arrays (already developed for JWST) as the programmable spectroscopic multi-slit mask, and HAWAII-2 2048×2048 HgCdTe detectors from Rockwell Scientific. Figure 10.1 shows the main scientific data paths for JEDI and key implementation features.

The use of SNe Ia is the best-established method to probe cosmic acceleration, and the method through which cosmic acceleration was discovered (Riess *et al.*, 1998; Perlmutter *et al.*, 1999). However, supernova spectroscopy is the "bottle-neck" for obtaining a large number of SN Ia events usable for cosmology; slits are needed to obtain sufficient signal-to-noise for SN Ia spectra. JEDI has the unique ability of simultaneously obtaining slit spectra for all objects in the wide field of view, thus providing a factor of 10 improvement in the efficiency of supernova spectroscopy. Because of its unique NIR wavelength coverage (0.8–4.2 microns), JEDI has the advantage of observing SNe Ia in the rest frame J band for the entire redshift range of $0 < z < 2$, where they are less affected by dust, and appear to be nearly perfect standard candles (Phillips *et al.*, 2006).

The great wealth of cosmological data from a mission like JEDI will not only illuminate the nature of cosmic acceleration, but will help us solve other fundamental problems in cosmology as well. These include the nature of dark matter, and the evolution of galaxies and large scale structure in the universe. Although JEDI was not selected by NASA for a JDEM concept study, it provides a useful reference in highlighting the powerful scientific results that can be obtained by the suitable choice of a suite of instruments.

Another example of an optimal dark energy space mission is SPACE (the SPectroscopic All-sky Cosmic Explorer), a mission concept proposed for ESA's *Cosmic-Vision 2015–2025* planning cycle (Robberto and Cimatti, 2009; Cimatti *et al.*, 2009). SPACE aims to produce the largest three-dimensional evolutionary map of the universe over the past 10 billion years by taking NIR spectra and measuring redshifts for more than half a billion galaxies at $0 < z < 2$ down to $H_{AB} \sim 23$ over almost the entire sky. In addition, SPACE will also perform a deep spectroscopic survey of

10.1 Designing the Optimal Dark Energy Experiment

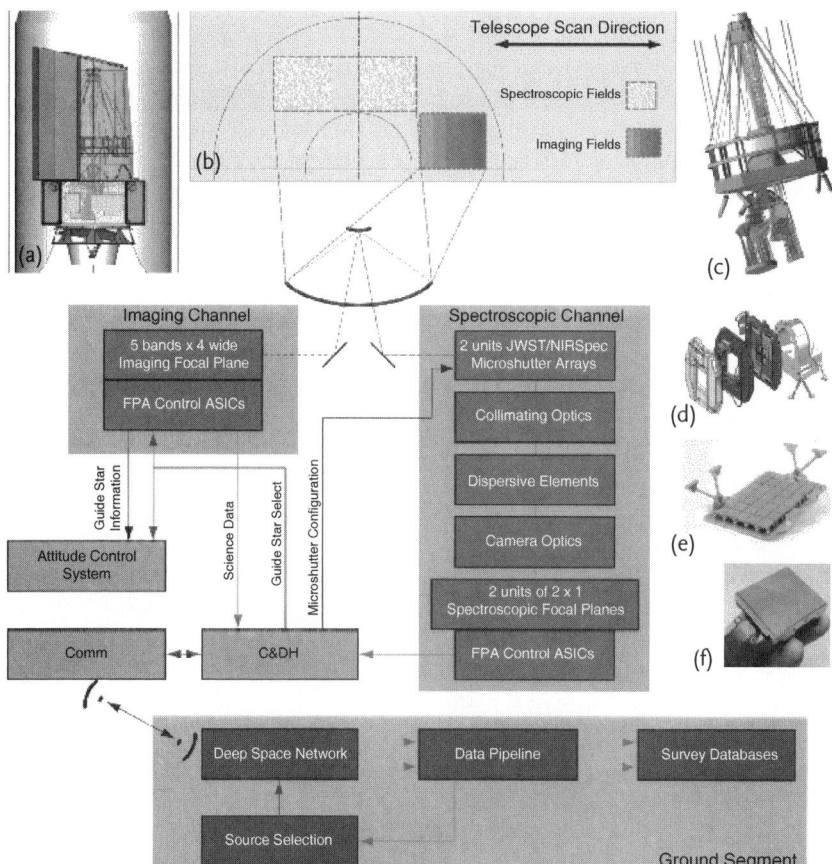

Figure 10.1 Main scientific data paths for JEDI and key implementation features (Cheng et al., 2006). The main body of the figure shows the data paths from the telescope through the ground processing. (a) The flight segment mounted in the fairing of a Delta-IV 4-m configuration. (b) The fields of view of the imaging and spectroscopic channels projected onto the sky. (c) A preliminary optical point design that demonstrates that the desired functions are packageable. (d) An exploded view of the JWST/NIRSpec microshutter array. This exact hardware is baselined for JEDI. Practical packaging constraints for this hardware cause the small horizontal gap between the two spectroscopic fields of view in panel (b). (e) A mechanical mockup of a 5 × 7 focal plane array built by Rockwell Scientific to demonstrate fabrication and alignment processes. (f) A single hybrid detector based on the HAWAII-2RG design that is being produced for three instruments on JWST.

millions of galaxies to $H_{AB} \sim 26$ and at $2 < z < 10+$ over a small field in the sky. These goals are only feasible for a space mission, since ground-based observations suffer ~ 500 times higher sky background (see e.g., Aldering (2001)).

SPACE achieves its main science objectives by using a 1.5 m diameter Ritchey–Chretien telescope with a field of view of $0.4 \, \text{deg}^2$, and a programmable spectroscopic slit-mask made of digital micromirror devices (DMDs). SPACE can obtain spectra of ≈ 6000 targets per pointing at a spectral resolution of $R \sim 400$, and perform diffraction-limited imaging with continuous wavelength coverage from 0.8 μm to 1.8 μm. Owing to the depth, redshift range, volume coverage and quality of its spectra, SPACE can reveal with unique sensitivity most of the fundamental cosmological signatures, including the power spectrum of matter density fluctuations and its turnover. SPACE can also place high accuracy constraints on the dark energy density and its evolution by measuring the evolution of $H(z)$ through baryonic acoustic oscillations, the distance-redshift relations of SNe Ia, the growth rate of cosmic large scale structure $f_g(z)$ (which also provides a test of gravity), and high-z galaxy clusters. The data sets from a mission like SPACE will represent a long lasting legacy for the whole astronomical community.

SPACE, along with DUNE (Refregier et al., 2009), was chosen by ESA for a concept study for a European dark energy mission in 2007. ESA has subsequently merged SPACE and DUNE into the proposed Euclid mission. If Euclid is selected in the next stages in competition with other proposed space missions, it will be launched in 2017.

10.2
Evaluating Dark Energy Experiments

In order to make useful comparisons of different dark energy experiments, it is important to choose the appropriate figure of merit (FoM) for dark energy constraints. Wang (2008a) proposed a relative generalized FoM given by

$$\text{FoM} = \frac{1}{\sqrt{\det \text{Cov}(f_1, f_2, f_3, \ldots)}}, \qquad (10.1)$$

where $\{f_i\}$ are the chosen set of dark energy parameters. For Gaussian distributed errors, FoM is the inverse of the N-dimensional volume enclosed by the 68.3% confidence level contours of all the parameters $\{f_1, f_2, \ldots, f_N\}$. This definition is a generalization of the FoM defined by the Dark Energy Task Force (DETF) (Albrecht et al., 2006), and has the advantage of being easy to calculate for either real or simulated data.

In order for this FoM to represent the dark energy constraints in an optimal manner, the dark energy parameters $\{f_i\}$ should have clear physical meaning, and be minimally correlated. Two simple choices for dark energy parametrization are (Wang, 2008a):

1. $w_X(a) = 3w_{0.5} - 2w_0 + 3(w_0 - w_{0.5})a$, where $(w_0, w_{0.5})$ are the values of $w_X(a)$ at $a = 1$ ($z = 0$) and $a = 2/3$ ($z = 0.5$).
2. $X(z) = \rho_X(z)/\rho_X(0)$ as a free function of z interpolated from its values at $z = 0.5, 1.0,$ and 1.5 ($X_{0.5}, X_{1.0},$ and $X_{1.5}$) for $0 < z < 1.5$, and $X(z > 1.5) = X_{1.5}$.

Parametrization (1) is simpler than and superior to the usual (w_0, w_a) parametrization (with $w_X(a) = w_0 + (1-a)w_a$), since $(w_0, w_{0.5})$ are always significantly less

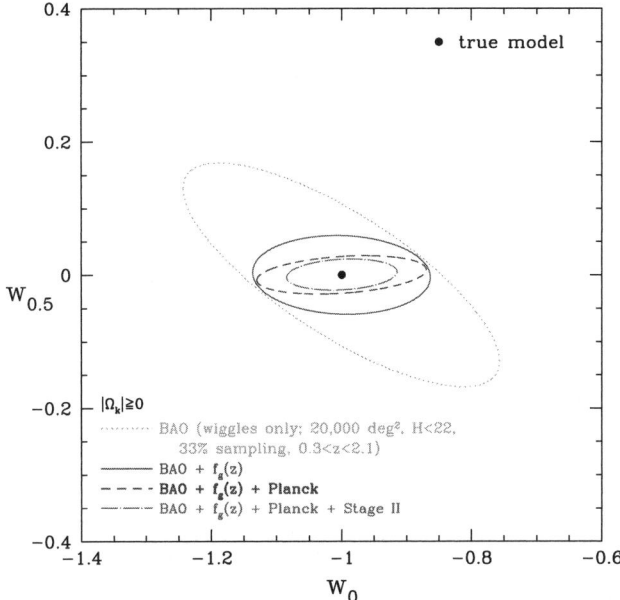

Figure 10.2 Joint 68.3% confidence level contours of $(w_0, w_{0.5})$ derived from the BAO scale measurements ("wiggles only") from a galaxy redshift survey, in combination with the growth rate $f_g(z)$ measurements from the same galaxy redshift survey, the constraints from Planck, and the DETF Stage II priors. The galaxy redshift survey covers 20 000 (deg)^2 to a depth of $H_{AB} = 22$, with 33% sampling, and a redshift range of $0.3 < z < 2.1$.

correlated than (w_0, w_a), as long as (Wang, 2008a)

$$\sigma^2(w_0) < \frac{2}{3}\left|\sigma^2(w_0 w_a)\right|. \tag{10.2}$$

Figures 10.2–10.3 show the joint 68.3% C.L. contours of $(w_0, w_{0.5})$ and (w_0, w_a) derived from the BAO scale measurements ("wiggles only") from a galaxy redshift survey, in combination with the growth rate $f_g(z)$ measurements from the same galaxy redshift survey, the constraints from Planck, and the DETF Stage II priors (expected constraints from ongoing dark energy projects) (Albrecht et al., 2006). The galaxy redshift survey covers 20 000 (deg)^2 to a depth of $H_{AB} = 22$, with 33% sampling, and a redshift range of $0.3 < z < 2.1$ (Cimatti et al., 2009). Clearly, Figure 10.2 is more informative in indicating how the various probes contribute to the joint dark energy constraints. It shows clearly that adding the growth rate $f_g(z)$ measurements to the BAO measurements breaks the degeneracy between w_0 and $w_{0.5}$, further adding of Planck priors mainly tightens the constraints on $w_{0.5}$, and further adding of DETF Stage II priors mainly tightens the constraints on w_0.

Note that for real data analyzed using the Markov Chain Monte Carlo (MCMC) method, the covariance matrices of $(w_0, w_{0.5})$ and (w_0, w_a) cannot be transformed into each other, since choosing $(w_0, w_{0.5})$ and choosing (w_0, w_a) as the base parameters correspond to different priors, if uniform priors are assumed for the base

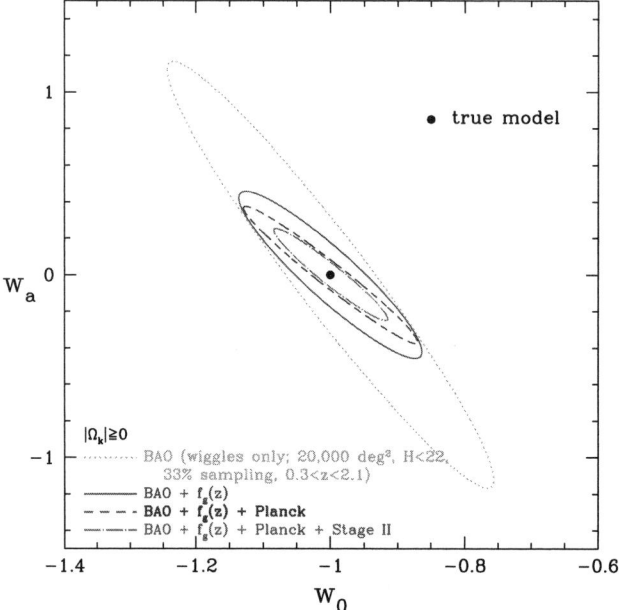

Figure 10.3 The same as Figure 10.2, but for (w_0, w_a).

parameters. For the Fisher matrix forecast, one can simply transform the covariance matrix of (w_0, w_a) into that of $(w_0, w_{0.5})$ by using (see Eq. (8.32)):

$$w_{0.5} = w_0 + \frac{w_a}{3} \,. \tag{10.3}$$

Parametrization (2) allows us to obtain minimal model-independent constraints on dark energy. The FoM for (w_0, w_a) and $(X_{0.5}, X_{1.0}, X_{1.5})$ for different experiments allow us to evaluate their ability to answer the first fundamental question about dark energy posed in Section 1.2: *Is dark energy density constant in cosmic time?*

A minimal way to evaluate the capability of a survey to test gravity is through its forecast measurement of the power law index γ of the growth rate $f_g(z)$:

$$f_g(z) = \left[\Omega_m(a)\right]^\gamma \,, \tag{10.4}$$

where $\Omega_m(a) \equiv 8\pi G \rho_m(a)/(3H^2)$ (see Eq. (1.32)). Lue, Scoccimarro, and Starkman (2004) pointed out that the value of γ allows one to differentiate between modified gravity and dark energy, since $f_g(z) \simeq \Omega_m(a)^{2/3}$ for DGP gravity models, while $f_g(z) \simeq \Omega_m(a)^{5/9}$ for a flat universe with a cosmological constant. The value of $f_g(z)$ does not vary significantly among dark energy models; Wang and Steinhardt (1998) found that $f_g(z) \simeq \Omega_m(a)^{6/11}$ for quintessence models. A high-precision fit to dark energy models is given by (Linder, 2005)

$$\gamma \simeq 0.55 + 0.05\left[1 + w_X(z=1)\right] \quad w_X > -1 \tag{10.5}$$

$$\gamma \simeq 0.55 + 0.02\left[1 + w_X(z=1)\right] \quad w_X < -1 \,. \tag{10.6}$$

A more rigorous and model-independent approach to testing gravity is to use the measurements of $f_g(z)$ and $H(z)$ together, as discussed in Section 5.6.3. If an alternative gravity model gives the same fit to the $H(z)$ data as a dark energy model, it will likely not fit the $f_g(z)$ data, and the usual χ^2 statistic can be used to compute the confidence level at which the alternative gravity model can be ruled out (Wang, 2008b). Thus we should also add two more parametrizations:

3. $f_g(z)$ as a free function of z measured in $\Delta z = 0.2$ bins for $0.2 \lesssim z \lesssim 2$.
4. $H(z)$ as a free function of z measured in $\Delta z = 0.2$ bins for $0.2 \lesssim z \lesssim 2$.

A modest baseline galaxy redshift survey from space can cover 20 000 (deg)^2 to a depth of $H_{AB} = 22$, with 33% sampling, and a redshift range of $0.3 < z < 2.1$ (Cimatti et al., 2009). Such a survey can provide measurements of the cosmic expansion history $H(z)$ and the growth rate $f_g(z)$ that are more than a factor of ten better than current measurements (see Figure 5.15).

In comparing the FoM for $f_g(z)$ and $H(z)$ for different dark energy experiments, we can rank their ability to answer the second fundamental question about dark energy posed in Section 1.2 (without assuming a modified gravity model): *Is gravity modified?*

10.3
Current Status and Future Prospects

There are a large number of dark energy surveys that are ongoing or have been proposed. Ongoing projects include Essence (http://www.ctio.noao.edu/~wsne/), Supernova Legacy Survey (SNLS) (http://www.cfht.hawaii.edu/SNLS/), Carnegie Supernova Project (CSP) (http://www.ociw.edu/csp/), ESO Visible and Infrared Survey Telescope for Astronomy (VISTA) Surveys (http://www.vista.ac.uk/), Panoramic Survey Telescope & Rapid Response System (Pan-STARRS) (http://pan-starrs.ifa.hawaii.edu/), and WiggleZ (http://wigglez.swin.edu.au/). Proposed near-term projects include Advanced Liquid-mirror Probe for Astrophysics, Cosmology and Asteroids (ALPACA) (http://www.astro.ubc.ca/LMT/alpaca/); Dark Energy Survey (DES) (http://www.darkenergysurvey.org/); Hobby-Eberly Telescope Dark Energy Experiment (HETDEX) (http://www.as.utexas.edu/hetdex/); and Sloan Digital Sky Survey (SDSS) III (http://www.sdss.org/). Proposed long-term projects include Large Synoptic Survey Telescope (LSST) (http://www.lsst.org/), Joint Dark Energy Mission (JDEM) (http://jdem.gsfc.nasa.gov/), Square Kilometre Array (SKA) (http://www.skatelescope.org/), and Euclid (http://sci.esa.int/science-e/www/area/index.cfm?fareaid=102).

The US Dark Energy Task Force (DETF) (Albrecht et al., 2006) has recommended an aggressive, multi-stage, multi-method program to explore dark energy as fully as possible. DETF recommended a dark energy program with multiple techniques at every stage, with at least one of these being a probe sensitive to the growth of cosmic structure in the form of galaxies and clusters of galaxies.

The DETF defined a figure of merit (FoM) that is the inverse of the area enclosed by the 95% confidence level error ellipse in the w_0–w_a plane (assuming a dark energy equation of state $w_X(a) = w_0 + w_a(1-a)$). DETF recommended that dark energy program in Stage III (near-term, medium-cost projects) should be designed to achieve at least a factor of 3 gain over Stage II (ongoing projects) in the FoM, and that a dark energy program in Stage IV (long-term, high-cost projects JDEM, LST, SKA) should be designed to achieve at least a factor of 10 gain over Stage II in the FoM.[13]

The DETF recommended continued research and development investment to optimize JDEM, LST, and SKA (Stage IV) to address remaining technical questions and systematic-error risks, and high priority for near-term projects to improve understanding of dominant systematic effects in dark energy measurements, and wherever possible, reduce them. The DETF recommended a coherent program of experiments designed to meet the goals and criteria it proposed.

The ESA-ESO Working Group on Fundamental Cosmology (Peacock et al., 2006) made specific recommendations to ESA and ESO. It recommended a wide-field optical and NIR imaging survey (suitable for weak lensing and cluster surveys) as a high priority, with ESA launching a satellite for high-resolution wide-field optical and NIR imaging, and ESO carrying out optical multi-color photometry, as well as a large spectroscopic survey (>100 000 redshifts over ~10 000 (deg)^2) to calibrate photometric redshifts. They also recommended that ESA-ESO secure access to an instrument with capability for massive multiplexed deep spectroscopy (several thousand simultaneous spectra over 1 (deg)^2) suitable for large galaxy redshift surveys, as well as conduct a supernova survey with multi-color imaging to extend existing samples of $z = 0.5 - 1$ SNe Ia by an order of magnitude, and improve the local sample of SNe Ia. They suggested the use of an European Extremely Large Telescope (ELT) to study SNe Ia at $z > 1$.

The US National Research Council's Committee on NASA's Beyond Einstein Program recently made the recommendation that NASA and DOE should proceed immediately with a competition to select a JDEM for a 2009 new start. They concluded that, "The broad mission goals in the Request for Proposal should be (1) to determine the properties of dark energy with high precision and (2) to enable a broad range of astronomical investigations. The committee encourages the Agencies to seek as wide a variety of mission concepts and partnerships as possible."

Euclid is the proposed European-led dark energy mission being studied for ESA's Cosmic Vision 2015–2025 program, and the result of merging two mission concepts, SPACE (the SPectroscopic All-sky Cosmic Explorer) (Cimatti et al., 2009) and DUNE (the Dark UNiverse Explorer) (Refregier et al., 2009). The synergy between JDEM and Euclid will be important for a strategically optimized approach to discovering the nature of cosmic acceleration.

In evaluating the various dark energy projects, it is critical to remember that the challenge to solving the cosmic acceleration mystery will not be the statistics of

---

13) In practice, it is far more convenient to use FoM $= 1/\sqrt{\det \mathrm{Cov}(w_0, w_a)}$ defined in Eq. (10.1), since the factor of change in FoM is the same as that of FoM$_\mathrm{DETF}$.

the data obtained, but the tight control of systematic effects inherent in the data. A combination of the most promising methods (as discussed in this book), each optimized by having its systematics minimized by design, provides the tightest control of systematics (Wang *et al.*, 2004; Crotts *et al.*, 2005; Cheng *et al.*, 2006). Ultimately, the discovery of the nature of cosmic acceleration will revolutionize our understanding of the universe.

References

Abdalla, F.B. et al. (2008) *MNRAS*, **387**, 969.
Abraham, R.G. et al. (2004) *AJ*, **127**, 2455.
Abrahamse, A. et al. (2008) *PRD*, **77**, 3503.
Adelberger, K.L. and Steidel, C.C. (2000) *ApJ*, **544**, 218.
Aguirre, A.N. (1999) *ApJ*, **525**, 583.
Aguirre, A.N. and Haiman, Z. (2000) *ApJ*, **532**, 28.
Alam, U. and Sahni, V. (2006) *PRD*, **73**, 084024.
Albrecht, A., Bernstein, B., Cahn, R., Freedman, W.L., Hewitt, J., Hu, W., Huth, J., Kamionkowski, M., Kolb, E.W., Knox, L., Mather, J.C., Staggs, S., and Suntzeff, N.B. *Report of the Dark Energy Task Force*, astro-ph/0609591.
Alcock, C. and Paczynski, B. (1979) *Nature*, **281**, 358.
Aldering, G. (2001) LBNL report number LBNL-51157.
Aldering, G. et al. arXiv:astro-ph/0405232.
Allen, S.W. et al. (2004) *MNRAS*, **353**, 457.
Allen, S.W. et al. (2008) *MNRAS*, **383**, 879.
Almeida, C., Baugh, C.M., and Lacey, C.G. (2007) *MNRAS*, **376**, 1711.
Almeida, C., Baugh, C.M., Wake, D.A., Lacey, C.G., Benson, A.J., Bower, R.G., and Pimbblet, K. (2008) *MNRAS*, **386**, 2145.
Alternative Gravity Theory references: see e.g. Sakharov, A.D. (1968) *Sov. Phys. Dokl.*, **12**, 1040; Jacobson, T. (1995) *Phys. Rev. Lett.*, **75**, 1260; Padmanabhan, T. (2002) *Mod. Phys. Lett. A*, **17**, 1147; (2004) *Int. Jour. Mod. Phys. D*, **13**, 2293–2298; (2003) **18**, 2903; (2004) *Class. Quan. Grav.*, **21**, 4485; Volovik, G.E. (2001) *Phys. Rept.*, **351**, 195; Visser, M. (2002) *Mod. Phys. Lett.*, **A17**, 977; Barcelo, C. et al. (2001) *Int. J. Mod. Phys.*, **D10**, 799; Volovik, G.E. gr-qc/0604062; Volovik, G.E. (2003) *The universe in a helium droplet*, Oxford University Press; Huang, C.-G., Sun, J.-R. gr-qc/0701078; Makela, J. gr-qc/0701128.
Amara, A. and Refregier, A. (2007) arXiv:0710.5171.
Amati, L. et al. (2002) *A & A*, **390**, 81.
Amendola, L., Polarski, D., and Tsujikawa, S. arXiv:astro-ph/0603703; arXiv:astro-ph/0605384.
Angulo, R. et al. (2005) *Mon. Not. Roy. Astron. Soc. Lett.*, **362**, L25–L29.
Angulo, R.E., Baugh, C.M., Frenk, C.S., and Lacey, C.G. (2008) *MNRAS*, **383**, 755.
Astier, P. et al. (2006) *A & A*, **447**, 31.
Bacon, D.J., Refregier, A.R., and Ellis, R.S. (2000) *MNRAS*, **318**, 625.
Bahcall, N.A., Fan, X., and Cen, R. (1997) *ApJ*, **485**, L53.
Bahcall, N.A., Ostriker, J.P., Perlmutter, S., and Steinhardt, P.J. (1999) *Science*, **284**, 1481.
Bailey, S. et al. (2009) arXiv:0905.0340.
Baker, J.G. (1969) *IEEE Trans. Aerosp. Electron. Syst. AES-5*, **261**.
Band, D. (1997) *ApJ*, **486**, 928.
Barber, A.J., Thomas, P.A., Couchman, H.M.P., and Fluke, C.J. (2000) *MNRAS*, **319**, 267.
Barber, A.J. (2003) private communication.

Barreiro, T., Bertolami, O., and Torres, P. (2008) *PRD*, **78**, 043530.

Barris, B.J. (2004) *ApJ*, **602**, 571.

Bartelmann, M. and Schneider, P. (2001) *Phys. Rept.*, **340**, 291–472.

Baugh C.M., Lacey C.G., Frenk C.S., Granato G.L., Silva L., Bressan A., Benson A.J., and Cole S. (2005) *MNRAS*, **356**, 1191.

Baugh, C.M. (2006) *Reports of Progress in Physics*, **69**, 3101.

Bean, R. and Dore, O. (2003) *PRD*, **68**, 023515.

Begelman, M.C., Blandford, R.D., and Rees, M.J. (1984) *Rev. Mod. Ph.*, **56**, 255.

Begelman, M.C. and Cioffi, D.F. (1989) *ApJ*, **345**, L21.

Bennett, C.L. et al. (2003) *Astrophys. J. Suppl.*, **148**, 1.

Benson, A.J., Cole, S., Frenk, C.S., Baugh, C.M., and Lacey, C.G. (2000) *MNRAS*, **311**, 793.

Bernardeau, F., Van Waerbeke, L., and Mellier, Y. (1997) *A & A*, **322**, 1.

Bernstein, G.M. and Jarvis, M. (2002) *Astron. J.*, **123**, 583.

Bernstein, G. and Jain, B. (2004) *ApJ*, **600**, 17.

Bertotti, B., Iess, L., and Tortora, P. (2003) *Nature*, **425**, 374.

Blake, C. and Glazebrook, K. (2003) *ApJ*, **594**, 665.

Blake, C. et al. (2006) *MNRAS*, **365**, 255.

Blandford, R.D. and Rees, M.J. (1974) *MNRAS*, **169**, 395.

Blandford, R.D., Saust, A.B., Brainerd, T.G., and Villumsen, J.V. (1991) *MNRAS*, **251**, 600.

Bloom, J.S., Frail, D.A., and Kulkarni, S.R. (2003) *ApJ*, **594**, 674.

Bode, P. et al. (2007) *ApJ*, **663**, 139.

Boehmer, C.G., Caldera-Cabral, G., Lazkoz, R., and Maartens, R. (2008) *PRD*, 78, 023505.

Bongard, S. et al. (2006) *ApJ*, **647**, 513.

Bordemann, M. and Hoppe, J. (1993) *Phys. Lett. B*, **317**, 315.

Borgani, S. et al. (2004) *MNRAS*, **348**, 1078.

Bouchet, F.R., Juszkiewicz, R., Colombi, S., and Pellat, R. (1992) *ApJ*, **394**, L5.

Boulade, O. et al. (2003) *Proceedings of the SPIE*, Vol. 4841, pp. 72–81.

Branch, D. (1981) *ApJ*, **248**, 1076.

Branch, D. and Tammann, G.A. (1992) *Annu. Rev. Astron. Astrophys.*, **30**, 359.

Branch, D., Perlmutter, S., Baron, E., and Nugent, P. (2001) astro-ph/0109070, Contribution to the *SNAP*, Supernova Acceleration Probe, Yellow Book (Snowmass 2001).

Branch, D. et al. (2004) *ApJ*, **606**, 413.

Branch, D. (2009) private communication.

Branch, D., Dang, L.C., and Baron, E. (2009) *PASP*, in press.

Bridle, S. and King, L. (2007) *NJPh*, **9**, 444.

Bridle, S. et al. (2009) *Annals of Applied Statistics*, **3**, 6.

Caldera-Cabral, G., Maartens, R., and Urena-Lopez, L.A. arXiv:0812.1827

Caldwell, R.R. (2002) *Phys. Lett. B*, **545**, 23–29.

Caldwell, R.R. and Kamionkowski, M. arXiv:0903.0866.

Cappellaro, E., Evans, R., and Turatto, M. (1999) *A&A*, **351**, 459.

Capozziello, S. and Salzano, V. arXiv:0902.0088v1 [astro-ph.CO]

Cardelli, J.A., Clayton, G.C., and Mathis, J.S. (1989) *ApJ*, **345**, 245.

Carroll, S.M., Hoffman, M., and Trodden, M. (2003) *PRD*, **68**, 3509.

Cash, W. (1979) *ApJ*, **228**, 939.

Catelan, P. and Moscardini, L. (1994) *ApJ*, **436**, 5.

Catelan, P., Lucchin, F., Matarrese, S., and Moscardini, L. (1995) *MNRAS*, **276**, 39.

Chaplygin, S. (1904) *Sci. Mem. Moscow Univ. Math. Phys.*, **21**, 1.

Charmousis, C., Gregory, R., Kaloper, N., and Padilla, A. (2006) *JHEP*, **0610**, 066.

Cheng, E., Wang, Y., Baron, E., Branch, D., Casertano, S., Crotts, A., Drosdat, H., Dubord, L., Egerman, R., Garnavich, P., Gulbransen, D., Kutyrev, A., MacKenty, J.W., Miles, J.W., Moustakas, L., Phillips, M., Roellig, T., Silverberg, R., Squires, G., Wheeler, J.C., Wright, E.L. (2006) *Proc. of SPIE*, Vol. **6265**, 626529.

Chevallier, M. and Polarski, D. (2001) *Int. J. Mod. Phys. D*, **10**, 213.

Chuh, T. et al. (2006) *Proceedings of the SPIE*, Vol. 6265, 62652.

Chung, D.J. and Freese, K. (2000) *PRD*, **61**, 023511.

Cimatti, A. et al. (the SPACE Science Team) (2009) *Experimental Astronomy*, **23**, 39.

Cline, J.M., Jeon, S., and Moore, G.D. (2004), *PRD*, **70**, 3543.

Cohn, J.D. (2006) *New Astron. Rev.*, **11**, 226.

Cole, S. et al. (2005) *MNRAS*, **362**, 505.

Conley, A. et al. (2007) *ApJ*, **664**, L13.

Connolly, A.J. et al. (1995), *AJ*, **110**, 2655.

Cooray, A., Holz, D., and Huterer, D. (2006) *ApJ*, **637**, L77.
Copeland, E.J., Liddle, A.R., and Wands, D. (1998) *PRD*, **57**, 4686.
Copeland, E.J., Sami, M., and Tsujikawa, S. (2006) *IJMPD*, **15**, 1753.
Crain, R.A., Eke, V.R., Frenk, C.S., Jenkins, A.J., McCarthy, I.G., Navarro, J.F., and Pearce, F.R. (2007) *MNRAS*, **377**, 41.
Crittenden, R.G., Natarajan, P., Pen, U.-L., and Theuns, T. (2001) *ApJ*, **559**, 552.
Crittenden, R.G., Natarajan, P., Pen, U.-L., and Theuns, T. (2002) *ApJ*, **568**, 20.
Crocce M. and Scoccimarro R. (2006) *PRD*, **73**, 063520.
Crocce M. and Scoccimarro R. (2008) *PRD*, **77**, 023533.
Croft, R. and Metzler, C. (2000) *Astrophys. J.*, **545**, 561.
Crotts, A *et al.* (2005) astro-ph/0507043.
Current Data Results references: see e.g. Barger, V., Gao, Y., Marfatia, D. astro-ph/0611775; Jassal, H.K., Bagla, J.S., Padmanabhan, T. astro-ph/0601389; Dick, J., Knox, L., Chu, M. (2006) *JCAP*, 0607, 001; Li, C., Holz, D.E., and Cooray, A. astro-ph/0611093; Liddle, A.R., Mukherjee, P., Parkinson, D., Wang, Y. (2006) *PRD*, **74**, 123506; Nesseris, S. and Perivolaropoulos, L. astro-ph/0612653; Wilson, K.M., Chen, G., Ratra, B. astro-ph/0602321; Xia, J.-Q. *et al.* (2006) *PRD*, **74**, 083521; Alam, U., Sahni, V. Starobinsky, A.A. (2007) *JCAP*, 0702, 011; Daly, R.A. *et al.* arXiv:0710.5112; Caldwell, R., Cooray, A., Melchiorri, A. astro-ph/0703375
da Angela, J. *et al.* astro-ph/0612401.
da Silva, A.C., Kay, S.T., Liddle, A.R., and Thomas, P.A. (2004) *MNRAS*, **348**, 1401.
Dahlen, T. *et al.* (2004) *ApJ*, **613**, 189.
Daly, R.A. (1990) *ApJ*, **355**, 416.
Daly, R.A. (1994) *ApJ*, **426**, 38.
Daly, R.A. (2009) private communication.
Daly, R.A. and Djorgovski, S.G. (2003) *ApJ*, **597**, 9.
Daly, R.A. and Guerra, E.J. (2002) *AJ*, **124**, 1831.
Daly, R.A. *et al.* (2009) *ApJ*, **691**, 1058.
Daly, R.A., Guerra, E.J., and Wan, L. (1998) *BAAS*, **30**, 843.
Dark Energy Reviews: see e.g. Padmanabhan, T. (2003) *Phys. Rep.*, **380**, 235; Peebles, P.J.,E., Ratra, B. (2003) *Rev. Mod. Phys.*, **75**, 55; Ruiz-Lapuente, P. (2007) *Class. Quantum. Grav.*, 24, 91; Ratra, B., Vogeley, M.S. (2007) arXiv:0706.1565; Frieman, J., Turner, M., Huterer, D. *ARAA*, in press, arXiv:0803.0982
Davis, M. and Huchra, J. (1982) *ApJ*, **254**, 437.
Davis, T.M. *et al.* (2007) *ApJ*, **666**, 716.
Davis, M. and Peebles, P.J.E. (1983) *ApJ*, **267**, 465.
Davis, M., Efstathiou, G., Frenk, C.S., and White, S.D.M. (1985) *ApJ*, **292**, 371.
de Bernardis, P. *et al.* (2000) *Nature*, **404**, 955.
De Felice, A., Mukherjee, P., and Wang, Y. (2008) *Phys. Rev. D*, **77**, 024017.
de Rham, C. *et al.* (2008) *Phys. Rev. Lett.*, **100**, 251603.
Dick, J., Knox, L., and Chu, M. (2006) *JCAP*, **0607**, 001.
Dodelson, S. (2003) *Modern Cosmology*, Academic Press.
Dodelson, S. and Vallinotto, A. (2005) astro-ph/0511086.
Dolney, D., Jain, B., and Takada, M. (2006) *MNRAS*, **366**, 884.
Dominguez, I., Höflich, P., and Straniero, O. (2001) *ApJ*, **557**, 279.
Dunkley, J. *et al.* (2008) arXiv:0803.0586.
Dunlop, J. *et al.* (1996) *Nature*, **381**, 581.
Dvali, G., Gabadadze, G., and Porrati, M. (2000) *PLB*, **485**, 208.
Dyer, C. and Roeder, R. (1973) *ApJ*, **180**, L31.
Eddington, A.S. (1957) *The mathematical theory of relativity*, Cambridge University Press.
Eke, V.R., Navarro, J.F., and Frenk, C.S. (1998) *ApJ*, **503**, 569.
Eisenstein, D. *et al.* (2005) *ApJ*, **633**, 560.
Eisenstein, D.J. *et al.* (2007) *ApJ*, **664**, 675.
Eisenstein, D. and Hu, W. (1998) *ApJ*, **496**, 605.
Eisenstein, D.J. and Hu, W. (1999) *ApJ*, **511**, 5; Hu, W., Eisenstein, D.J. and Tegmark, M. (1998) *PRL*, **80**, 5255.
Eisenstein, D.J., Seo, H., Sirko, E., and Spergel, D. (2006b) *ApJ*, **664**, 675.
Ettori S., Dolag K., Borgani S., and Murante G. (2006) *MNRAS*, **365**, 1021.
Evrard, A.E., Metzler, C.A., and Navarro, J.F. (1996) *ApJ*, **469**, 494.
Fairbairn, M. and Goobar, A. (2006) *Phys. Lett. B*, **642**, 432–435.
Feldman, H.A., Kaiser, N., and Peacock, J.A. (1994) *ApJ*, **426**, 23.

Fenimore, E.E. and Ramirez-Ruiz, E. arXiv:astro-ph/0004176.
Figueroa, D.G., Verde, L., and Jimenez, R. arXiv:0807.0039v1 [astro-ph].
Filippenko, A. (1997) *Annu. Rev. Astron. Astrophys.*, **35**, 309.
Fixsen, D.J. (1996) *ApJ*, **473**, 576.
Freedman, W.L. et al. (2001) *ApJ*, **553**, 47.
Freese, K. and Lewis, M. (2002) *Phys. Lett. B*, **540**, 1.
Fry, J.N. and Gaztanaga, E. (1993) *ApJ*, **413**, 447.
Fu, L. et al. (2008) *A & A*, **479**, 9.
Gardner, J.P. et al. (2006) *Space Sci. Rev.*, **123**, 485.
Garnavich, P.M. et al. (2004) *ApJ*, **613**, 1120.
Garriga, J., Linde, A., and Vilenkin, A. (2004) *Phys. Rev. D*, **69**, 063521.
Gaztanaga, E., Cabre, A., and Hui, L. (2008) arXiv:0807.3551.
Geach, J.E. et al. (2009) MNRAS in press.
Gerardy, C.L. et al. (2004) *ApJ*, **607**, 391.
Ghezzi, C.R. et al. (2004) *MNRAS*, **348**, 451.
Ghirlanda, G., Ghisellini, G., and Lazzati, D. (2004) *ApJ*, **616**, 331.
Giavalisco, M. et al. (2004) *ApJ*, **600**, L103.
Goldhaber, G. et al. (2001) *ApJ*, **558**, 359.
Gondolo, P. and Freese, K. (2003) *Phys. Rev. D*, **68**, 063509, hep-ph/0211397.
GRB and cosmology references: Xu, D., Dai, Z.G., and Liang, E.W. (2005) *ApJ*, **633**, 603; Nava, L., Ghisellini, G., Ghirlanda, G., Tavecchio, F., and Firmani, C. (2006), *A&A*, **450**, 471; Qi, S., Wang, F.-Y., and Lu, T. (2008) *A&A*, **483**, 49–55.
Guerra, E.J. and Daly, R.A. (1998) *ApJ*, **493**, 536.
Guerra, E.J., Daly, R.A., and Wan, L. (2000) *ApJ*, **544**, 659.
Gunn, J.E. et al. (1998) *AJ*, **116**, 3040.
Gunn, J.E. et al. (2006) *AJ*, **131**, 2332.
Guy, J., Astier, P., Nobili, S., Regnault, N., and Pain, R. (2005) *A & A*, **443**, 781.
Guy, J. et al. (2007) *A & A*, **466**, 11.
Guzik, J. and Bernstein, G. (2005) *PRD*, **72**, 3503.
Guzzo, L. et al. (2008) *Nature*, **451**, 541.
Hachinger, S. et al. (2008) *MNRAS*, **389**, 1087.
Hachisu, I., Kato, M., and Nomoto, K. (2008) *ApJ*, **683**, L127.
Haiman, Z., Mohr, J.J., and Holder, G.P. (2001) *ApJ*, **553**, 545.
Hallman, E.J., Motl, P.M., Burns, J.O., and Norman, M.L. (2006) *ApJ*, **648**, 852.
Hamilton, A.J.S. (1992) *ApJ*, **385**, L5.
Hamilton, A.J.S. (1998) *The Evolving Universe*, (ed. D. Hamilton), Kluwer Academic, p. 185–275, astro-ph/9708102.
Hamuy, M. et al. (1996) *AJ*, **112**, 2398.
Hardin, D. et al. (2000), *A & A*, **362**, 419.
Hawkins, E. et al. (2003) *MNRAS*, **346**, 78.
Heavens, A.F., Matarrese, S., and Verde, L. (1998) *MNRAS*, **301**, 797.
Hetterscheidt, M. et al. (2007) *A & A*, **468**, 859.
Heymans, C. et al. (2006) *MNRAS*, **368**, 1323.
HgCdTe references: Norton, P. (2002) *Opto-Electronics Review*, **10** (3), 159; Rogalski, A. (2005) *Rep. Prog. Phys.*, **68**, 2267.
Hicken, M. et al. (2007) *ApJ*, **669**, 17.
Hirata, C.M. and Seljak, U. (2004) *Phys. Rev. D*, **70**, 063526.
Hirata, C.M., Mandelbaum, R., Ishak, M., and Seljak, U. (2007) *MNRAS*, **381**, 1197.
Hirata, C.M. (2009) private communication.
Hockney, R.W. and Eastwood, J.W. (1988) *Computer simulation using particles*, Bristol, Hilger.
Hoekstra, H. (2007) *MNRAS*, **379**, 317.
Hoekstra, H., Franx, M., Kuijken, K., and Squires, G. (1998) *ApJ*, **504**, 636.
Hoekstra, H., Yee, H., and Gladders, M.D. (2002) *ApJ*, **577**, 595.
Hoekstra, H. et al. (2006) *ApJ*, **647**, 116.
Hoekstra, H. and Jain, B. (2008) *Annual Review of Nuclear and Particle Science*, **58**, 99.
Höflich, P. and Khokhlov, A. (1996) *ApJ*, **457**, 500.
Höflich, P., Wheeler, J.C., and Thielemann, F.K. (1998) *ApJ*, **495**, 617.
Hogg, D.W., Baldry, I.K., Blanton, M.R., and Eisenstein, D.J. arXiv:astro-ph/0210394.
Hoppe, J. hep-th/9311059.
Howell, D.A. et al. (2006) *Nature*, **443**, 308.
Hsiao, E.Y., Conley, A., Howell, D.A., Sullivan, M., Pritchet, C.J., Carlberg, R.G., Nugent, P.E., and Phillips, M.M. (2007) *ApJ*, **663**, 1187.
Hu, W. and Sugiyama, N. (1996) *ApJ*, **471**, 542.
Hu, W., Spergel, D.N., and White, M. (1997) *PRD*, **55**, 3288.
Huterer, D. et al. (2006) *MNRAS*, **366**, 101.
Hutsi, G. (2006) *Astron. & Astrophys.*, **449**, 891.
Jackiw, R. and Polychronakos, A.P. (2000) *Phys. Rev. D*, **62**, 085019.

Jain, B. and Taylor, A. (2003) *PRL*, **91**, 141302.
Jenkins, A. et al. (2001) *MNRAS*, **321**, 372.
Jeong, D. and Komatsu, E. (2006) *ApJ*, **651**, 619.
Jha, S. et al. (2006) *AJ*, **131**, 527.
Jha, S., Riess, A.G., and Kirshner, R.P. (2007) *ApJ*, **659**, 122.
Jimenez, R. and Loeb, A. (2002) *ApJ*, **573**, 37.
Jimenez, R., Verde, L., Treu, T., and Stern, D. (2003) *ApJ*, **593**, 622.
Jones, L.R. et al. (1998) *ApJ*, **495**, 100.
Kaiser, N. (1986) *MNRAS*, **222**, 323; (1991) *ApJ*, **383**, 104.
Kaiser, N. (1987) *MNRAS*, **227**, 1.
Kaiser, N. (1998) *ApJ*, **498**, 26.
Kaiser, N., Squires, G., and Broadhurst, T. (1995) *ApJ*, **449**, 460.
Kaiser, N., Wilson, G., and Luppino, G.A. (2000) arXiv:astro-ph/0003338.
Kallosh, R. et al. (2003) *JCAP*, **0310**, 015.
Kaloper, N. and Sorbo, L. (2006) *JCAP*, **0604**, 007.
Kamenshchik, A.Y., Moschella, U., and Pasquier, V. (2001) *Phys. Lett. B*, **511**, 265.
Kantowski, R. (1998) *ApJ*, **507**, 483.
Kasen, D. et al. (2003) *ApJ*, **593**, 788.
Kasen, D., Nugent, P., Thomas, R.C., and Wang, L. (2004) *ApJ*, **610**, 876.
Kasen, D. (2006) *ApJ*, **649**, 939.
Kasen, D. and Woosley, S.E. (2007) *ApJ*, **656**, 661.
Kauffmann, G., Nusser, A., and Steinmetz, M. (1997) *MNRAS*, **286**, 795.
Kay, S.T., Thomas, P.A., Jenkins, A., and Pearce, F.R. (2004) **355**, 1091.
Kendall, M.G. and Stuart, A. (1969) *The Advanced Theory of Statistics*, Vol. II, Griffin, London.
Kessler, R. et al. (2009) ApJS, **185**, 32.
Kim, A., Goobar, A., and Perlmutter, S. (1996) *PASP*, **108**, 190.
Knop, R.A. (2003) *ApJ*, **598**, 102.
Knox, L., Song, Y.S., and Tyson, J.A. (2006) *PRD*, **74**, 023512.
Koehler, R.S., Schuecker, P., and Gebhardt, K. (2007) *A&A*, **462**, 7.
Kolb, E.W. and Turner, M.S. (1990) *The Early Universe*, Addison-Wesley.
Komatsu, E. et al. (2009) *ApJS*, **180**, 330.
Korsch, D. (1977) *Appl. Opt.*, **16**, 2074–2077.
Kowalski, M. et al. (2008) *Astrophys. J*, **686**, 749.
Koyama, K. arXiv:0709.2399

Kravtsov, A.V., Nagai, D., and Vikhlinin, A.A. (2005) *ApJ*, **625**, 588.
Kravtsov, A.V., Vikhlinin, A., and Nagai, D. (2006) *ApJ*, 650, 128.
Krisciunas, K. et al. (2004a) *AJ*, **127**, 1664.
Krisciunas, K. et al. (2004b) *AJ*, **128**, 3034.
Krisciunas, K. et al. (2007) *AJ*, **133**, 58.
Krisciunas, K., Phillips, M.M., and Suntzeff, N.B. (2004) *ApJ*, **602**, L81.
Kuo, C.L. et al. (2004) *Astrophys. J.*, **600**, 32.
Landy, S.D. and Szalay, A.S. (1993) *ApJ*, **388**, 310.
Lanzetta, K.M., Yahil, A., and Fernandez-Soto, A. (1996), *Nature*, **381**, 759.
Lentz, E., Baron, E., Branch, D., Hauschildt, P.H., and Nugent, P. (2000) *ApJ*, **530**, 966.
Lewis, A. and Bridle, S. (2002) *Phys. Rev. D*, **66**, 103511.
Lewis, A., Challinor, A., and Lasenby, A. (2000) *ApJ*, **538**, 473; http://camb.info/.
Li, L.-X. (2007) *MNRAS*, **374**, L20.
Liang, N., Xiao, W.K., Liu, Y., and Zhang, S.N. (2008) *ApJ*, in press arXiv:0802.4262.
Lima, M. and Hu, W. (2004) *PRD*, **70**, 043504.
Limber, D.N. (1953) *ApJ*, **117**, 134.
Limmongkol, S., Owen, R.E., Siegmund, W.A., and Hull, C.L. (1993) *ASP Conference Series*, Vol. 37, Fiber Optics in Astronomy II, (BYU: Provo, UT), 127.
Lin, Y.-T. and Mohr, J.J. (2004) *ApJ*, **617**, 879.
Linde, A. (1987) *Three hundred years of gravitation*, (eds S.W. Hawking and W. Israel), Cambridge Univ. Press, p. 604
Linder, E.V. (2005) *PRD*, **72**, 3529.
Loeb, A. (2007) *JCAP*, **3**, 1.
Loose, M. et al. (2002) *Proc. SPIE*, Vol. **4850**, 867; (2002) *Proc. SPIE*, Vol. **4841**, 782.
Lue, A., Scoccimarro, R., and Starkman, G.D. (2004) *PRD*, **69**, 124015.
Lue, A. (2006) *Physics Report*, **423**, 1.
Luppino, G.A. and Kaiser, N. (1997) *ApJ*, **475**, 20.
Lupton, R. (1993) *Statistics in Theory and Practice*, Princeton University Press.
Ma, Z., Hu, W., and Huterer, D. (2006) *ApJ*, **636**, 21.
Ma, C.-P. and Bertschinger, E. (1995) *ApJ*, **455**, 7.
Ma, C.P., Caldwell, R.R., Bode, P., and Wang, L. (1999) *ApJ*, **521**, L1.
MacKenty, J.W. and Stiavelli, M. (2000) Imaging the Universe in Three Dimensions, in *Proceedings from ASP Conference*,

Vol. 195, (eds W. van Breugel and J. Bland-Hawthorn), p. 443.
MacKenty, J.W. et al. (2006) *SPIE*, **6269**, 37.
Madau, P., Pozzetti, L., and Dickinson, M.E. (1998) *ApJ*, **498**, 106.
Majumdar, S. and Mohr, J.J. (2003) *ApJ*, **585**, 603.
Majumdar, S. and Mohr, J.J. (2004) *ApJ*, **613**, 41.
Mandelbaum, R. et al. (2006) *MNRAS* **367**, 611.
Mannucci, F., Della Valle, M., Panagia, N., Cappellaro, E., Cresci, G., Maiolino, R., Petrosian, A., and Turatto, M. (2005) *A & A*, **433**, 807.
Mannucci, F., Della Valle, M., and Panagia, N. (2006) *MNRAS*, **370**, 773.
Mantz, A., Allen, S.W., Ebeling, H., and Rapetti, D. (2008) *MNRAS*, **387**, 1179.
Marietta, E., Burrows, A., and Fryxell, B. (2000) *ApJS*, **128**, 615.
Marinoni, C. et al. (2005) *A & A*, **442**, 801.
Marion, G.H. et al. (2006) *ApJ*, **645**, 1392.
Marion, G.H. et al. (2009) *ApJ*, in press.
Martineau, P. and Brandenberger, R. astro-ph/0510523
Massey, R. and Refregier, A. (2005) *MNRAS*, **363**, 197.
Massey, R. et al. (2007) *ApJS*, **172**, 239.
Matarrese, S., Verde, L., and Heavens, A.F. (1997) *MNRAS*, **290**, 651.
Mathiesen, B., Evrard, A.E., and Mohr, J.J. (1999) *ApJ*, **520**, L21.
Mathiesen, B.F. and Evrard, A.E. (2001) *ApJ*, **546**, 100.
Matsubara, T. (2004) *ApJ*, **615**, 573.
Matsubara T. (2008) *PRD*, **77**, 063530.
Mazzali, P.A. et al. (2005) *ApJ*, **623**, 37.
McCarthy, P.J. et al. (2004) *ApJL*, **614**, L9.
Meszaros, P. (2006) *Rep. Prog. Phys.*, **69**, 2259.
Miknaitis, G. et al. (2007) *ApJ*, **666**, 674.
Milani, A., Vokrouhlický, D., Villani, D., Bonanno, C., and Rossi, A. (2002) *PRD*, **66**, 082001.
Miller, L. et al. (2007) *MNRAS*, **382**, 315.
Milne, P.A., The, L.-S., and Leising, M.D. (2001) *ApJ*, **559**, 1019.
Milton, K.A., Kantowski, R., Kao, C., and Wang, Y. (2001) **16**, 2281.
Mobasher, B., Rowan-Robinson, M., Georgakakis, A., and Eaton, N. (1996) *MNRAS*, **282**, L7.

Modified Gravity (references with more details): Uzan, J.-P. and Bernardeau, F. (2001) *Phys. Rev.*, **D64**, 083004; Stabenau, H.F. and Jain, B. (2006) *PRD*, **74**, 084007; Heavens, A.F., Kitching, T.D., and Verde, L. astro-ph/0703191; Uzan, J.-P. (2007) *Gen. Relat. Grav.*, **39**, 307; Zhang, P., Liguori, M., Bean, R., and Dodelson, S. (2007) *Phys. Rev. Lett.*, **99**, 141302
Modified Gravity Models references: see e.g. Sahni, V. and Habib, S. (1998) *PRL*, **81**, 1766; Parker, L. and Raval, A. (1999) *PRD*, **60**, 063512; Deffayet, C. (2001) *Phys. Lett. B* **502**, 199; Onemli, V.K. and Woodard, R.P. (2004) *PRD*, **70**, 107301.
Mohr, J.J., Mathiesen, B., and Evrard, A.E. (1999) *ApJ*, **517**, 627.
Molnar, S.M. et al. (2004) *ApJ*, **601**, 22.
Moseley et al. (2000) *ASP Conf.*, **207**, 262.
Motl, P.M., Hallman, E.J., Burns, J.O., and Norman, M.L. (2005) *ApJ*, **623**, L63.
MSA: http://www.jwst.nasa.gov/microshutters.html.
Mukhanov, V. (2005) *Physical Foundations of Cosmology*, Cambridge University Press.
Mukhanov, V.F., Feldman, H.A., and Brandenberger, R.H. (1992) *Physics Reports*, **215**, 203.
Mukherjee, P. and Wang, Y. (2003) *ApJ*, **598**, 779; *ApJ*, **599**, 1.
Munshi, D. and Valageas, P. (2006) astro-ph/0601683.
Nagai, D. (2006) *ApJ*, **650**, 538.
Nagai D., Vikhlinin A., and Kravtsov, A.V. (2007) *ApJ*, **655**, 98.
Nagashima M., Lacey C.G., Okamoto T., Baugh C.M., Frenk C.S., and Cole S. (2005) *MNRAS*, **363**, L31.
Nagashima M., Lacey C.G., Baugh C.M., Frenk C.S., and Cole, S. (2005) *MNRAS* **358**, 1247.
Nakajima, R. and Bernstein, G. (2007) *Astron. J.*, **133**, 1763.
Nakar, E. and Piran, T. (2005) *MNRAS*, **360**, L73.
Neil, R.M. (1993) ftp://ftp.cs.utoronto.ca/pub/~radford/review.ps.gz
Nolan, L., Dunlop, J., Jimenez, R., and Heavens, A.F. (2003) *MNRAS*, **341**, 464.
Nomoto, K. (1982) *ApJ*, **253**, 798.
Norris, J.P., Marani, G.F., and Bonnell, J.T. (2000) *ApJ*, **534**, 248.
Nugent, P. et al. (1997) *ApJ*, **485**, 812.

Nugent, P., Kim, A., and Perlmutter, S. (2002) *PASP*, **114**, 803.
Oemler, A. Jr. and Tinsley, B.M. (1979) *AJ*, **84**, 9850.
O'Dea, C.P. *et al.* (2009) *A & A*, **494**, 471.
Oguri, M. and Takahashi, K. (2006) *Phys. Rev. D*, **73**, 123002.
Okumura, T. *et al.* (2008) *ApJ*, **676**, 889.
Orsi, A. *et al.* (2009) in preparation.
Paczynski, B. (1995) *PASP*, **107**, 1167.
Padmanabhan, N., White, M., and Cohn, J.D. arXiv:0812.2905.
Padmanabhan, T. arXiv:0807.2356, Invited article to appear in Advanced Science Letters Special Issue on Quantum Gravity, Cosmology and Black holes (ed. M. Bojowald)
Page, L. *et al.* (2003) *ApJS*, **148**, 233.
Panaitescu, A. and Kumar, P. (2002) *ApJ*, **571**, 779.
Paul, M. (1935) *Rev. Opt.*, **14**, 13.
Peacock, J.A. (1999) *Cosmological Physics*, Cambridge University Press.
Peacock, J.A. *et al.* (2001) *Nature*, **410**, 169.
Peacock, J.A., Schneider, P., Efstathiou, G., Ellis, J.R., Leibundgut, B., Lilly, S.J., and Mellier, Y. (2006) *Report by the ESA-ESO Working Group on Fundamental Cosmology*, arXiv:astro-ph/0610906.
Pearson, T.J. *et al.* (2003) *Astrophys. J.*, **591**, 556.
Peebles, P.J.E. (1980) *The Large Scale Structure of the Universe*, Princeton Univ. Press, Princeton, NJ.
Peebles, P.J.E. (1993) *Principles of Physical Cosmology*, Princeton University Press.
Pen, U.-L., Van Waerbeke, L., and Mellier, Y. (2002) *ApJ*, **567**, 31.
Percival, W.J. *et al.* (2007) *MNRAS*, **381**, 1053.
Percival, W.J. and White, M. (2009) *MNRAS*, **393**, 297.
Perlmutter, S. *et al.* (1999) *ApJ*, **517**, 565.
Phantom Models references: see e.g. Chen, C.M., Gal'tsov, D.V., and Gutperle, M. (2002) *Phys. Rev. D*, **66**, 024043; Townsend, P.K. and R.Wohlfarth, M.N. (2003) *Phys. Rev. Lett.*, **91**, 061302; Ohta, N. (2003) *Phys. Lett. B*, **558**, 213; (2003) *Phys. Rev. Lett.*, **91**, 061303; (2003) *Prog. Theor. Phys.*, **110**, 269; (2005) *Int. J. Mod. Phys. A*, **20**, 1; Roy, S. (2003) *Phys. Lett. B*, **567**, 322.
Phillips, M.M. (1993) *ApJ*, **413**, L105.

Phillips, M.M. *et al.* (1999) *AJ*, **118**, 1766.
Phillips, M.M., Garnavich, P., and Wang, Y. *et al.* (2006) astro-ph/0606691, Proc. of SPIE, Vol. 6265, 626569.
Podariu, S., Daly, R.A., Mory, M.P., and Ratra, B. (2003) *ApJ*, **584**, 577.
Poole, G.B. *et al.* (2007) *MNRAS*, **380**, 437.
Premadi, P., Martel, H., Matzner, R., and Futamase, T. (2001) *ApJ*, **135**, 7.
Press, W.H. and Schechter, P. (1974) *ApJ*, **187**, 425.
Press, W.H., Teukolsky, S.A., Vettering, W.T., and Flannery, B.P. (2007) *Numerical Recipes 3rd Edition*, Cambridge University Press, Cambridge.
Pskovskii, Y.P. (1977) *Soviet Astron.*, **21**, 675.
Puschell, J.J., Owen, F.N., and Laing, R.A. (1982) *ApJ*, **257**, L57.
Quintessence Models references: see e.g. Freese, K., Adams, F.C., Frieman, J.A., and Mottola, E. (1987) *Nucl. Phys.*, **B287**, 797; Peebles, P.J.E. and Ratra, B. (1988) *ApJL*, **325**, 17; Wetterich, C. (1988) *Nucl. Phys.*, **B302**, 668; Frieman, J.A., Hill, C.T., Stebbin, A., and Waga, I. (1995) *PRL*, **75**, 2077; Caldwell, R., Dave, R., and Steinhardt, P.J. (1998) *PRL*, **80**, 1582; Wang, L. and Steinhardt, P.J. (1998) *ApJ*, **508**, 483
Rapetti, D., Allen, S.W., and Mantz, A. (2008) *MNRAS*, **388**, 1265.
Refregier, A. (2003a) *MNRAS*, **338**, 35.
Refregier, A. *et al.* (2009) *Experimental Astronomy*, **23**, 17.
Refregier, A. and Bacon, D. (2003) *MNRAS*, **338**, 48.
Reiprich, T.H. and Böhringer, H. (2002) *ApJ*, **567**, 716.
Reynaud, S. and Jaekel, M. (2007) *Notes of a lecture given during the International School of Physics Enrico Fermi on Atom Optics and Space Physics*, Varenna, July 2007, arXiv:0801.3407v1 [gr-qc].
Ricker, P.M. and Sarazin, C.L. (2001) *ApJ*, **561**, 621.
Riess, A.G. (2000) *PASP*, **112**, 1284.
Riess, A.G. *et al.* (1998) *ApJ*, 504, 935.
Riess, A.G. *et al.* (1998) *Astron. J.*, **116**, 1009.
Riess, A.G. *et al.* (1999) *AJ*, **117**, 707.
Riess, A.G. *et al.* (2004) *ApJ*, **607**, 665.
Riess, A.G. *et al.* (2007) *ApJ*, **659**, 98.
Riess, A.G., Press, W.H., and Kirshner, R.P. (1995) *ApJ*, **438**, L17.

Rines, K., Diaferio, A., and Natarajan, P. (2007) *ApJ*, **657**, 183.

Robberto, M. and Cimatti, A. for the SPACE Science Team, arXiv:0710.3970; http://urania.bo.astro.it/cimatti/space/.

Ross, N.P. et al. (2007) *MNRAS*, **381**, 573.

Sahni, V. and Starobinsky, A. (2006) *IJMPD*, **15**, 2105.

Sanchez, A.G., Baugh, C.M., and Angulo, R. (2008) *MNRAS*, **390**, 1470.

Samushia, L. and Ratra, B. arXiv:0806.2835v1 [astro-ph].

Sandvik, H.B., Tegmark, M., Zaldarriaga, M., and Waga, I. (2004) *Phys. Rev. D*, **69**, 123524.

Sawicki, M.J., Lin, H., and Yee, H.K.C. (1997) *AJ*, **113**, 1.

Scannapieco, E. and Bildsten, L. (2005) *ApJ*, **629**, 85S.

Schaefer, B.E. (2002) *Gamma-Ray Bursts: The Brightest Explosions in the Universe*, Harvard.

Schaefer, B.E. (2003) *ApJ*, **583**, L71.

Schaefer, B.E. (2004) *ApJ*, **602**, 306.

Schaefer, B.E. (2007) *ApJ*, **660**, 16.

Schaefer, B.E. and Collazzi, A.C. (2007) *ApJ*, **656**, L53.

Scheuer, P.A.G. (1974) *MNRAS*, **166**, 513.

Schneider, P., Ehlers, J., and Falco, E. (1992) *Gravitational Lenses*, Springer-Verlag, Berlin.

Schneider, P., Van Waerbeke, L., and Mellier, Y. (2002) *A & A*, **389**, 729.

Schuecker, P. et al. (2003) *A & A*, **398**, 867.

Scoccimarro, R. (2000) *ApJ*, **544**, 597.

Seitz, C. and Schneider, P. (1997) *A & A*, **318**, 687.

Seljak, U. and Zaldarriaga, M. (1996) *ApJ*, **469**, 437; http://www.cfa.harvard.edu/~mzaldarr/CMBFAST/cmbfast.html.

Semboloni, E. et al. (2006) *A & A*, **452**, 51.

Seo, H. and Eisenstein, D.J. (2003) *ApJ*, **598**, 720.

Seo, H. and Eisenstein, D.J. (2005) *ApJ*, **633**, 575.

Seo, H. and Eisenstein, D.J. (2007) *ApJ*, **665**, 14.

Seo, H. et al. (2008) arXiv:0805.0117.

Sheldon, E.S. et al. (2004) *AJ*, **127**, 2544.

Sigad, Y., Branchini, E., and Dekel, A. (2000) *ApJ*, **540**, 62S.

Simon, J., Verde, L., and Jimenez, R. (2005) *PRD*, **71**, 123001.

Smith, R.E. et al. (2003) *MNRAS*, **341**, 1311.

Smith, R.E., Scoccimarro, R., and Sheth, R.K. astro-ph/0703620.

Smith, R.E., Scoccimarro, R., and Sheth, R.K. (2008) *PRD*, **77**, 043525.

Somerville, R., Primack, J.R., and Faber, S.M. (2001) *MNRAS*, **320**, 289.

Song, Y.-S., Sawicki, I., and Hu, W. (2007) *PRD*, **75**, 064003.

Sorokina, E.I. and Blinnikov, S.I. (2000) *Astron. Lett.*, **26**, 67.

Spergel, D.N. et al. (2007) *ApJS*, **170**, 377.

Spinrad, H. et al. (1997) *ApJ*, **484**, 581.

Springel, V. et al. (2005) *Nature*, **435**, 629.

Starobinsky, A.A. (1980) *Phys. Lett. B*, **91**, 99.

Stebbins, A. (1996) arXiv:astro-ph/9609149.

Steidel, C.C. et al. (1999) *ApJ*, **519**, 1.

Strolger, L.-G. et al. (2004) *ApJ*, **613**, 200.

Sumiyoshi, M. et al. arXiv:0902.2064.

Sun, M. et al. (2009) *ApJ*, **693**, 1142.

Sunyaev, R.A. and Zeldovich, Y.B. (1972) *Comments on Astrophysics and Space Physics*, **4**, 173.

Takada, M. and Jain, B. (2004) *MNRAS*, **348**, 897.

Tanvir, N.R. et al. (2009) arXiv:0906.1577.

Tegmark, M. (1997) *PRL*, **79**, 3806.

Tegmark, M. (2002) *Phys. Rev.*, **D66**, 103507.

Tegmark, M. et al. (2004) *ApJ*, **606**, 702.

Tegmark, M. et al. (2006) *Phys Rev. D*, **74**, 123507.

The Sloan Digital Sky Survey Project Book, http://www.astro.princeton.edu/PBOOK/.

Thomas, R.C., Branch, D., Baron, E., Nomoto, K., Li, W., and Filippenko, A.V. (2004) *ApJ*, **601**, 219.

Tonry, J.L. et al. (2003) *ApJ*, **594**, 1.

Totani, T., Morokuma, T., Oda, T., Doi, M., and Yasuda, N. (2008) *PASJ*, **60**, 1327.

Treu, T. et al. (1999) *MNRAS*, **308**, 1037.

Treu, T. et al. (2001) *MNRAS*, **326**, 221.

Treu, T. et al. (2002) *ApJL*, **564**, L13

Tsamis, N.C. and Woodard, R.P. (1995) *Ann. Phys.*, **238**, 1.

Uenishi, T., Nomoto, K., and Hachisu, I. (2003) *ApJ*, **595**, 1094.

Vale, C. and White, M. (2003) *ApJ*, **592**, 699,

van Waerbeke, L. et al. (2000) *Astron. Astrophys.*, **358**, 30.

Valageas, P. (2000) *A & A*, **354**, 767; *A & A*, **356**, 771.

Ventimiglia, D.A., Voit, G.M., Donahue, M., and Ameglio, S. (2008) *ApJ*, **685**, 118.

Verde, L. et al. (2002) *MNRAS*, **335**, 432.

Verde, L., Heavens, A.F., Matarrese, S., and Moscardini, L. (1998) *MNRAS*, **300**, 747.

Vikhlinin, A. et al. (2003) *ApJ*, **590**, 15.

Vikhlinin, A., Kravtsov, A., Forman, W., Jones, C., Markevitch, M., Murray, S.S., and Van Speybroeck, L. (2006) *ApJ*, **640**, 691.

Vikhlinin, A. et al. (2009a) *ApJ*, **692**, 1033.

Vikhlinin, A. et al. (2009b) *ApJ*, **692**, 1060.

Wambsganss, J., Cen, R., Xu, G., and Ostriker, J.P. (1997) *ApJ*, **475**, L81.

Wan, L. and Daly, R.A. (1998) *ApJS*, **115**, 141

Wan, L. and Daly, R.A. (1998) *ApJ*, **499**, 614.

Wang, L. and Steinhardt, P.J. (1998) *ApJ*, **508**, 483.

Wang, L. and Wheeler, J.C. (2008) *ARA & A*, **46**, 433.

Wang, L. et al. (2003) *ApJ*, **591**, 1110.

Wang, L. et al. (2006) *ApJ*, **641**, 50.

Wang, L., Baade, D., and Patat, F. (2007) *Science*, **315**, 212.

Wang, Y. (1990) *Phys. Rev. D*, **42**, 2541.

Wang, Y. (1999) *ApJ*, **525**, 651.

Wang, Y. (2000a) *ApJ*, **536**, 531.

Wang, Y. (2000b) *ApJ*, **531**, 676, astro-ph/9806185.

Wang, Y. (2005) *JCAP*, **0503**, 005.

Wang, Y. (2006) *ApJ*, **647**, 1.

Wang, Y. (2008a) *Phys. Rev. D*, **77**, 123525.

Wang, Y. (2008b) *JCAP*, **0805**, 021.

Wang, Y. (2008c) *PRD*, **78**, 123532.

Wang, Y. arXiv:0904.2218.

Wang, Y. and Freese, K. (2006) *Phys. Lett.*, **B632**, 449.

Wang, Y. and Garnavich, P. (2001) *ApJ*, **552**, 445.

Wang, Y. and Hall, N. (2008) *MNRAS*, **389**, 489.

Wang, Y. and Lovelace, G. (2001) *ApJ Letter*, **562**, L115.

Wang, Y. and Mukherjee, P. (2004) *ApJ*, **606**, 654.

Wang, Y. and Mukherjee, P. (2007) *PRD*, **76**, 103533.

Wang, Y. and Tegmark, M. (2004) *Phys. Rev. Lett.*, **92**, 241302.

Wang, Y. and Tegmark, M. (2005) *Phys. Rev. D*, **71**, 103513.

Wang, Y. et al. (2004) *BAAS*, **36** (5), 1560.

Wang, Y., Bahcall, N., and Turner, E.L. (1998) *AJ*, **116**, 2081.

Wang, Y., Freese, K., Gondolo, P., and Lewis, M. (2003) *ApJ*, **594**, 25–32.

Wang, Y., Holz, D.E., and Munshi, D. (2002) *ApJ*, **572**, L15.

Wang, Y., Kratochvil, J.M., Linde, A., and Shmakova, M. (2004) *JCAP*, **12**, 006.

Wang, Y., Spergel, D.N., and Strauss, M.A. (1999) *ApJ*, **510**, 20.

Wang, Y., Tenbarge, J., and Fleshman, B. (2003) *ApJ*, **624**, 46 (2005).

Weinberg, S. (1972) *Gravitation and Cosmology: Principles and Applications of the General Theory of Relativity*, John Wiley & Sons.

Weinberg, S. (2008) *Cosmology*, Oxford University Press.

Weymann, R.J., Storrie-Lombardi, L.J., Sawicki, M., and Brunner, R.J. (eds) (1999) *Photometric Redshifts and High Redshift Galaxies*, ASP Conference Series, Vol. 191.

Wheeler, J.C. (2003) AAPT/AJP Resource Letter. *Am. J. Phys.*, **71**.

Wheeler, J.C. private communication.

White, M. (2005) *Astropart. Phys.*, **24**, 334.

White, S.D.M. and Frenk, C.S. (1991) *ApJ*, **379**, 52.

White, M., Song, Y.-S., and Percival, W.J. (2009) *MNRAS*, in press; arXiv:0810.1518.

Will, C.M. (2006) *The Confrontation between General Relativity and Experiment*, Living Rev. Relativity 9, http://www.livingreviews.org/lrr-2006-3.

Williams, J.G., Turyshev, S.G., and Boggs, D.H. (2004) *Physical Review Letters*, **93**, 261101.

Wittman, D.M., Tyson, J.A., Kirkmand, D., Dell'Antonio, I., and Bernstein, G. (2000) *Nature*, **405**, 143.

Wittman, D.M. (2009) arXiv:0905.0892.

Wood-Vasey, W.M. et al. (2008), *ApJ*, **689**, 377.

Woodard, R.P. (2005) *Proceedings for the 3rd Aegean Summer School*, Chios, 26 September–1 October 2005. arXiv:astro-ph/0601672

Woosley, S.E., Kasen, D., Blinnikov, S., and Sorokina, E. (2007) *ApJ*, **662**, 487.

Yoon, S.C. and Langer, N. (2002) *A & A*, **419**, 623.

Zhang, P. and Corasaniti, P.S. (2007) *ApJ*, **657**, 71.

Zhang, J., Hui, L., and Stebbins, A. (2005) *ApJ*, **635**, 806.

Zhang, Y.-Y. et al. (2006) *A&A*, **456**, 55.

Zwicky, F. (1933) *Physica Acta*, **6**, 110.

Index

a

affine connection   16
affine parameter   67
Alcock–Paczynski test   134 ff
angular diameter distance   2, 92 ff, 135
aperture-mass, variance   142 ff
assymetric measurement errors   188
astigmatism   205, 208
average color correction law   75
average spectral sequence   75

b

Baker–Paul three-mirror telescope   206
BAO   223
– data analysis techniques   103 ff
– density-field reconstruction   99 ff
– Fisher matrix   119
– forecast   119
– full $P(k)$ method   122 ff
– galaxy correlation function measurement   114 ff
– galaxy power spectrum measurement   104 ff
– nonlinearity parameter   121
– redshift-space distortions   99 ff
– scale-dependent bias   101 ff
– wiggles only method   120 ff
BAO in power spectra   95 ff
BAO in redshift-space correlation function   94
BAO scale, calibration   92 ff
BAO scale, radial and transverse directions   96
BAO scale, spherically averaged   94 ff
BAO, damping scale   109 ff, 112 ff
BAO, emission line galaxies   111 ff
BAO, galaxy correlation function measurement
– mitigation of systematic effects   116 ff

BAO, galaxy power spectrum measurement
– mitigation of systematic effects   108
BAO, nonlinear effects   98 ff
BAO, observational results   93 ff
BAO, red galaxies   110 ff
BAO, scale parameter   103 ff, 109 ff, 117 ff
BAO, systematic effects   97 ff
baryon acoustic oscillations (BAO)   91 ff, 219
baryon/photon ratio   92
Bayes' theorem   42
bias factor   101, 122, 128 ff, 141

c

CAMB   105, 108
Cardassian model   46 ff
CFHTLS   142 ff
Chandra   166, 172
Chandrasekhar limit *see* Chandrasekhar mass
Chandrasekhar mass   56
Chaplygin gas   38
cluster abundance *see* cluster mass function
cluster gas mass fraction   166 ff, 179 ff
cluster gas-to-mass ratio *see* cluster gas mass fraction
cluster gass mass fraction   182
cluster mass   166
cluster mass function   164 ff
– cosmological parameter constraints   175 ff
– estimation   172 ff
– measured   174
clusters, hydrostatic mass measurements   166 ff, 170
CMB shift parameters   33 ff
CMBFAST   105, 108
coma   205, 208
comoving distance   2, 51, 79, 191
comoving distance and redshift relation   18

Dark Energy. Yun Wang
Copyright © 2010 WILEY-VCH Verlag GmbH & Co. KGaA, Weinheim
ISBN: 978-3-527-40941-9

comoving number density   119
comoving sound horizon   33
comoving wavelength   2
comoving wavenumber   21
complex ellipticity   139, 149
complex lensing shear   138
conformal Newtonian gauge   23 ff
conformal time   22
conformal transformation   44
conservation of energy and momentum   13
continuity equation   19
convergence power spectrum   141 ff
convolution   116
cosmic acceleration   1 ff
– current observational evidence   8 ff
– models   35 ff
cosmic far infrared background   71
cosmic microwave background (CMB)   33 ff
cosmic scale factor   1 ff, 92
cosmic time and redshift relation   18
cosmological constant   3 ff, 47 ff, 224
cosmological N-body simulations
   97 ff, 116, 165 ff, 168 ff, 171 ff, 182
– galaxy modeling   97 ff
cosmological perturbations   18
– generalized case   22
– nonrelativistic case   18
cosmological redshift   2 ff
covariance matrix   223 ff
Cramér-Rao inequality   32
critical density   3
curvature constant   1
curvature scalar   15
curvature tensor   16

d

dark current   211
dark energy   9
– density   11
– density function   17
– equation of state   10 ff, 11
– fundamental questions   10, 224 ff
dark energy density   164
dark energy density function   79, 195
dark energy equation of state
   144, 178 ff, 194 ff
dark energy experiments
– current status   225
– figure of merit (FoM)   222 ff
– future prospects   219 ff
– JEDI (Joint Efficient Dark-energy
   Investigation)   219 ff
– optimization   219 ff

dark energy model   35
– doomsday model   40 ff
dark energy parametrization   222 ff
dark energy perturbation   27
dark energy space mission
– DUNE   222, 226
– Euclid   226
– JDEM   226
– SPACE (the SPectroscopic All-sky Cosmic
   Explorer)   220, 220 ff, 226
Dark Energy Task Force (DETF)   223, 225
– figure of merit (FoM)   226
dark matter haloes   100
deceleration parameter   4
deflection angle   135
deflection potential   137 ff
density field Poisson sample   106
dewiggled power spectrum   112 ff
2dFGRS   127, 131
DGP gravity model   12 ff, 30, 45 ff, 132, 224
– characteristic length scale   46
– dark energy model equivalent   13
digital micromirror devices (DMDs)
   214 ff, 221
Dirac delta function   129
distance modulus   2 ff, 78, 80
distance-redshift relation   4 ff
drag epoch   91, 93 ff
Dyer–Roeder distance   67

e

ecliptic poles   62
Einstein frame   44
Einstein radius, angular   137
Einstein ring   137
Einstein tensor   15
Einstein's equation   15 ff
ellipticity, observed   140
energy-momentum tensor   16
equation of state   4, 17
ESA-ESO   226
Euler equation   19
expansion history of the universe *see* Hubble
   parameter
extended radio galaxies (ERG)   199
– systematic effects   200 ff
extinction   74
– host galaxy dust   62 ff
– Milky Way   62

f

$f(R)$ gravity models   44 ff
fast Fourier transform   130
Fermat's principle   136 ff

field curvature 205, 208
field distortion 205
figure of merit (FoM) 222
finger of God effect 100 ff, 125
Fisher matrix 32, 119 ff, 157
fit to straight line 187
FKP method 130
flux averaging 5, 83
flux averaging, recipe 78
flux statistics, SNe Ia 77
Fourier transform 104, 116, 128
Friedmann equation 16 ff
future prospects 219 ff
future SN Ia surveys, forecast 83 ff

g
galactic extinction 73
galaxy bispectrum 129 ff
galaxy cluster 163 ff
– abundance 163
– future surveys 183
– likelihood analysis 175 ff
– mass 166 ff
– mass and temperature relation 169 ff
– mass and Y_X relation 171 ff
– radius 166
– relaxed 170
– standard candles 163
– systematics 182 ff
– unrelaxed 170
galaxy-convergence power spectrum 141
galaxy correlation function 125 ff
– measurement 114 ff
– minimum-variance estimator 115 ff
galaxy density distribution 115
galaxy density perturbation 128
galaxy ellipticity 139 ff
galaxy-galaxy power spectrum 141
galaxy haloes see dark matter haloes
galaxy number density 123 ff, 160
galaxy power spectrum 141
galaxy power spectrum measurement 104
– FKP method 105 ff
galaxy redshift survey 91 ff, 223 ff
– linear growth rate 124 ff
– redshift space distortions 124 ff
galaxy redshift survey, effective volume 119
galaxy, large scale flows 124
galaxy, pairwise peculiar velocity dispersion 128
galaxy, small scale random motion 127
galaxy, small scale random velocity 124
gamma-ray bursts (GRBs) 185 ff

gauge-invariant perturbation equations 23
Gaussian filter 109
general relativity
– parametrized post-Newtonian approach 202
– solar system test 202
geometric weak lensing method
– linear scaling 160 ff
– off-linear scaling 160 ff
gray dust 71
GRBs
– assymetric measurement errors 188
– bulk Lorentz factor of the jet 186
– calibration relations 186 ff
– distance ratio 190 ff
– impact on dark energy constraints 193 ff
– Malmquist bias 196
– minimum rise time 186
– model-independent distance measurements 190 ff
– peak energy 186
– systematic errors 188, 195 ff
– time lag 186
– total burst energy in the gamma rays 186
– variability 186
– weak lensing 196
GREAT08 152
growing mode 21
growth history of cosmic large scale structure see linear growth rate

h
halo mass function 164 ff
halofit 116
Heaviside step function 143
HgCdTe 210 ff
Hubble constant 3, 105, 122, 191
Hubble diagram 4 ff, 54
Hubble parameter 3, 91 ff, 132 ff, 141, 225
– old passive galaxies 196 ff
– uncorrected estimate 79 ff

i
ICM temperature 169
ideal fluid 3
impact parameter 135
Indium Antimonide (InSb) 210
intra-cluster medium (ICM) 166
intrinsic alignments 156
intrinsic shear stress 30
intrinsic SN Ia color variation 63
inverse metric tensor 15
IRMOS 216

j

Jacobian, lensing mapping   138 ff
Joint Dark Energy Mission (JDEM)
   84 ff, 226
Joint Efficient Dark-energy Investigation
   (JEDI)   84 ff, 219
Jordan frame   44
JWST   214 ff
JWST NIRCam   210
JWST telescope   209

k

K-correction   64, 73, 75, 84 ff, 173
kernel function   130
Korsch three-mirror anastigmatic telescope
   see three-mirror anastigmat (TMA)
KPNO   216

l

late-time integrated Sachs–Wolfe (ISW)
   effect   29
lens equation   135 ff
lensing convergence   138
lensing potentialsee deflection potential
lensing shear   138 ff
lensing shear, reduced   138 ff
light curve-fitting technique   54
likelihood   109
likelihood analysis   175
likelihood function   175 ff
Limber's approximation   140
linear growth factor   141, 164
linear growth rate   12, 132 ff, 223 ff
– power-law index γ 224
linear regime   125 ff
longitudinal gaugesee Newtonian gauge
LST   226
luminosity distance   2 ff, 51, 79

m

magnification   139
magnification, SNe Ia   76
magnitude   2
magnitude statistics, SNe Ia   77
Malmquist bias   196
marginalization over H_0 78 ff
Markov Chain Monte Carlo (MCMC)
   32 ff, 223
massless neutrino species   92
matter   17
matter and radiation equality   92
matter density fluctuations   164
matter density perturbation
   11, 18 ff, 104, 114, 128

matter power spectrum
   68, 100, 104 ff, 141, 164
matter transfer function   13, 104 ff, 122
mean extinction ratio   63
Mercury Cadmium Telluride (HgCdTe) see
   HgGdTe
metric perturbations   22 ff
– modification to gravity   30
metric tensor   15 ff
micro-electro-mechanical-systems (MEMS)
   214
microshutter arrays (MSAs)   214 ff
minimum convergence   66 ff
minimum magnification   66
minimum-variance weighting   115
mock catalogs   128
model-independent constraints   31
– dark energy   224
modified gravity   9 ff
– solar system test   203
modified gravity models   44 ff
modified gravity, $f_g(z)$ test   132 ff
modified gravity, $G(z)$ test   158 ff
M_{tot}-T_X correlation   167
multiple-object spectroscopic masks
– fibers   213
– micro-electro-mechanical-systems
   (MEMS)   214
multiplicative shear bias   148

n

near infrared (NIR)   63
nearby SNe Ia   54
NIR detectors   210 ff
noise vector   80
nonlinear effects   123
nonlinear regime   126 ff
normal SNe Ia   51, 56

o

old passive galaxies   196
– Hubble parameter   197 ff
– sample selection   197
– systematic effects   198
optical axis   135

p

pairwise velocity dispersion   100
peculiar gravitational acceleration   21
peculiar gravitational potential   19
peculiar velocities   124
perfect fluid   16
phantom field   37
photometric redshifts   154 ff

photon-decoupling epoch   92
photon-decoupling surface   33
physical wavelength   2
Planck mass   46
point spread function (PSF)   148 ff
Poisson equation   19
Poisson statistics   175
Press–Schechter model   164
primordial matter power spectrum   104
priors   223
progenitor population drift   72
PSF anisotropy   149 ff
PSF correction
– current status   152 ff
– Fourier space   153
– impact on dark energy constraints   154
– KSB+ method   149 ff
– shapelet method   151 ff
Pskovskii–Branch–Phillips (PBP) relation see SN Ia peak luminosity, correlation with light curve width

q
quintessence   36 ff, 224
– PNGB   39

r
radiation   17
radiation energy density   92
radio galaxies   198 ff
– distance measurement   199 ff
– extended radio galaxies (ERG)   199
real-space power spectrum   125
redshift-space distortions   122
redshift-space power spectrum   125, 129
redshift uncertainties, damping factor   121
reduced convergence   68
reduced shear   140
reference spectrum   108
renormalized perturbation theory (RPT)   112 ff
– mode-coupling   118
Ricci tensor   15
Riemann–Christoffel curvature tensor   16
Robertson–Walker metric   1, 16

s
scalar perturbations   22
scale-dependent bias   101, 113
Schwarzschild radius   135
SDSS   94 ff, 205 ff, 214
seeing   148
seeing correction   150
self-similar theory   169

shapelet expansion   151
shear correlation function   142 ff
shear polarizability tensor   150
shear power spectrum   141
SIDECAR ASIC   211 ff
σ_8 177 ff
Silk damping scale   121
simulated random galaxy catalog   115
SKA   226
smear polarizability tensor   150
smoothing angle   68
smoothly distributed matter, mass-fraction   66 ff
SN Ia color   55
SN Ia data analysis   73 ff
SN Ia light curve fitting
– MLCS method   73 ff
– SALT method   74 ff
SN Ia light curves and spectra inhomogeneity   52
SN Ia peak luminosity
– correlation with light curve width   53 ff
– intrinsic scatter   53
SN Ia photometry calibration   61
SN Ia progenitors
– double-degenerate   60 ff
– single-degenerate   60 ff
SNe Ia
– delay time distribution   59 ff
– possible causes of observational diversity   56
– rate   57 ff
– ultra-deep survey   55
SNe Ia, flux-averaging analysis   76 ff
SNe Ia, late-time light curves   88 ff
SNe Ia, NIR light curves   86 ff
SNe Ia, NIR spectra   86
SNe Ia, optimized observations   86 ff
SNe Ia, spectral luminosity indicators   87
SNe Ia, systematic uncertainties   61 ff
SNe Ia, weak lensing   66 ff, 84 ff
solar system test
– future prospects   203
– general relativity   202
– modified gravity   203
sound speed   21, 92
source ellipticity distribution, transformation   140
spectral energy distribution (SED)   64
spherical aberration   205, 208
standard candles   4, 61
star-formation rate   59 ff
STEP   150 ff, 152 ff

stray light suppression 208
sub-Chandrasekhar-mass white dwarf 51
Subaru/XMM-Newton Deep Survey 59
super-Chandrasekhar-mass white dwarf 51
supernova peak luminosity, evolution 72 ff
surface mass density, dimensionless *see*
  lensing convergence
survey selection function 115
synchronous gauge 24 ff
– conversion to conformal Newtonian gauge 26
synthetic catalog 105 ff

t

tangential shear 141
telescope 205 ff
– astigmatism 205
– Baker–Paul three-mirror telescope 206
– coma 205
– field curvature 205
– field distortion 205
– Korsch three-mirror anastigmatic telescope 206
– Ritchey–Chrétien telescope 205
– Schmidt telescope 205
– spherical aberration 205
three-mirror anastigmat (TMA) 206, 209
tidal gravitational forces 156
time delay 136
– geometric 136
– potential 136
time dilation 52, 73
Type Ia supernovae
– apparent peak brightness 4
– difference in analysis techniques 7
– intrinsic peak luminosity 4
Type Ia supernovae (SNe Ia) 51 ff, 219 ff

u

unbiased estimator, lensing shear 152
uniform priors 223

v

vacuum energy *see* cosmological constant
variance, minimazation 82
volume-averaged distance 95

w

weak lensing 135 ff, 219 ff
– angular correlation functions 140 ff
– future prospects 157 ff
– geometric method *see* geometric weak lensing method
– observational results 142
– systematics 147 ff
weak lensing amplification, universal probability distribution function (UPDF) 68, 85
weak lensing of SNe Ia, signatures 70
weak lensing systematics, parametrization 148
weight function 105 ff, 112
window function 82, 106
window function lensing 138
WL
– B mode 142 ff
– basic concepts 135 ff
– calibration bias 150 ff, 152 ff
– E and B decomposition 141
– E mode 142 ff
– higher-order statistics 141
– residual offset 150 ff, 152 ff
– self-calibration 154
– window functions 141

x

X-ray flux 173
X-ray luminosity 173
X-ray spectral temperature 169
XMM-Newton 166